W0230596

DEMOCRATIZING FOREST GOVERNANCE IN INDIA

edited by

SHARACHCHANDRA LELE
AJIT MENON

OXFORD
UNIVERSITY PRESS

OXFORD
UNIVERSITY PRESS

Oxford University Press is a department of the University of Oxford.
It furthers the University's objective of excellence in research, scholarship,
and education by publishing worldwide. Oxford is a registered trademark of
Oxford University Press in the UK and in certain other countries

Published in India by
Oxford University Press
YMCA Library Building, 1 Jai Singh Road, New Delhi 110 001, India

© Oxford University Press 2014
The copyright to Chapter 5 of this volume vests with its authors

The moral rights of the authors have been asserted

First Edition published in 2014

All rights reserved. No part of this publication may be reproduced, stored in
a retrieval system, or transmitted, in any form or by any means, without the
prior permission in writing of Oxford University Press, or as expressly permitted
by law, by licence, or under terms agreed with the appropriate reprographics
rights organization. Enquiries concerning reproduction outside the scope of the
above should be sent to the Rights Department, Oxford University Press, at the
address above

You must not circulate this work in any other form
and you must impose this same condition on any acquirer

ISBN-13: 978-0-19-809912-3
ISBN-10: 0-19-809912-6

Typeset in Dante MT Std 10.5/13
by The Graphics Solution, New Delhi 110 092

Contents

Tables, Figures, and Boxes

Tables

Figures

Boxes

Preface

The idea for this book was sparked by conversations we had with Vasant Saberwal at the Ford Foundation on the idea of 'thinking beyond Joint Forest Management'. The Ford Foundation had provided support to Winrock International India for research on JFM, and it was agreed that a set of forward-looking papers compiled into a book would be a fitting ending to this research programme. Accordingly, a set of papers were commissioned and a workshop to discuss these papers with a wider set of academics, foresters, and activists was organized with financial and logistic support from Winrock. Subsequently, although Winrock was unable to continue its involvement in this effort, we have carried the idea towards completion in book form.

We are grateful to our colleagues in Winrock International India, especially Mamta Borgoyary, Sharmistha Bose, and Regina Hansda for their help in organizing the workshop. Designated commentators, including Amita Baviskar, Kanchan Chopra, D. Suryakumari, K.D. Singh, B.M.S Rathore, Manoj Pattanaik, Sanjay Kumar, Ravi Rebbapragada, Usha Ramanathan, K.B. Thampi, Manoj Mishra, and Nandini Sundar played an extremely valuable role by providing a counterpoint to the papers in different sessions. We are also grateful to the other participants in the workshop who made it an intense and useful discussion. Financial support from the Ford Foundation (via Winrock and via ATREE) enabled us to commission the papers, hold the workshop, and also keep the price of the book reasonable. Also important was the moral support and intellectual stimulus provided by Vasant through a rather protracted process. Two anonymous referees provided us valuable suggestions for further moulding the contributions. Finally, thanks also to Meena Venkataraman for helping with the final, and to H. Usha and Sowmyashree, M.V. for secretarial support, and to Sowmyashree, M.V. for map-making support.

Abbreviations

ACF	Assistant Conservator of Forests
AP	Andhra Pradesh
APO	Annual Plan of Operation
APCFMP	Andhra Pradesh Community Forestry Management Project
APFD	Andhra Pradesh Forest Department
APTWD	Andhra Pradesh Tribal Welfare Department
ATREE	Ashoka Trust for Research in Ecology and the Environment
BoR	Board of Revenue
BLR Act	Bihar Land Reforms Act
BRT	Biligiri Rangaswamy Temple Wildlife Sanctuary
CA	Compensatory Afforestation
CAF	Compensatory Afforestation Fund
CAMPA	Compensatory Afforestation Fund Management and Planning Authority
CBD	Convention on Biological Diversity
CBNRM	community-based natural resource management
CCA	Community Conserved Area
CEC	Central Empowered Committee
CESS	Centre for Economic and Social Studies
CFR	community forest rights
CIFOR	Center for International Forestry Research
CNTA	Chhota Nagpur Tenancy Act
CoP	Conference of Parties
CPR	common property resource
CPLR	common property land resources
CRZ	Coastal Regulation Zone
CSD	Campaign for Survival and Dignity

CSE	Centre for Science and Environment
CSO	civil society organization
CT/WH	Critical Tiger/Wildlife Habitat
DCF	Deputy Conservator of Forests
DDS	Deccan Development Society
DFID	Department for International Development
DFO	Divisional Forest Officer
DPAP	Drought Prone Areas Programme
DRDA	District Rural Development Agency
EGoM	Empowered Group of Ministers
EIA	Environmental Impact Assessment
ERRP	Economic Rehabilitation of Rural Poor
FAC	Forest Advisory Committee
FCA	Forest Conservation Act, 1980
FD	Forest Department
FDA	Forest Development Agency
FES	Foundation for Ecological Security
FoC	Future of Conservation
FPC	Forest Protection Committee
FRA	Forest Rights Act
FRC	Forest Rights Committee
FSI	Forest Survey of India
FSO	Forest Settlement Officer
GO	Government Order
GoI	Government of India
GoM	Group of Ministers
GoO	Government of Odisha
GP	gram panchayat
GS	gram sabha
Ha	hectare
HPPCL	Himachal Pradesh Power Corporation Limited
IA	Interlocutory Application
IBA	Important Bird Area
CCA	Community Conserved Area
ICCAs	Indigenous Peoples' and Local Community Conserved Areas and Territories
ICFRE	Indian Council of Forestry Research and Education

IIED	International Institute for Environment and Development
IFA	Indian Forest Act
IFAD	International Fund for Agriculture Development
ITDA	Integrated Tribal Development Authority
IUCN	International Union for Conservation of Nature
JBIC	Japanese Bank for International Cooperation
JFM	Joint Forest Management
JFMC	Joint Forest Management Committee
JFPM	Joint Forest Planning and Management
JICA	Japan International Cooperation Agency
JJJA	Jungle Jameen Jan Andolan
KAS	Kisan Adivasi Sanghatan
LAA	Land Acquisition Act
LSM	Lok Sangharsh Morcha
MFP	minor forest produce
MoC	Ministry of Coal
MoEF	Ministry of Environment and Forests
MoPR	Ministry of Panchayati Raj
MoTA	Ministry of Tribal Affairs
MoU	Memorandum of Understanding
MPSEZL	Mundra Port and Special Economic Zone Limited
MZPSG	Mining Zone Peoples' Solidarity Group
NALA	Non-Agricultural Lands Assessment Act, 1963
NALCO	National Aluminium Company Limited
NAP	National Afforestation Programme
NBSAP	National Biodiversity Strategy and Action Plan
NBWL	National Board for Wildlife
NCA	National Commission on Agriculture
NEP	National Environment Policy
NGO	non-governmental organization
NGT	National Green Tribunal
NPV	net present value
NREGS	National Rural Employment Guarantee Scheme
NSSO	National Sample Survey Organisation
NTCA	National Tiger Conservation Authority
NTFP	non-timber forest product

ODA	Official Development Assistance
OMC	Odisha Mining Corporation
ORSAC	Odisha Space Applications Centre
OTFD	Other Traditional Forest Dweller
OTDP	Orissa Tribal Development Project
PA	protected area
PCCF	Principal Chief Conservator of Forests
PDS	public distribution system
PESA	Panchayats (Extension to the Scheduled Areas) Act, 1996
PF	Protected Forest
PIL	public interest litigation
PMO	Prime Minister's Office
PPSS	Posco Pratirodh Sangram Samiti
PoWPA	Programme of Work on Protected Areas
PPP	Private Protected Forests
REDD+	Reduced Emissions from Deforestation and Forest Degradation Plus
RF	Reserved Forest
RFA	recorded forest area
RTI	Right to Information
SAIL	Steel Authority of India
SC	Scheduled Caste
SEZ	Special Economic Zone
SPTA	Santhal Pargana Tenancy Act
ST	Scheduled Tribe
TERI	Tata Energy Research Institute
TPCG	Technical and Policy Core Group
VFCs	Village forest committees
VSS	Van Samrakshan Samitis
WII	Wildlife Institute of India
WLPA	Wild Life Protection Act, 1972
WLS	Wildlife Sanctuary
WPC	World Parks Congress
WWF	World Wildlife Fund

SHARACHCHANDRA LELE
AJIT MENON

Introduction

Forest Governance beyond Joint Forest Management, Godavarman, and Tigers

Over the past decade or so, India's forests have reached a crossroads. Even as the official monitoring agency claims that the forest cover in the country has been steady at about 21 per cent of the geographical area (Forest Survey of India 2011), tigers and other biota are disappearing from these forests (Tiger Task Force 2005) and forests are also being increasingly converted for developmental purposes (Das 2012), thereby displacing forest-dwelling communities (Sahu 2008). Even as the government claims that the official participatory forest management programme is a huge success (Singh *et al.* 2011), communities are beginning to claim forest management rights under a new law in the teeth of fierce opposition from foresters (Joint Committee 2010). Even as new concepts, laws, and procedures are being framed, the judiciary continues to micro-manage forest-related decision-making in a variety of ways (Rosencranz and Lele 2008). Even as the state pushes forward with an overall agenda of growth-through-industrialization, its committees are proposing a minimum support price programme for non-timber forest products (MoPR 2011). It appears that the forest sector is undergoing a churning as never before.

A useful entry point to understanding some of the key tensions here is to ask the question, 'what is a forest?', an innocuous-sounding question that actually cuts quite deep. Most foresters and ecologists think of a 'forest' as a well-defined physical entity, but they do not agree on what this is: foresters include single-species plantations of teak or pine or even exotics such as eucalyptus, while ecologists think of pristine treeland with multiple natural species. The country's official forest monitoring agency (Forest Survey of India) counts even areca nut and coffee plantations in its estimates of forest cover (Lele 2012). And legally, even pure grassland and snow-capped peaks have been classified as 'Reserve Forest' in many parts of the country (Sarin 2003a). The Supreme Court insisted on following the 'dictionary meaning' in the famous Godavarman case (Anonymous 2006; Lele 2007), but contradicted itself by allowing legally defined forest land (including grassland) to also be included. An attempt by the Ministry of Environment and Forests (MoEF) to clarify the issue went nowhere (Anonymous 2007).

The reason why the definition matters is because the word forest is an emotive one, and in the public mind, there is still a simplistic equation between forests and environmental benefits to society. But in the churning that has taken place over the last two decades, this simplistic equation is increasingly being questioned. It is being gradually acknowledged that a forest is not a simple ecological entity, but a complex socio-ecological construct in which different forest management systems (and non-forest systems) provide different mixes of benefits to different stakeholders located at varying distances from the forest, and that stakes are ecologically and socially determined.

From this constructivist understanding of forests, it is obvious that forest policy and management will always be a political process (though informed by ecology) because it involves balancing and prioritizing between different benefits and beneficiaries of different systems of managing forests, and also balancing between forests and non-forests, because even non-forest land-uses benefit someone. The discourse has indeed shifted from the earlier focus on technical forestry and the intermediate focus on participatory management to a richer engagement with the idea of forest governance. The questions have shifted from 'how to conserve forests' or 'how to afforest wasteland' to 'who are the stakeholders relevant to the forest question', 'who should have the primary say', 'what are the rights of non-local stakeholders', and 'what should be the process for converting to non-forest land'. Questions of forest rights, responsibilities, regulatory structures, transparency, and accountability have increasingly become central to the discourse, be it in the context of forests managed for local use, forests managed for wildlife conservation, or forests being converted to other uses. The debates have expanded from the technical quality of silvicultural practices to questions of democracy in all aspects of forest governance.

This book seeks to highlight and contribute to this shift in the discourse by capturing the multiple dimensions of the idea of democratic governance of forests in India. The chapters in this book review developments over the last decade or so, developments that have made the forestry debate so much more complex than it was in the 1980s or even the 1990s. The chapters are broadly grouped into four sets: the three dimensions of forest governance (governing local use, governing wildlife conservation, and governing conversion to non-forest purposes) and the wider socio-economic context—how it poses challenges to the idea of democratic governance. In the remaining part of this chapter, after a brief overview of the historical debate, we discuss what new dimensions have emerged in forest policy, management, law, and science in recent times, how these constitute important aspects of the current debate that require further investigation, and why the concept of democratic forest governance provides a useful umbrella for characterizing this exploration. We then provide an overview of the book and summarize the key points emerging from individual chapters.

Forestry Debates in India: Old Threads and New Dimensions

Forest policy and consequent forest-related debates in India till the 1990s are easily divided into two major phases. The first phase starts with the colonial period but extends for two decades into the post-colonial period. The colonial period, well documented by a number of scholars, was characterized by a British forest policy that was, notwithstanding its ebbs and flows, its regional variations, and its internal disagreements, essentially focused on revenue maximization and industrial production (Guha 1989; Guha and Gadgil 1989) while also being a part of colonial state-making (Sivaramakrishnan 1999) and the project of 'civilizing nature' (Philip 2004). The structures it created were those of a technocracy and the main mechanisms for management were establishing areas of exclusive state control and controlling the extraction and trade of commercially valuable forest products. The policy met with active resistance in pockets—sometimes violent, as in Uttarakhand (Guha 1989) and Jharkhand, and in some places peaceful, as in the erstwhile North Canara district (Nadkarni *et al.* 1989). The rationale behind wresting control was also contested intellectually through petitions, court cases, and public statements arguing that forests were essential for local livelihoods and that local communities had customary rights over these forests. This protest and resistance wrested localized concessions, such as the Van Panchayats in Kumaon (Agrawal 1999), the *mundari-khuntkattidar* settlement in Jharkhand (Upadhya 2005), and the *soppi-nabetta* privileges granted in North Canara (Saberwal and Lele 2004) and elsewhere in the Western Ghats of Karnataka (Srinidhi and Lele 2001). Nevertheless, the overall picture that emerges from this period is of a fairly successful policy of taking control of large fractions of the forested landscape, of excluding or limiting local use, and of reorienting forest ecologies towards revenue-generating models (Guha and Gadgil 1989).

This policy continued and even expanded in the first two decades of postcolonial India. More lands were brought under the category of Reserve Forests, and silviculture in state forests was even more emphatically aimed at meeting industrial needs of a modernizing India. There was no rethinking of colonial forest policy and structures; indeed, there was no new forest policy at all, just a small statement tucked into the National Agricultural Policy of 1952. Remarkably, this period was not even marked by major debates or protests, as the newly independent state focused attention elsewhere.

It was only with the emergence of Chipko in the 1970s in the Garhwal Himalayas that the debate was sparked afresh, and entered its second phase. This phase was marked not only by protests on the ground—protests that extended beyond Garhwal to Jharkhand (Anonymous 1979; Das 1991)—but it also eventually led to increasing attention to forest and environmental issues at the national level, both in academia and in policy circles. The forest question received renewed attention and the forest debate was re-opened.

The debate in this second phase represented both continuity and change with respect to the first phase in ways that are important to understand. One dimension of the debate that seemed common between the colonial and post-1970 periods, as Guha (2012) has noted, was the question of the role of the state vis-à-vis local communities in the use and management of forests. But while the protest and resistance in the first phase was followed by the granting of specific rights in pockets, the state response in the second phase was different. The response to Chipko was not the granting of rights to the Garhwalis but the banning of tree felling above a certain altitude—a blanket ban that actually caused further hardship to villagers. The protests in Jharkhand, which were triggered by state encroachments on common lands as well as teak-oriented silviculture that was antithetical to the local dependence on sal (*Shorea robusta*), met with bullets (Corbridge and Jewitt 1997) and in turn set off a chain of violence. Elsewhere, the forest bureaucracy attempted to solve the problem of local needs by launching a series of so-called Social Forestry projects in the 1980s that took a technical, rather than institutional, approach. These projects attempted to increase tree production on non-forest land (typically grazing land with the revenue department) to address what they saw as a fuelwood crisis, without addressing issues of rights or control.

At the same time, a quieter approach was being tried by foresters in the Bengal countryside. Foresters, fed up with their perennial conflict with villagers over grazing and firewood collection, began to strike deals with village groups to protect plantations of sal trees in return for a share in the pole harvest (Chatterji 1996). Eventually, through a convergence of various factors, the concept of 'participatory management' entered the rhetoric of forest policy (Government of India 1988). This eventually led to the now well-known June 1990 circular by the central government to states asking them to involve local communities in the regeneration of degraded forests, a circular that triggered what is today called the joint forest management (JFM) programme. Today, more than two decades after the first official JFM programmes were launched, the central government claims that a third of the country's forest estate is under participatory management. With massive funds for implementation, numerous studies, and many conferences and workshops that focus on analysing its performance, JFM appears to have become the dominant paradigm for resolving the state–people conflict over the use and management of forests as forests (Sundar, Jeffery, and Thin 2001).

A second thread or dimension to the debate that had emerged in the colonial period but underwent significant changes and broadening in the post-1970s was the question of whether and how to control the conversion of forests to other land-uses. During the colonial period, it took the form of agriculture versus forestry, about whether and how much forest land should be given out for cultivation. Expanding cultivation was one of the goals of the overall colonial economic policy, since it increased land revenue. While foresters objected, the labelling of forests as 'wasteland' reflected the colonial preference for agriculture as the best or most beneficial land-use (Whitehead 2010). The debates that occurred between the British revenue and forest departments were strong but nevertheless controlled and contained. The post-independence period, however, saw much greater tension building up on this question: provincial governments, rather than imposing political inconvenient land reforms, found it easier to hand out common lands (pastures and forests) to the landless. The national focus on 'Grow More Food', driven by an expanding population, helped this process.

The 1970s, however, were seeing the emergence of broader 'environmental' concerns in a variety of ways. One theme was supra-local environmental benefits of forests, and the idea that Himalayan deforestation was causing widespread flooding in the Gangetic plains (Myers 1986). The Chipko movement was interpreted as communities fighting to protect nature rather than communities fighting for their right to use forests (Agarwal *et al.* 1982). Consequently, state responses to Chipko aimed at halting tree felling and then regulating forest conversion. First, the government put a moratorium on the felling of trees above 1000m altitude and 30-degree slopes in the eight hill districts of present-day Uttarakhand. Then, forestry was transferred from the state list to the joint subject list, allowing the centre a greater say in forest management. The third and most significant response was the passing of the Forest Conservation Act 1980 (FCA) that required states to obtain approval from the central government for the conversion of any forest land to non-forest activities such as mining, dams, or agriculture, and also for their forest working plans. The Act, at least initially, put a brake on populist handing out of forest land in the states, which had reached an astounding 150,000 ha per year in the 1970s (Damodaran and Engel

2003, p. 6). The ambit of the FCA was further expanded by the Supreme Court via the Godavarman case, an expansion that was welcomed by conservationists (Dutta and Yadav 2005).

A third dimension has emerged since the 1970s that was almost entirely missing in the colonial period, namely, wildlife and biodiversity conservation goals in forest policy. With the rapid decline of wildlife worldwide in the twentieth century, including in India, wildlife conservation concerns emerged strongly in the early 1970s, albeit largely in a top-down manner and with erstwhile sport hunters turning conservationists (Rangarajan 1996). These concerns manifested themselves first as Project Tiger, followed by the enactment of the Wild Life Preservation Act 1972 (WLPA). The WLPA not only banned hunting of all threatened species, but was also used to rapidly create a network of 'inviolate areas' that increased from 0.5 per cent of the country's land area in 1969 to 5 per cent in 2001. The central MoEF, created in 1980, began to provide substantial financial support for these activities. Thus, the first two decades after Chipko not only saw some new approaches by the state to dealing with the age-old question of how to reconcile production forestry with local needs, but also some attempts to slow down the rate of forest conversion and the emergence of a new goal for forestry, namely wildlife and biodiversity conservation. The National Forest Policy document of 1988 appeared to capture this expanded set of concerns and revised priorities quite well. It categorically rejected revenue-generation as the prime objective of forest management, instead giving priority to environmental stability and meeting local needs. It also articulated the need for people's involvement in forest regeneration. And the central order on participatory forest management followed immediately in June 1990. At this juncture, it appeared that forest management in the country was about to enter a new era of inclusiveness, conservation-orientation, and better regulation of developmental pressures.

Path-breaking Initiatives or Quicksand?

The promise of the 1990s has, however, not panned out. On the contrary, forestry has become an enormously contentious sector. First, after an initial honeymoon, the shortcomings of the JFM programme soon became apparent to many academics and civil society groups

(Lele 2001; Sarin 2003b; Sundar 2001), even though others, supported financially by donors and the central government, continued to work on tinkering within its framework. The programme has become particularly untenable in light of the passing of the Scheduled Tribes and Other Traditional Forest Dwellers (Recognition of Forest Rights) Act, 2006 (hereinafter Forest Rights Act or FRA). But the government and the forest departments continue to promote the JFM programme while making vague promises of 'revamped JFM committees' (as in the Green India Mission document: MOEF 2010); promises that have not been fulfilled subsequently. The question of local forest management for meeting local needs seems to have reached an impasse.

The WLPA was always a blunt top-down instrument. It temporarily arrested the dramatic decline in tiger, elephant, and other mega-faunal populations that had occurred in earlier decades. But at the same time its non-consultative and non-participatory nature created a new form of state–people conflict around forests. Creation of inviolate spaces in the Indian context meant the inevitable displacement of (or threat of displacement) millions of forest-dwelling and forest-dependent peoples, and sparked a long-drawn battle that continues till today (Lasgorceix and Kothari 2009). Moreover, the non-transparent and unaccountable functioning of the forest bureaucracy has meant that even the blinkered pursuit of wildlife conservation at all costs has failed to meet its goals, as witnessed by the disappearance of the tiger from Sariska National Park (Johari 2007). The setting up of the Tiger Task Force in 2004 and its multi-dimensional recommendations in 2005 created a ray of hope that has been rapidly extinguished by the bureaucratic manner in which the National Tiger Conservation Authority has functioned since its creation. Yet conservationists by and large continue to be locked in visceral battles with tribal[1] rights activists in a false debate over 'tiger versus tribal' (see articles in *Seminar*, issue 552, 2005). Thus, between the social costs of, and hence agitation against, exclusionary conservation and the failures of bureaucratic conservation, it is not clear what the future holds for conservation in India.

Like the WLPA, the FCA was a blunt instrument, introducing paperwork without clarifying criteria or setting up transparent and effective monitoring and implementation procedures. Moreover, it generally reinforced central authority over the states, in turn creating the impression that the states themselves had nothing to gain by forest conservation.

The intervention of an activist judiciary in the forest sector has created a bigger mess. The initial orders passed by the Supreme Court in 1996 in the now famous Godavarman case drew much approbation from many quarters, as it appeared the Court had spotted flagrant violations and taken emergency action, while also bringing about more consistency in the application of the FCA. Over the years, however, in what has now become the longest-running case with possibly the highest number of interlocutory applications and interim orders in judicial history, the Godavarman case has become a vehicle for unbridled and unjustifiable judicial overreach (Rosencranz *et al.* 2007; Rosencranz and Lele 2008). As a consequence, the process of regulating developmental pressures on forests has become extraordinarily confusing without necessarily leading to a reduction in the conversion rate; in fact, the last decade has seen a huge rise in forests being handed over for dams, mining, roads, and other projects (Das 2012).

From Forest Management to Forest Governance

A major shift in the forestry discourse took place with the passing of the FRA by the Indian parliament in 2006. Although the formulation of this Act began simply as a response to the problem of historical 'encroachments' in forest land (and, therefore, seemed more related to land rights rather than forest rights), it eventually incorporated elements that seek to reform (if not revamp) rights and procedures across all the three dimensions discussed above: forest-use rights, conservation-linked displacement, and forest conversion.[2] In each of these dimensions, the Act sought to not just give voice or rights to local communities but also to push the process towards clearer articulation of objectives and criteria.[3] Consciously or otherwise, this Act has forced a shift in the forestry debate from forest management to forest governance.

How does forest governance differ from forest management? There are several ways of describing this difference. One is as what Schlager and Ostrom (1992) call a shift from the 'operational choice' level to 'constitutional choice' level. For instance, earlier critiques of eucalyptus-based social forestry in Karnataka (Shiva *et al.* 1981) were quickly reduced to silvicultural questions of whether eucalyptus should be used in plantations or not—a question that, though important, prevented the larger debate on who should decide on species choice in forestry programmes or whether common lands should be covered with tree plantations at all.

Another way of characterizing the difference is at the normative level, where the focus shifts from efficiency in achieving single pre-defined goals to fairness and due process in balancing between multiple conflicting goals or interests. As mentioned above, the National Forest Policy 1988 included a long wish list of objectives (priorities) that included environmental balance, biodiversity conservation, soil and water conservation, provision of local needs and supra-local national needs. These goals, however, are potentially conflicting (Lele 2011; Lele and Srinidhi 1998). For example, the creation of inviolate tiger reserves will almost certainly prevent local communities' harvesting of non-timber forest produce and grazing activities, thereby also affecting their agriculture. The interests of timber producers do not quite coincide with the interests of graziers. Similarly, the conversion of forest land for non-forest industrial purposes could result in both the alienation of local communities and the undermining of wider ecosystem benefits. A major, perhaps *the* major, task in forest governance is to identify and implement institutional arrangements and processes that will transparently and fairly balance between these conflicting objectives.

At a third level, from an single agency-centric approach, the focus shifts to the roles to be played by the local citizens, the judiciary, civil society groups, and also multiple layers of government (Sampford 2002). Finally, in the context of environmental governance, scales of ecological processes and their linkages become important points for discussion (Lele 1996; Murphree 2000; Sikor *et al.* 2010). Thus, terms such as rights, responsibilities, regulatory mechanisms, subsidiarity, transparency, and accountability become central to the discourse on forest governance.

None of the above can be separated from the social, political, or economic context in which forests are embedded. If forests are an arena of competing interests, then the governance of forests involves the exercise of power by different agencies. This power may be defined, directed, and circumscribed by rules. These rules may also be interpreted and implemented in particular ways depending upon the larger political economy and cultural norms of the agents. This power is also built upon and contested by knowledge claims. Furthermore, the contestation ebbs and flows depending upon the changes in the level and nature of interest, changes that may be driven by larger economic processes.

As the discourse shifts from asking how the forest department should manage forests to how society should govern forests and how different actors within it should or could play different roles in this governance system, the social context of these actors, their economic interests, their abilities to mobilize and legitimize knowledge become as important as the formal rules that govern their interactions.

While the term 'governance' has a largely descriptive connotation, the term 'democratic governance' has a clear normative thrust. The contributors to this book believe that there is a strong need to democratize the process of forest (and overall environmental) governance in India. It is important to note here that we are not using the term governance in the manner in which it is used by advocates of 'good governance', where the focus is limited to the goal of accountability and efficiency of public administration, or 'sound development management' or even the broader view that includes respect for human rights and the existence of democratically elected governments (Leftwich 1994: 371–2). Such conceptualizations of good governance are limited because they either see good governance as simply effective administration, or take for granted that in democracies, a plurality of interests are adequately addressed in practice.

We prefer to phrase the societal goal as *democratic* governance, which involves deepening the democratic process in multiple ways: spatial decentralization, devolution of actual power to lower tiers, and (perhaps most important) the functioning of all tiers and all arms (political, executive, and judicial) in ways that are democratic, transparent, and accountable. It means the tuning of institutions of governance to the socio-ecological context in ways that enable the participation of the weakest and the feasibility of addressing environmental sustainability and justice goals. This would involve a far deeper and wider set of changes, changes that will come out only through contestation and struggle. This book is about the promise of, progress towards, and challenges in achieving such democratic governance in the forest context in India.

Overview of the Book

Several books that have emerged within the last decade have dealt with some of the issues or dimensions identified above. Some have focused on assessing the JFM programme within a fairly narrow framework (Bhattacharya *et al.* 2010; Ravindranath and Sudha 2004), while others (Sundar *et al.* 2001) have sought to unpack the larger political economy of this programme. Some have looked at the politics of wildlife and biodiversity conservation from the perspectives of the politics of knowledge (Shahabuddin and Rangarajan 2007) or of state–society interaction (Chhatre and Saberwal 2006). We felt, however, that it was necessary to address all the major dimensions of forest governance. With this in mind, we have brought together in this book a set of original contributions on the problems and potential of democratic forest governance in the country today. The key issues explored relate to the shift from so-called participatory management to rights-based bottom-up governance, the challenge of integrating conservation concerns into bottom-up governance, the even more complex question of balancing between forests and non-forests and the role to be played by the judiciary in that process, and the larger questions of agrarian structures, social inequities, and shifting dependencies that are ever-present but often unaddressed in the forestry discourse. Our aim is to reach out to a broad cross-section of readers who might be interested in questions related to forest governance, management, and policy in India and south Asia at large, including activists, bureaucrats, academics, NGO persons, and the wider public.

The contributions were commissioned from a group of activist scholars and scholarly activists who at some level share a common understanding of the forest question: that forests are social constructs and how they get defined depends upon which of multiple interests prevail, that a conflict in forest management is often a conflict over both the goals and the structures needed to achieve those goals, that political economy perspectives are as important as institutional ones, and that forest governance has to be placed in a larger context for enabling a richer analysis of possibilities and constraints.

At the same time, the contributors differ in their understanding of the state's role and reach. While much of the activist literature in the forest debate is extremely critical of the state and the forest department, much of the recent academic literature on the forest question in India, both historical and more contemporary, has focused on the complexity of state-making, emphasizing that state territorialization is neither writ large nor completely controlled by the bureaucracy. This latter literature has focused more on the limits of state power, the segmented nature of state interests and the ways in which 'civil' society has resisted and negotiated spaces for itself (Cederlof 2008; Sivaramakrishnan 1999). The perspectives of the contributors to this book vary within this range of positions vis-à-vis the state. While they recognize that processes of state-making are complex and context-specific, they do not want to lose sight of the fact that 'bad' laws and policy can have severe adverse impacts on local communities (even if they resist or seek to resist them) and that 'good' laws and policy can open up opportunities for these same communities while, of course, not guaranteeing good outcomes for all. Thus, there are several levels of analyses (both across and within chapters)—one that is aimed more at a critical appraisal of specific policies and laws, one focused more on the conceptual and process-related dilemmas or stumbling blocks that arise from these policies and laws in specific sectors or regional contexts, and one looking at the politics of a pro-capital state but also, in some cases, the politics of the local.

The chapters are organized into four major sections. The first section deals with governing local use of forests. It begins with Lele's meta-analysis of the JFM programme, based on a rigorous normative and analytical framework and drawing upon a number of published studies. Lele points out that different assessments of JFM reach different conclusions largely because of the different normative frameworks used to make such assessments. He argues that JFM should be assessed by its 'jointness' and hence the space it offers to forest-dependent communities to determine the nature of forest governance. From his meta-analysis, he concludes that JFM is not joint at all and, in fact, is merely a means through which the forest bureaucracy is able to fulfil its supra-local priorities. The limits to JFM, Lele argues, cannot be sought in explanations of poor implementation or even design flaws from within the institutional perspective from which JFM is supposed to have emerged. He suggests that the success of JFM is itself its biggest failure, as it addresses a very different goal from that of democratic or even administrative decentralization.

The second chapter by Sagari R. Ramdas examines the limitations of current paradigms of local management from the point of view of the livestock sector. Ramdas provides a detailed historical analysis of policies regarding common lands in general and forest legislation in particular in terms of their adverse impact on livestock-rearers in Andhra

Pradesh. JFM, rather than offering fresh possibilities to livestock-rearers, is found to be an extension of centralized forms of forest management that not only ignore but stymie customary access to the commons for livestock and disrupt customary relationships between pastoral and more sedentary agricultural communities. Ramdas argues that the FRA offers new possibilities to grazing communities, highlighting how its potential is as yet under-utilized. She concludes by warning that competing policies such as REDD+ may result in the FRA itself being marginalized.

The third chapter in this section is by Madhu Sarin, in which she focuses on the genesis and potential of the FRA. The chapter begins with a detailed critique of conventional forest law and policy and the manner in which the state has created 'national' forests under state control with unifunctional ecological priorities. Sarin, like Ramdas, argues that the creation of such forests has undermined more inclusive regional laws that recognized local claims and complexities to land, including forests, but also points out how the overall forest reservation process was greatly faulty in not recognizing the historical presence, occupation, and use of many lands by tribal communities in central India. The labelling of these communities as 'encroachers' then led to persistent conflict, which reached flashpoint due to an overzealous interpretation of Supreme Court orders. Sarin explains how the FRA is not only the means through which local communities can claim rights but also a means to democratize forest governance. She ends on an optimistic note, describing recent events in which several hundreds of villages have received community forest rights. But, like Ramdas in the previous chapter, Sarin reminds us, that the FRA cannot be viewed in isolation. New threats to democratic forest governance exist in terms of the overly active intervention of the Courts to promote exclusive conservation and the government's commitments to international agreements on climate change.

The second section of the volume deals with the governance of biodiversity-rich areas, areas that are traditionally considered to be 'conservation-priority' areas. Nitin D. Rai's chapter presents a strong critique of conventional conservation thinking at two levels. At one level, using examples from recent ecological research in India and elsewhere, he shows how the assumptions that underpin conventional conservation policies, namely, that the presence of people is harmful to conservation goals, that 'inviolate areas' are essential to conservation, or that use of forests by local communities is the biggest driver of degradation, are all based upon an outdated science of equilibrial ecology when, in fact instability, complexity, and uncertainty are the norm. Hence he argues that adaptive, locally specific management that draws upon local knowledge and involves local communities has to be the approach. At another level, Rai argues that the fact that such knowledge and such management practices are not being adopted reflects not so much the lack of evolution in scientific thinking as the selective use of science by forest departments whose agenda is first and foremost to control the landscape. The consequences in the form of displacement of local (mostly tribal) communities and conflict have been enormous.

In this backdrop, Neema Pathak Broome, Shiba Desor, Ashish Kothari, and Arshiya Bose suggest that recent changes in legislation, both the amendment to the Wildlife Act and the passing of the FRA include potentially significant improvements to conservation policy, in the form of the processes and requirements for identification of Critical Tiger/Wildlife Habitats (CTH/CWH), in the stringent processes laid down before communities can be displaced, and in the possibility (under the FRA) of communities getting forest rights before displacement is considered. The authors assess the progress made in the actual implementation of these provisions, which sadly, is quite inadequate. Notwithstanding these implementational challenges, basic limitations in these legal provisions and also larger economic forces ranged against conservation, the authors remain hopeful that conservation can become more people-friendly and democratic in India.

The third section deals with the question of conversion or diversion of forests to non-forest (that is, agricultural or developmental) purposes. Kanchi Kohli and Manju Menon provide a wide-ranging critique of the manner in which the tension between forests and development is currently being regulated, especially by the bureaucracy. The authors argue that the Forest Conservation Act (FCA) fails to articulate a collective public intent about what conservation should mean and hence in practice is simply a means through which the state 'regulates' non-forest use of forest lands, and in fact, justifies particular conversions in the name of development. Furthermore, they illustrate how such conversion has been marketized and monetized through the adoption of the principle of net present value.

Shomona Khanna presents an insider's view and detailed critique of the judicial process behind the Godavarman case. Khanna suggests that the Godavarman judgment has further centralized forest management by empowering a special bench of the Supreme Court to decide upon matters of forest conservation and conversion and the forest bureaucracy to implement such decisions. Though the judiciary was initially seen as a possible last resort to protect people's rights in the face of an insensitive bureaucracy, Khanna highlights how the judiciary has by and large supported the executive's centralized and exclusionary vision of forest governance. Added to this, she argues that the Godavarman judgement has redefined the boundaries of forests in such a way as to deny people's historical claims to forests.

Cutting across these three dimensions of forest governance (local use, conservation, and conversion) are a wider set of challenges that have not received much attention in past discussions of the forest question. These wider issues are taken up in the last section of the book. On the one hand, the question of land rights of the forest-users (for habitation and cultivation) is more complicated than even the FRA would have us believe. Shifting cultivation is the extreme example of this complication. The chapters by Kundan Kumar and Dhrupad Choudhury discuss the enormous complexities surrounding the phenomenon of shifting cultivation. Kumar provides a detailed account of how forest law in India has marginalized shifting cultivation and rendered it 'illegal'. He illustrates, through a case study of Orissa, how land used by shifting cultivators has been captured by state line departments for tree plantations and even coffee plantations. Though he sees the FRA as a possible opportunity for shifting cultivators to claim their rights, he argues that since the Act does not specifically mention shifting cultivation rights, it is likely that claims to this land will be denied by the forest bureaucracy.

Choudhury's commentary suggests that the lack of mention of shifting cultivation rights within the FRA explains why the FRA has not been met with much enthusiasm in the Northeast. He details how 'traditional' village councils in the Northeast regulated shifting cultivation, ensuring that questions of intra-village equity were addressed, and how state and district level legislation have progressively undermined such traditional legislation. Law, Choudhury argues, has also privileged sedentary cultivation over shifting cultivation. The FRA, he suggests, has not reversed this bias.

On the other hand, Prakash Kashwan and Viren Lobo's critique of the FRA points to the 'internal' dimensions of the governance problem, namely, the acute challenge posed by intra- and inter-village inequities of assets and power. The authors go one step further than others have done in this book by suggesting that the FRA and those advocating it have tended to essentialize the community and community rights: the FRA, they argue, has both winners and losers. Yet despite their concerns, the authors highlight how social movements/NGOs can address these issues of intra-village inequity in the process of campaigning for the FRA.

Finally, Ajit Menon, Viren Lobo, and Sharachchandra Lele suggest the need to relook at the forest dependence question given the changing, and perhaps declining, nature of forest dependence amongst local communities. They highlight a number of points: first, that dependence on the forested commons should be contextualized within the wider dependence on land-based common property resources; second, that while dependence on the forested commons is significant, forest-based livelihoods might not be adequate in and of themselves; and third, that there are increasing 'external' pressures on the commons that threaten livelihoods. They conclude that it is these external pressures that are chiefly responsible for declining dependence and, therefore, local governance of the commons should not be 'justified' on the basis of economic dependence alone (nor denied on the basis of a decline in such dependence) but rather should stand on normative principles of democratizing the governance of common property.

Together, these 11 chapters provide a multi-dimensional exploration of the forest policy debate in India. Of course, given the complexity of the forest question and the diversity of socio-ecological conditions in the country, several nuances and variations remain unexplored. For instance, the currently popular idea of 'payments for ecosystem services' has not found a place in this volume (but see Lele 2013). Similarly, the forest rights questions are distinctly different in the Western Ghats, where individualized forest rights on the one hand (see Srinidhi and Lele 2001), and large swathes of coffee and tea plantations on the other, pose enormous challenges to the idea of bottom-up democratic governance of forests. But perhaps the greatest challenge facing researchers who engage with the question of democratic forest governance in India is the need to deepen and widen the concept itself, to go beyond the strong critique of forest governance as it has unfolded so far into realms of what the shape of forest governance could be, what possibilities of forest-based livelihoods might be worked out, and what kinds of action- and advocacy-relevant knowledge will be required to support this process. We return to some of these issues in a brief epilogue at the end of the book.

Notes

1. The terms 'tribal', 'adivasi', and 'indigenous' are used interchangeably throughout this volume.

2. The FRA does not explicitly talk about forest conversion, but the intent was visible enough that the Ministry of Environment and Forests passed an order in July 2009 requiring community consent for forest conversion in areas where rights under FRA had been granted (MoEF 2009).

3. For instance, whereas the WLPA gives the state the power to declare any area as a wildlife sanctuary or national park and then proceeds towards evicting people residing within it, the FRA requires the state to define criteria for Critical Wildlife Habitat, demonstrate that a particular site fits those criteria, and further demonstrate that co-existence of local communities with wildlife is not possible, before moving towards a more people-friendly resettlement process.

References

Agarwal, A., R. Chopra, and K. Sharma. 1982. *The State of India's Environment: The First Citizen's Report*. Delhi: Centre for Science and Environment.

Agrawal, R. 1999. 'Van Panchayats in Uttarakhand', *Economic and Political Weekly*, 34 (39): 2779–81.

Anonymous. 1979. 'Exploitation, Protest and Repression', *Economic and Political Weekly*, 14 (22): 940–3.

———. 2006. 'How Will you Define a Forest? *Daily News & Analysis*, 13 February; http://www.dnaindia.com/india/report-how-will-you-define-a-forest-1012733.

———. 2007. 'At Stake', *Down To Earth*, June 1–15; http://www.downtoearth.org.in/print/6057.

Bhattacharya, P., L. Pradhan, and G. Yadav. 2010. 'Joint Forest Management in India: Experiences of Two Decades', *Resources, Conservation and Recycling*, 54 (8): 469–80.

Cederlof, G. 2008. *Landscapes and the Law: Environmental Politics, Regional Histories, and Contests over Nature*. Ranikhet: Permanent Black.

Ravindranath, N.H. and P. Sudha. 2004. *Joint Forest Management in India: Spread, Performance and Impact*. Hyderabad: Universities Press.

Rosencranz, A., E. Boenig, and B. Dutta. 2007. 'The *Godavarman* Case: The Indian Supreme Court's Breach of Constitutional Boundaries in Managing India's Forests', *ELR News & Analysis*, 37 (1): 10032–42.

Rosencranz, A. and S. Lele. 2008. 'Supreme Court and India's Forests', *Economic and Political Weekly*, 43 (5): 11–14.

Saberwal, V. and S. Lele. 2004. 'Locating Local Elites in Negotiating Access to Forests: Havik Brahmins and the Colonial State 1860–1920', *Studies in History*, 20 (2): 273–303.

Sahu, G. 2008. 'Mining in the Niyamgiri Hills and Tribal Rights', *Economic and Political Weekly*, 43 (15):19–21.

Sampford, Charles. 2002. 'Environmental Governance for Biodiversity', *Environmental Science & Policy*, 5 (1): 79–90.

Sarin, M. 2003a. 'Conserving Forests: Trees Hide Woods', in Anonymous (ed.), *The Hindu Survey of Environment 2003*, pp. 111–15. Chennai: *The Hindu*.

———. 2003b. 'Forest Conservation is Too Complex an Issue to be Resolved by Executive Fiat'. *Down To Earth*, 15 July: 36–40; http://www.downtoearth.org.in/node/13145.

Schlager, E. and E. Ostrom. 1992. 'Property-rights Regimes and Natural Resources: A Conceptual Analysis', *Land Economics*, 68 (3): 249–62.

Shahabuddin, G. and M. Rangarajan. 2007. *Making Conservation Work: Securing Biodiversity in This New Century*. New Delhi: Permanent Black.

Shiva, V., H.C. Sharatchandra, and J. Bandyopadhyay. 1981. *Social, Economic, and Ecological Impact of Social Forestry in Kolar*. Bangalore: Indian Institute of Management.

Sikor, T., J. Stahl, T. Enters, J.C. Ribot, N. Singh, W.D. Sunderlin, and L. Wollenberg. 2010. 'REDD-plus, Forest People's Rights and Nested Climate Governance', *Global Environmental Change*, 20 (3): 423–5.

Singh, V.R.R., Deepak Mishra, and V.K. Dhawan. 2011. *Status of Joint Forest Management in India (as on June 2011)*. Dehra Dun: Forest Research Institute.

Sivaramakrishnan, K. 1999. *Modern Forests: Statemaking and Environmental Change in Colonial Eastern India*. New Delhi: Oxford University Press.

Srinidhi, A.S. and S. Lele. 2001. 'Forest Tenure Regimes in the Karnataka Western Ghats: A Compendium'. Bangalore: Institute for Social and Economic Change.

Sundar, N. 2001. 'Is Devolution Democratisation?', Delhi: Institute of Economic Growth.

Sundar, N., R. Jeffery, and N. Thin. 2001. *Branching Out: Joint Forest Management in India*. New Delhi: Oxford University Press.

Tiger Task Force. 2005. 'Joining the Dots: Report of the Tiger Task Force', New Delhi: Ministry of Environment and Forests, Government of India.

Upadhya, C. 2005. 'Community Rights in Land in Jharkhand', *Economic and Political Weekly*, 40 (41): 4435–8.

Whitehead, J. 2010. 'John Locke and the Governance of India's Landscape: The Category of Wasteland in Colonial Revenue and Forest Legislation', *Economic and Political Weekly*, 45 (50): 83–93.

GOVERNING FORESTS FOR LOCAL USE

SHARACHCHANDRA LELE

What is Wrong with Joint Forest Management?

Responses to the question of what is wrong with Joint Forest Management (JFM) come in very different, sometimes opposing, forms. Ask the question to foresters, and their response is 'nothing much'. If on a public platform, they praise the concept of participatory management and the JFM programme that is its manifestation, and list all of its achievements: 1.1 million committees protecting 22.9 million hectares, and regenerating large swathes of the landscape (for example, chapters in Singh *et al.* 2011). In private, they might admit that many JFM committees are not functional, but they put all that down to implementational failures that are part and parcel of all public programmes. Ask the question to one set of researchers and non-government organizations (NGOs), and their response is likely to be 'there are a few second-generation issues that need to be sorted out' (Saigal 2000), and they launch into how the JFM orders need to be fine-tuned, funding agencies need to be more alert, the central government needs to monitor more closely, NGOs need to be more involved, local capacities need to be built further, and so on (Ravindranath and Sudha 2004). But a much smaller group of researchers, along with a much larger group of activists, answer that 'everything is wrong', and they point to almost the same features as the others, but in a critical manner: that JFM is based on orders and not the law, that it is heavily funded thereby creating perverse incentives, that it is not joint management at all, or that joint management is not necessary, what is necessary is community forest rights as in the Forest Rights Act (FRA) 2006.

Why is it that understandings and assessments of the same programme, implemented on a large scale for two decades now, should differ so radically from each other? Is it because people are referring to different regions? Or different periods? Is it because some studies are more rigorous than others? That some studies have better data or larger samples than others? Even if statements by activists were set aside for being based on anecdotal data or ideological positions, and those of foresters were discounted for being self-serving, there remains a vast academic literature on JFM that refuses to converge. While exchanges in the policy arena have been sharp and increasingly bitter with the passing of the FRA 2006 (see, for example, Campaign for Survival and Dignity (CSD) 2010), there has been a singular lack of academic engagement with the question of how to look at JFM, on why similar findings lead to very different recommendations.

This chapter is an attempt to provide some of this clarity by deriving more logical benchmarks against which to measure JFM, particularly the idea of 'jointness' that is central to JFM. I show how differences in normative stance and theoretical assumptions might underpin different assessments, thereby leading to divergent conclusions, even with similar data. I do so by presenting findings from a range of major studies and evaluations of JFM and then juxtaposing and integrating their interpretations.

The chapter is organized as follows. I begin with a brief review of the prehistory of JFM, the rationale for its emergence, and the nature of its spread. I then pose the question that must be answered before any assessment of JFM is carried out, namely, what are the goals of JFM and what are the specific notions of 'meeting local needs' or 'participation' embedded in the JFM enterprise? I also present an overview of the various assessments that I use as a basis for my analysis and the conclusions I reach. Finally, I discuss the reasons for the trends observed and divergent interpretations of similar trends. While a number of implementation and design issues emerge clearly, the continued acceptance of JFM as a desirable and largely successful concept appears to be driven by hidden normative positions, that is, the *real* goals of JFM. This pushes us towards a political economy explanation based on the interests that are tied to the JFM concept at various levels. Broadening the normative position is, therefore, central to any meaningful discussion on local forest governance.

Prehistory, Emergence, and Spread of JFM

The history of pre- and post-colonial forest policy in India has been documented exhaustively by scholars (Gadgil *et al.* 1983; Guha and Gadgil 1989; Pathak 1997; Rangarajan 1996). In essence, colonial forest policy started with the assumption of the state's eminent domain over the forested landscape, initiated a process of large-scale reservation of hitherto locally used and managed forests, and focused on maximizing timber production and revenue from these and other forests. This led to significant conflict between tribal groups and peasants on the one hand and the colonial state on the other over the rights of local communities to meet their domestic and livelihood needs from forests. After independence, unfortunately, the policy hardly changed: instead of the colonial state, the focus (officially enunciated in the National Forest Policy of 1952) was on using forests for the needs of national and industrial development. As a result, the conflict re-emerged in the form of the (now world-famous) Chipko movement in Uttarakhand in the 1970s and other protests elsewhere. The passing of the Forest Conservation Act in 1980 changed nothing, as it sought to limit forest conversion but did not address the question of local needs, rights, or management.

It was only in 1988 that the government actually revised the goals of forest policy in the country (Government of India [GoI] 1988). In a landmark shift from the colonial approach, it rejected the idea of 'forests as a source of revenue' and prioritized 'preservation/restoration of ecological balance', 'meeting the requirements of fuel-wood, fodder, minor forest produce and small timber of the rural and tribal populations', 'meeting essential national needs', and 'creating a massive people's movement for achieving these objectives' as the core goals of forest policy.[1] The policy went on to state that 'a primary task of all agencies responsible for forest management... should be to associate the tribal people closely in the protection, regeneration and development of forests'.

Following this policy shift, the GoI in June 1990 issued a circular to all the states on 'involvement of village communities... in the regeneration of degraded forest lands' (GoI 1990). The circular stated that 'the requirements of firewood, fodder and small timber... of the tribals and other villagers... are to be *treated as first charge* on forest produce' (emphasis mine). Although it did not use the term 'joint forest management', it was clearly drawing upon the JFM experiment in Arabari and the term JFM quickly became popular. The core idea was the formation of village-level user groups who would enter into a contractual arrangement with the Forest Department (FD) to protect and regenerate a degraded forest patch in return for the usufruct and a share in the final (commercial) harvest.

Over the next decade, different states issued orders to implement JFM programmes. In 1998, the central Ministry of Environment and Forests created a 'JFM cell' for guiding the implementation of JFM across the country, and this cell issued a fresh set of guidelines in 2000 (GoI 2000a) and modifications in 2002 (GoI 2002). JFM has now been adopted by all states and by the union territory of Andaman and Nicobar Islands as the official approach to forest management. As of 2011, there were about 1,18,000 JFM committees across the country protecting about 22.93 million hectares of forests—approximately a third of the land with the FDs in the country (Bahuguna 2011).

One major factor responsible for this rapid spread was the enormous financial support that state FDs received for JFM implementation from international aid agencies. The British Department for International Development (DFID) and the World Bank were the earlier sponsors, supporting projects in Karnataka and Maharashtra, respectively, starting in 1992. The World Bank eventually ended up supporting JFM in eight states with loans worth $528 million, and the Japanese Bank for International Cooperation (JBIC) has now emerged as the biggest lender for JFM-type programmes, having invested a total of $1 billion across four states.[2] Following the launch of the National Afforestation Programme (NAP) in 2002, which had an annual afforestation target of 3 million ha, the GoI began to fund JFM from its own funds, and created the institution of Forest Development Agencies (FDAs) to channel the funds to the JFM committees. The central government has pumped in more than $400 million into the programme till date.[3] After the NAP ended, funding for JFM continued to flow from the Green India Mission (Ministry of Environment and Forests (MOEF) 2010) and Compensatory Afforestation Planning and Management Authority

(CAMPA).[4] Although the mission document mentions 'revamping of JFM Committees' and also funding the gram sabhas recognized under the FRA 2006, in practice only JFM committees have been funded.

Understanding the Goals of JFM

It is clear from the June 1990 circular that the primary objective of the JFM programme was to 'involve local communities' in 'regenerating degraded forests'. But it is also clear from the language that the purpose of regenerating such forests was to meet 'local requirements of firewood, fodder, small timber, etc.', over which these communities had 'first charge'. Subsequent state-level government orders vary somewhat in their wording, but the core objectives remain similar. However, these simple phrases hide complex possibilities. It is, therefore, useful to step back and discuss some of the nuances in these ideas.

Participation

The involvement of local communities or their participation is a broad concept that means many things to many people (Cohen and Uphoff 1980; Midgley 1986). In its narrowest sense, participation is a fairly state-centric and limited concept, standing for getting local people to collaborate in or cooperate with a state-initiated programme, with almost no role for the local community in setting the agenda of the programme (what Cohen and Uphoff call 'participation in implementation'). Here, people's participation need not even mean community participation, as individuals collaborating with the state is also seen as participation: for instance, when individual farmers agree to have their individual farm plots 'treated' under a government watershed development programme (Joy *et al.* 2004).[5] But even here, participation cannot be equated with engaging in wage labour—people must volunteer some labour or resource contribution for long-term gain.

Intermediate approaches to participation allow a greater role for the local community in planning or post-implementation management of the resource. This in turn implies the setting up of a collective action institution at the local level. Yet, the word participation suggests a continued state presence, and indeed, *joint* management (or co-management, a term more popular in the West) is the appropriate term for this situation. Thus, the 'jointness' is both between members of the community and between the community and the state agency (Lele 1998). These collective action institutions will typically be user groups organized around a particular function or sectoral activity, and may, therefore, be exclusive to some extent, that is, not involving all citizens in a particular location. The relationship with the state is a longer term one, with some responsibilities transferred, but with significant residual powers with the state, including the power to recognize or de-recognize the local institution and to allocate the resource in question. Hence, this approach is also referred to as 'administrative de-concentration' or 'administrative decentralization', where tasks are decentralized but not much authority (Larson and Ribot 2005; Ramakrishnan *et al.* 2002).

At the other end of the spectrum is what is called 'political devolution',[6] 'political decentralization' or 'democratic decentralization' (Ribot 2002), that refers to a political process going beyond administrative decentralization and involving transfer of some of the state's authority to lower rungs for enabling self-governance within certain domains. Participation taking place here is of citizens in decision-making within this domain, with a fair degree of autonomy, that is, with no day-to-day presence of the higher organs of the state, even if the latter set some broad limits to the authority that the local body can exercise.[7]

Clearly, these different versions are underpinned by different normative ideas, that is, notions of *why* the involvement of local communities should be attempted in the first place. Typically, the reasons given for involving local communities cover a whole range:

- it is efficient and cost-effective, because if local people are using resources anyway then it is easier to regulate resource use through them and solve the problem of free-ridership through their collective action,
- it can better meet local needs and even sustainability goals (because local communities understand local needs, and have better local ecological knowledge, and because giving them a role increases their sense of ownership), or
- people have a fundamental right to local self-governance, including a right to manage their local environment.[8]

The first two arguments for local involvement are purely *instrumental* ones, in which participation is simply seen as a means to an end, and the end is determined by the state/implementer. If these are the motivations, then the form of participation is also likely to be of the limited kind: simple participation or joint management. The third argument is more expansive, and would result in participation being conceived of in more radical, devolutionary terms. Note also that the instrumental arguments work only if certain conditions are met (local people being more knowledgeable or being more environment-friendly) whereas the 'self-governance as a right' argument does not depend upon people's interest, preparedness, or special skills.

Local Requirements

In much of the discourse around forest management, it is taken for granted that rural communities in south Asia are heavily dependent on surrounding forests and common lands for firewood, grazing and small timber. Although the dependence is generally quite high, it is important to tease apart various levels or layers to this idea of 'requirements' that need to be met through decentralized management.

First, local needs go much beyond firewood, grazing, and small timber. Depending upon the region, forest condition and culture, rural households may extract food products, medicinal plants, bamboo, other non-timber forest products, agricultural inputs, as well as timber for house construction. Clearly, these cannot all be met from the same kind of vegetation; indeed grassland, shrubland, or heavily pruned treeland may be essential parts of a landscape that meets local needs. Second, these needs vary not just by region, but also within villages by type of occupation (animal husbandry, farming, forest product collection, and so on), economic condition (those who can afford to purchase alternatives and those who cannot), and even gender roles.

Third, in the official discourse, there is an obsession with 'bona fide' needs which in turn are equated with 'subsistence' needs. But although many forest products (such as firewood) are needed by rural households for domestic activities (cooking, shelter), many other products are used in production activities (such as the use of mulch, manure, dung produced by animals grazing in forests, fencing material and implements in agriculture) that generate income or are sold directly to generate income (such as commercially valuable forest products). And although its scale may have increased in recent times, making a living by selling

forest products either directly or indirectly is hardly a recent phenomenon—trade in forest products and in agriculture has been going on
for centuries. In other words, if 'viable livelihood' is a local need and a
right, then local needs include the need to harvest products for income-
generating activities, directly or indirectly.

A related point is that many products whose extraction is not seen as
part of 'bona fide needs' of local communities are those whose ownership was appropriated by the colonial state, such as timber. Timber-
based rural livelihoods have been common not just in other parts of
the tropical world, including Latin America, and recently have become
possible in Nepal, but have also been part of local livelihoods in northeastern India (that did not share a colonial history with peninsular
India).

In short, the concept of local needs is multi-layered and defining
it involves taking normative positions about what is need and whose
needs are to be prioritized. Any attempt to meet local needs must then
provide space for proper articulation of needs of different groups and
for negotiating what is a 'legitimate' need.

Regeneration and Non-degradation

Regeneration of degraded forests sounds like an obvious and straightforward concept, and is treated by the FD as equivalent to creating forest
plantations or planting up grassland. But there is a substantial literature
which points out that 'degradation' is a social construct (Dove 1992;
Lele 1994) and that there is no simple linear relationship between tree
cover and ecological balance, wildlife conservation, soil or water conservation benefits, or production of locally useful products (Chhattre
and Agrawal 2009; Lele *et al.* 2011; Lele and Srinivasan 2012; Putz and
Redford 2010; Rodríguez *et al.* 2006). It follows that sustainable forest
management (or forest regeneration) does not refer to one particular
form of forest (for example, pristine forest) or silvicultural practice (for
example, complete human exclusion), but rather to a range of practices
that ensure the forest's ability to continuously generate different mixes
of products, services, and intangible benefits. And there is ample room
to negotiate between different mixes within the bounds imposed by
sustainability or 'non-degradation'. Moreover, given that in the past,
forest management objectives were explicitly tuned to meeting state

revenue or industrial needs, any regeneration that also has addressing local needs as one of its objectives must deviate significantly from past silvicultural practices that focused on felling and planting.

Reading Goals from the Orders

In light of the above ambiguities and nuances, any attempt to assess JFM must begin with understanding the goals of the programme, that is, which particular interpretations of local needs and forest regeneration, and particular ideas of and rationales for participation it adopted. The empirical assessment must then proceed to evaluate achievements or progress with respect to those particular goals, not variants that we may normatively prefer. This is not to say that normative choices made should not be questioned, but rather to separate normative debates from empirical ones, and to use the appropriate benchmarks for the latter.

Having said this, understanding the goals is not easy. Government circulars tend to be worded in ways that are vague and slippery for an analyst. Goals and means are lumped together,[9] symptoms and root causes are listed together, phrases are repeated rather than clarified, and of course, there is some diversity across the orders of different states in terms of the emphasis they put. Nevertheless, some clear threads are visible.

First, the approach to participation is clearly an instrumental one. JFM programmes have never claimed that they are about fully decentralizing forest management to local communities as a matter of their right. The word 'governance' is never used, nor is devolution. The focus is on somehow getting local communities to work or collaborate with the FD in the management of a resource to reduce its degradation. But it is recognized that such collaboration will only happen if people are given not just a role in implementation but a voice in planning. The adjective 'joint' thus applies to both planning and management.[10] And all government orders specify in detail how jointness is to be ensured: requests for forming the JFM committee must ideally come from the local community, the committee must be elected by the community but will have a forest official as its secretary (and usually other government officials in it also), the committee must then (with the participation of other villagers) create a 'micro-plan' for the protection and management of the forest patch assigned to it, and it must implement this plan with the support of the FD. Economic returns from such forest management are to be shared between the community and the department.

Second, meeting local needs is clearly a goal but there is ambivalence about what that means. On the one hand, there is much talk about only meeting 'bonafide' needs. On the other hand, there is a clear incentive being offered to grow timber and other commercially valuable products. On the whole, it appears that subsistence needs and some indirect livelihood needs (grazing and collection and sale of non-timber forest products (NTFPs) are recognized, while major income-generation (from the sale of timber) is a bonus.

Third, degradation and regeneration are defined in terms of tree canopy cover. Less than 25 per cent of canopy cover is equated with degraded forest, while 100 per cent of canopy cover was treated as prime forest and kept outside the purview of JFM till some limited changes were introduced in 2000. This simplified idea of degradation and regeneration is at odds with the objective of fulfilling certain local needs, such as grass for livestock or leafy matter for manure or fodder or leaf plate-making. And while it leaves open the possibility that some tree-based local needs could be addressed, it also leaves open the possibility of continuing past practices of softwood and timber-oriented forestry.

Finally, neither the National Forest Policy nor the 1990 circular refer to equity or social justice goals, except by emphasizing the needs of 'tribal' communities. Most JFM orders use the term 'rural community', not 'rural poor'. Nevertheless, most orders impose various conditions on how the general body is to be constituted and how various marginalized social groups have to be given representation on the managing committee. Some orders also explicitly say that the needs of the poorer and landless sections must be taken into consideration in forest planning. This suggests that ensuring some level of equity *within* the community was a goal or a concern. Moreover, it is reasonable to expect that, in a country committed to social justice, any state-funded programme will at least not aggravate inequalities or produce adverse social impacts.

Thus, it is reasonable to deduce that JFM is officially about using a participatory or 'joint' process of forest planning and management involving the village community and the department so as to meet local needs and achieve forest regeneration. It would, therefore, be fair to assess a JFM committee or programme only in terms of how much it contributed to regeneration of natural forest (to meet the wider environmental goals), how much it contributed to meeting local livelihood needs sustainably, whether it was at least equity-neutral, and whether all this was achieved through a participatory mode, that is, some basic level of joint planning and management.

Primacy of the Joint Process

A crucial conceptual point to be noted at this juncture is that these four dimensions are not on par. Joint planning and management is a process, while the others are outcomes. Even more important, joint planning and management is a *pre-requisite* for any outcome to be called the outcome of JFM, otherwise the situation would be indistinguishable from any ordinary afforestation project. This implies that any JFM programme must be first judged on whether it was at least minimally 'jointly' planned and managed *before* judging its outcomes in terms of meeting local needs fairly or regenerating or protecting forests. In other words, it would be conceptually incorrect to say that one evaluated a JFM programme and found (say) regeneration high but participation low, because this would mean that the programme had failed to qualify as a 'joint' management programme at the outset (see criteria outlined in Lele *et al.* 2005, chapter 2). The following analysis reflects this thinking.

Available Studies

The JFM programme has spawned a vast body of literature, with studies of varying focus, spread, and quality.[11] I have selected a sample of studies that has several characteristics. First, I have chosen three different types of studies: state-wide or multi-state rapid assessments (typically by external evaluators), multi-state detailed studies (by academics), and state-level detailed studies (by academics). Second, I have chosen studies that occurred at different points in time, ranging from 1997 to 2008. Third, the study must have focused on all the above dimensions, rather than studies that have looked at only one dimension (such as gender in Correa 1996; or broader equity issues in Kumar 2002). Fourth, the study must have a relatively clear methodology for data collection and analysis. Finally, as foresters often complain of a negative bias amongst academics, I have specifically included a study conducted by the Indian Council for Forest Research and Education (Environmental Impact Assessment (EIA) Division 2008), an agency controlled and peopled largely by foresters themselves. A list of the studies used, including their characteristics in terms of regions covered, sample size, and period of study, is given in Table 1.1. When discussing these studies in the next section, for the sake of brevity, I will refer to them by their number in square brackets.

Findings About JFM Process and Outcomes

What do these studies tell us about JFM? To what extent do their basic empirical findings converge and where do they diverge or contradict each other? I explore these questions at two levels. First, how participatory or joint has the process been? Second, where the process has met some basic standard of jointness, what it has resulted in? I also refer to other studies where they substantiate or nuance particular points.

Is the Process 'Joint' at All?

The level of information that even these studies provide on the extent of jointness in the process of setting up village forest committees (VFCs),[12] micro-planning and subsequent protection and management is variable. Part of this variation is due to differing frameworks: some studies (numbered 3, 6, 7 in Table 1.1) look at the process as only that of community participation rather than joint management, thereby evaluating only what the community did but not what the FD delivered vis-à-vis its commitments. Nevertheless, some features emerge quite clearly across all of the studies.

First, it must be noted that only a very small fraction of the VFCs are community-initiated (mostly in parts of Gujarat and Odisha). A significant fraction is NGO-initiated, but the vast majority across all states are FD-initiated. This suggests a shift from an open-ended invitation to the community to a target-driven programme implementation mode. Second, only in a small fraction of cases was the process laid down for initiation, planning, and management followed with some degree of rigour. In a large fraction (sometimes majority) of the cases, problems such as committees being constituted too quickly with

TABLE 1.1 Studies on JFM used in this overview

No.	Study	Regions	Sample size	Method	Period
1.	Saxena et al. (1997)	Karnataka (two districts)	Varies	Rapid evaluation	1997
2.	Kumar et al. (1999)	Madhya Pradesh, Andhra Pradesh	Varies	Rapid evaluation	1999
3.	EIA Division (2008)	All India	182 FDAs, 600 villages	Survey method	2007–8
4.	Sundar et al. (2001)	Gujarat, Madhya Pradesh, Andhra Pradesh, Odisha (two districts each)	16 villages	In-depth case studies, and other qualitative methods	Mainly 1995–1997
5.	Springate-Baginski and Blaikie (2007) and studies therein led by Banerjee (Bengal), Reddy (AP), and Sarap (Odisha)	West Bengal, Odisha, Andhra Pradesh (sample districts)	12–14 villages in each state	Survey method + interviews	2005
6.	Sudha et al. (2004a)	Karnataka (10 districts)	495 villages in 10 districts	Survey method	2000–2
7.	Rao et al. (2004a)	Andhra Pradesh (four divisions)	167 villages	Survey method	1999–2002
8.	Lele et al. (2005)	Eastern plains of Karnataka (eight districts)	28 rapid surveys, Four village case studies	Rapid assessment + case study	2001–2

Source: Author.

minimal consultation amongst villagers, cut-and-paste micro-plans, and delayed signing of agreements with the FDs are reported (1, 2, 4, 5, 8). This is corroborated by several other studies by Anonymous (1999) and Ravindranath and Sudha (2004). In West Bengal, even community involvement in protection was higher (Dutta *et al.* 2005), the process of micro-planning was still largely perfunctory (Banerjee in 5). An extreme case was some of the districts in the Karnataka eastern plains, where in a large number of villages, committees were formed *after* plantation activity had already taken place (8). Even the Indian Council of Forestry Research and Education (ICFRE) assessment (3) admits that 'a comprehensive micro-planning of village assets based on participatory principles… keeping in view the requirements of local communities has been found wanting in most of the committees' (3, p. 21). Some senior foresters also admit that the control of the agenda still rests with the FD (Verma 2008).

When it comes to actual involvement in protection and continued management of the resource, the vast majority of cases show such involvement either not occurring even at the outset (3)[13] or dropping dramatically after an initial (usually project-funded) phase (2, 6, 7). This is not to say that protection stopped everywhere: typically, community protection has continued where it was self-initiated (such as in Gujarat and Odisha) or not set up with funding to begin with (West Bengal). While 'institutionalization' even in simplistic terms seems to have happened only in 50–60 per cent of cases (Roy and Roy 2008), study 6 reports that general body meetings were not held in 60 per cent of cases in Karnataka. In some places where committees take their job of protection seriously, they do not get back-up from the FD (Banerjee in 5, 8).

Where Process was Followed Somewhat, What did it Result in?

Although all of the chosen studies look at the four dimensions identified above, I must note that most of them do not look at outcomes contingent on process. That is, they do not distinguish between outcomes obtained under the lack of participation and outcomes obtained after some minimal participation or jointness was achieved. Keeping this in mind, I summarize below the findings on outcomes from these studies.

Forest Regeneration

Most studies report some improvement in forest conditions, but this finding needs to be interpreted carefully. At the outset, note that 'forest cover' and 'tree cover' are taken as synonymous by most studies/ assessments, although some go on to look at species composition as well (for example, 6, 7). Also note that new plantations are typically included when talking about whether forest status has improved or not, when it is rather a truism that a newly planted area, protected by an FD-paid guard, will necessarily show increases in tree cover in that initial period (for example, 6, 7). In the West Bengal case, however, part of the improvement has come about due to rapid regeneration of sal forests from rootstock. Interestingly, however, plantations still form a major component of JFM areas in West Bengal and they are not necessarily of local species (Banerjee 2007 in [5]). In Andhra Pradesh, apart from new (often eucalyptus) plantations, the 'improvement in forest condition' is in the form of reclaiming of land that was previously 'encroached' under *podu* (shifting cultivation) or settled agriculture (Reddy *et al.* 2007 in [5]). In cases such as in the tribal areas of Gujarat and Odisha, where community-led protection has preceded the formal JFM committees by several years; the forests have actually improved due to natural regeneration (Ravindranath *et al.* 2000, which is a companion volume of 6, 7). Finally, some of the detailed studies (4, 5) report that protection of one degraded patch may have occurred at the expense of degradation of other (unprotected) patches of standing forest. Overall, given that a majority of cases do not meet minimum participation standards, JFM begins to look indistinguishable from any conventional plantation-based afforestation programme.

Meeting Local Livelihood Needs Sustainably and Substantially

The gains in terms of local livelihood needs have been much more mixed. While the increased availability of fuelwood is reported in many studies, they also uniformly report decreases in fodder availability (or accessibility) due to the closure of JFM areas for planting and tree growth in general (4, 5, 6, 7). In a few cases such as West Bengal, the income from NTFPs has increased significantly (Banerjee in 5), but by and large gains from NTFPs have been much smaller than anticipated. In terms of a share in the final sale of timber/softwood, most villages had not reached the stage of final harvest at the time of the studies (confirmed by 3). But even in West Bengal, where JFM began very early and harvests have taken place, the returns to the individual households have been fairly paltry, even when actually paid out (Banerjee in 5, also Dutta *et al.* 2004). In other locations, the FD has tried to entice villagers by offering them a share in the harvest of pre-existing plantations (1, 8), a move that has generated quick returns first time around but cannot really be seen as an indicator of JFPM outcomes. Wage employment has increased in most places, but virtually in all cases it corresponds to the initial and funded period of JFM, when funds are spent on planting and other activities (3, Banerjee and Reddy in 5, also Corbridge and Jewitt 1997).

Equity Impacts

As clarified earlier, poverty alleviation or reducing socio-economic equities is not the stated objective of JFM, but one would look for JFM to not aggravate existing inequities. Unfortunately, the pro-poor impact of JFM has been limited, and in many cases grossly negative. Poorer households that depend upon the commons for firewood and grazing have suffered the most from the closure of forest areas brought under JFM (4, 5 ,8). This is the direct consequence of the focus being on planting trees per se rather than understanding and meeting local needs. In several cases, nomadic communities and even villagers from neighbouring villages who might have had some customary rights have been left out. The eviction of tribal podu cultivators and other 'encroachers' under the name of JFM has been a major issue in Andhra Pradesh, Madhya Pradesh, and some other states.[14] Other studies show that women have often had to face the brunt of forest closure (Sarin 1997; Sarin *et al.* 2003). As even the ICFRE report acknowledges, 'in majority of the cases, JFMCs are either dysfunctional or captured by local elites' (3, p. 27).

Summary

In short, we find that across the spectrum of evaluations and studies, the vast majority of cases do not even meet the basic criteria of being a joint planning and management exercise. Thus, the increased tree cover that is observed in most cases is not due to JFM itself, but to an afforestation programme that funds planting and protection for the initial years, and whose long-term impact on tree cover (or 'forest regeneration') is quite uncertain. The goals of meeting local needs have been achieved partially where some participation has occurred, but even here some needs such as grazing have been left out or actively suppressed.

Fifteen years ago, the World Bank's own evaluation of the Social Forestry project it had funded in Karnataka concluded that

> The [Social Forestry] project achieved its physical targets.... The project, however, did not alleviate fuelwood shortages; the poverty groups which provided an important part of the rationale for the project, were addressed only marginally, effective community participation in project implementation was not attained; and institutional development of the implementing agency to improve its capacity to work collaboratively with villagers, failed to occur.... Project sustainability is rated as unlikely on social and institutional grounds. (Agriculture Operations Division 1993)

It is indeed tragic that, in spite of spending nearly $2 billion from bilateral and multi-lateral grants and loans and taxpayer money, and the notional formation of more than 1 lakh VFCs, 'effective community participation' and 'sustainability' continue to be a mirage.

Why the Shortcomings?

When I initiated this meta-analysis using the literature on JFM outcomes, I expected to find a diverse, if not conflicting, picture. However, even though the specific criteria and indicators used vary, and there are clearly some regional variations (or variations depending upon whether JFM was initiated by communities or by the state), the broad findings across different studies are surprisingly similar at the level of the quality of the process and the outcomes. However, the interpretation of these findings in terms of what has caused the shortfall in JFM quality and performance are quite varied. This is where the role of a theoretical framework[15] becomes crucial.

Unfortunately, there has been a tendency in the literature on JFM, particularly the semi-academic one, to not employ explicit theoretical frameworks. The pitfalls of this can be illustrated by comparing two studies from among our sample itself: studies 6 and 8. Both studies are on JFM in Karnataka, and study 6 contains all the districts covered by

8. Both studies found that people's participation in actual forest protection in the JFM villages was very low: study 6 found that in 63 per cent of the sample villages, the forest department was protecting the JFM plantation on its own even after the first three years and in 15 per cent of the villages, there was no protection happening at all. Study 8 found that only three out of 28 villages were actually protecting their forest, 11 others were 'cooperating' with the FD's protection efforts, and the rest were simply not functional. And yet, while study 8 concluded that JFM was simply not happening in the majority of the cases, and eventually located the problem in the FD's lack of interest and motivation, study 6 concluded that 'overall, JFPM has enabled people to share the responsibility of forest management'. The reason for this difference lies in the frameworks: while study 8 lays down a clear framework starting with participation as the core objective of the JFM process and factors in the implementation, design, and community level that might inhibit JFM, study 6 does not outline any explanatory scheme, and in practice examines only some factors in the JFM orders that might play a role in influencing its quality, without examining the role of the FD. This is true of most of the semi-academic literature on JFM, including study 3. The consequence is that interpretations and recommendations that emerge are in the form of laundry lists, without touching upon core issues or assessing the big picture. Thus, it is necessary to outline an overarching framework that would coherently link different kinds of explanations or levels of explanation, before we interpret the common findings across our sample of studies.

A Theoretical Framework for Understanding JFM-related Findings

To the extent that the idea of JFM has emerged from the literature on collective action and common property resource (CPR) management that emerged around the late 1980s, one finds that a large fraction of academic studies on JFM treat it as an example of collective management of a CPR and hence seek to explain JFM outcomes in terms of categories of broad variables well accepted in this literature—namely, the nature of the resource (size, mobility, technology, divisibility), the decision-making arrangements (collective and constitutional choice mechanisms, operational rules, and external recognition), and the socio-economic context (number and homogeneity of resource users, dependency on the resource, prior collective action experience) (Arnold and Stewart 1991; see also Agrawal 2001, for more details). In particular, since the nature of the resource is fixed (forest), the focus has been on the rules and the context. Further, most of the rules are also fixed and hence many studies of community forestry focus largely on how characteristics of the households and communities shape participation in collective action (for example, Agarwal 2010; Agrawal and Gupta 2005; Behera 2009; Dutta *et al.* 2005; Lise 2000; Naik 1997).

Closer examination, however, would suggest that this framework is inadequate. First, as already pointed out, one has to distinguish between situations that are only notionally JFM attempts in the official record and those where some collaborative activity has been seriously attempted. The question of why JFM, where seriously attempted, succeeded or failed (in whatever sense) has to be separated from why JFM was not seriously attempted in most places. Second, and most important, JFM (where seriously attempted) is about 'joint' or collaborative management, not about community management. It is a partnership in which the FD's presence looms large all the time. Moreover, while the rules allow communities to apply for JFM, in practice the vast majority of cases correspond to FD-initiated JFM. Therefore, both the structures (the JFM rules) and the attitudes and actions of the FD officials become crucial factors that shape the participatory space and influence both whether communities collaborate with the FD and what the outcome of the collaboration is.[16] Third, political economists would point out that JFM is an attempt to modify the allocation of authority over forests, which has historically been with the FDs. Therefore, even if it is an officially adopted programme, there is reason to expect that the implementing agency (being the FD itself) or its members would be very ambivalent about the programme and may not do a sincere job of implementing it.

Corresponding to the above, explanations for the poor quality of the JFM process, that is, the absence of genuinely collaborative management that we observed above, pull in two different directions. One perspective would assume that the lead implementing agency is seriously attempting to implement it, and would focus on the *institutional design* of JFM factors and *contextual factors* that may lead to inadequate engagement by the community or adverse outcomes in spite of community participation. Another perspective would focus on the *political economy* of JFM, examining the assumption of serious implementation itself.[17] We will now discuss these two strands of explanation.

The Design of JFM: Insights from CPR Theory

As explained earlier, JFM involves two kinds of jointness or collaborations: within the community and between the community and the FD. The main theoretical framework for explaining whether or when individuals in a community might collaborate with each other in resource management is a CPR theory (Ostrom 1990). In simplified terms, this theory says that the reason for forest degradation is located in the de facto open-access nature of the property rights regime, and therefore, a clear demarcation of resource use areas and clear, substantial and secure assignment of rights to a small, homogeneous group of resource users who interact on a daily face-to-face basis will address this problem. An examination of the JFM orders and programmes shows that even these basic tenets of CPR theory are violated.[18]

Mismatch Between Resource Use Areas and JFM Areas

Across the board, with a few exceptions, JFM programmes made only part of the forest use area available for collaborative management. One reason was clearly the FD's limited focus on 'degraded' forest, which almost implied that one has to degrade one's forest before it can be brought under proper management. Even after the central government revised its guidelines in 2000 to allow dense forests to be brought under JFM in some cases, and even though some state governments (such as Karnataka) passed new orders also enabling this, in practice hardly any dense forests have been brought under JFM to date (the only exception being Madhya Pradesh, which created a different scheme for dense forests). Another reason was the firm belief that JFM meant creating new plantations (as in earlier afforestation programmes) and so only that area which could be planted up (depending upon resources available) was often made the focus of JFM. Whatever the reason, the mismatch between use area and management area was ubiquitous (see 4, 5, 8).

Insubstantial Rights

The core idea of JFM was to give greater rights on the forest resource to the community and link it to the responsibility of protection. These enhanced rights included full rights to firewood and fodder for meeting local needs, full rights to economic returns from NTFPs, and a share in the final timber harvest. They also included the right to plan forest use and silviculture, and the right to catch and punish offenders from within and outside the village. But in practice, these rights given were only partial, confusing, and inadequately backed up in practice. The lack of involvement in silvicultural decisions has already been mentioned. Further, local firewood needs were defined in a complicated manner (1, 2) and grazing rights were de facto taken away by the focus on planting up open areas (1, 2, 4, 5, 6, 7, 8). Moreover, rights to commercially valuable NTFPs had been leased out to contractors or taken over by the state under its policy of nationalization (Lele *et al.* 2010), and most states refused to cancel the leases or de-nationalize the products (1, 2, 8). Finally, profits from timber harvest have hardly been shared (3, 4, 5). Similarly, as described earlier, the right to exclude others could not be exercised without the support of the FDs, which was missing in many cases. All of these factors taken together mean that the de facto rights assigned to the community were inadequate (see also Behera and Engel 2006a).

Insecure Rights

If a community is to invest in collective protection and management of a forest, a process that yields benefits only after several years, it requires a secure tenure on the resource. But JFM functioned, and continues to function, on the basis of executive orders, which do not have the sanctity and security of a statute. The orders can be easily modified, overturned or withdrawn by the executive. Moreover, the orders only enable the signing of a Memorandum of Understanding (MoU) between the community and the FD. The tenure of the MoU was only five years to begin with, creating enormous insecurity. And the MoU and the JFM orders themselves give lopsided powers to the FD to cancel the agreement at any time (see 8 for details).

In this context, it should be pointed out that the mainstream literature has confused tenurial security with the legal status of the VFCs (for example, Damodaran and Engel 2003: 26; Ravindranath *et al.* 2004: 331). The central guidelines of 2000 (GoI 2000a) suggest that these institutions be registered under the Societies Act. But the point is that JFM is about a group of individuals having secure and exclusive access to a forest patch, and so the security that is needed is about resource access, not just about the group's identity or status. Registering the VFC as a society does not in any way give its members assured access to the forest! Only an arrangement such as that followed in the case of Van Panchayats in Uttarakhand (GoUP 1931) or Village Forests in the Indian Forest Act makes sense, whereby a patch of forest is identified and its users constituted into an body with exclusive rights to that patch, and the legal notification spells out both the body and the patch to which it has been given rights.

Large, Non-homogeneous Groups

One supposed advantage of the JFM programme was that it did not specify who the general body of users would be, giving room for groups of different size and location to form their own committees. This would have matched the theoretical prescription that the groups be as homogeneous and small as feasible. Unfortunately, most programmes insisted on forming one committee per revenue village, not recognizing the diversity within villages and the large size of many revenue villages also. Again, subsequent revisions specifically allowed formation of groups at the level of hamlets, but this provision was generally not implemented (1, 4, 5, 8).

Funding-based Implementation

From day one, JFM has always been implemented as yet another government programme, dependent upon funding for its functioning (1–8). This is at odds with the core idea of re-assigning rights and responsibilities. First, such reassignment is a one-time task that hardly requires any funds. Funds may at most be required for retraining FD staff and spreading awareness about JFM. Second, budgeting funds for tree planting a priori biases the entire effort away from community-driven planning (which in most cases must provide for grazing and fodder) towards conventional silvicultural choices, undermining the whole idea of joint planning. Nevertheless, all JFM implementation has been done only in projects or programmes where substantial funds have been provided specifically for such planting. The impact of such fund-driven implementation also is that communities then become dependent upon it, and genuine participation (driven by interest in resource regeneration) does not emerge.

An extension of the funding-based programme logic has been the idea of forming FDAs on the lines of the District Rural Development Agencies (DRDAs), and channelling central funds through them (GoI 2000b). There are several problems with this approach. On the one hand, the idea of DRDAs itself is outmoded (if not illegal), when the idea of panchayati raj has been accepted. The DRDAs are supposed to be dissolved and all development funds should be routed through the zilla panchayats, which are now the third tier of governance. On the other hand, the government also claims that the FDAs are district-level federations of JFM committees, when the FDA structure defies all notions of federations. The key officer-bearers are not elected by the representatives of the JFM committees at all, but are forest officials ex-officio (the conservator as chairman and the division forest officer as secretary). A large number of other government officials are ex-officio members of the executive body of the FDA. The general body of the FDA consists of member-secretaries of all the VFCs under it, who are again forest officials, in addition to one women member from each VFC. Not surprisingly, even the ICFRE report (3) acknowledges that 'the writ of the FDAs runs large over JFMCs'.[19]

Beyond CPR Theory: The Core Problem of FD-community Jointness

While CPR theory provides insights into why communities may not be willing to or able to collectively manage forests under JFM, the question remains as to whether and when the other kind of jointness, namely, between the community and the FD, is needed and will work. This jointness is characterized by the FD being involved in and having control of all operational decisions through its representative (typically a section forester) who is the ex-officio secretary of the VFC in all states (except Gujarat).[20]

There is no theoretical support for this extreme form of jointness. The entire literature on decentralization is about 'transferring' certain tasks and functions to lower, more suitable levels. Executing tasks requires some functional autonomy, even if it is circumscribed. For instance, JFM does not grant the local community the right to alienate forest land or hunt wildlife species. Ensuring that local communities do not transgress such boundaries does not require that the forester be present in all VFC meetings. Indeed, requiring such presence undermines the very purpose of even administrative de-concentration, which is to reduce the burden on the FD. The universal complaint across all regions is that the foresters do not have time to conduct VFC meetings (1, 2, 4, 5, 8).

Transferring the operational role to the local community and leaving a regulatory role for the FD would have been the obvious and logical form of devolution (see Joint Committee 2010, chapter 8; Lele 1999, 2004). This idea has support in the literature of co-management, where the principle of subsidiarity is enunciated (Murphree 2000). But the JFM structure does not follow this form.

Furthermore, while the FD has the power to dissolve the VFC any time, the VFC has no recourse if the FD fails to keep its side of the bargain. Again, this arrangement defies all notions of collaboration. A truly collaborative arrangement would require that disputes within the collaboration must be solved with third-party arbitration, not unilaterally. But the entire approach to jointness in JFM is lopsided: initiated by the FD, funded by the state through the FD or FDA, the area allotted by the FD, a micro-plan drafted by the FD official and approved by the FD, the prerogative to punish or not punish offenders caught by the villagers also with the FD, de-recognition in the hands of the FD, and so on. As a retired principal chief conservator of forests (PCCF) from Gujarat put it,

> the main reason of degradation is conceded to be bureaucratic control and JFM is expected to change this and bring forests to people. [But] the control of JFM is in the hands of the FD. Government is willing to entrust protection of forests to people but [the] orders do not make serious efforts to trust communities with control functions. (Verma 2008)

The Persistence of Lacunae—The Political Economy of JFM

The shortcomings in JFM at the implementation level have been pointed out by a very large number of studies. The shortcomings at the design level, both in terms of CPR theory and also the lopsidedness in the FD-community relationship, have also been highlighted by several major studies, including donor-supported evaluations. For instance, as far back as 1997, it was pointed out that inadequate area, lack of clear entitlement to forest produce, and excessive control of the FD were serious problems in JFM in Karnataka (1), and specific recommendations for amending the JFM guidelines were made. Similarly, the World Bank's evaluation of JFM in Andhra Pradesh and Madhya Pradesh concluded that for the majority of the FD staff, understanding of participation does not go beyond 'we (FD) manage and direct, you (FPC members) participate' kind of collaboration and that 'the reluctance to delegate decision-making power to the communities remains' (2, pp. 64–5). Several other studies (4, 5, 8) also make clear and strong recommendations for change. Several senior foresters themselves have called for major changes (Kumar 2002; Rangachari and Mukherji 2000; Verma 2008).

But the response of the state has been a limited one. Many of the recommendations of the revised JFM guidelines of 2000 and 2002 have been hardly implemented. For example, few instances of JFM having been extended to good forests can be found outside the state of Madhya Pradesh, which had devised its own guidelines earlier. Moreover, these revised central guidelines only addressed some of the issues flagged above, and did so inadequately. For instance, the focus on registering VFCs under the Societies Act distracts from the absence of statutorily supported access and management rights over well-defined pieces of land. Many activists and scholars have been pressing for the past decade or so for the activation of the Village Forest chapter of the Indian Forest Act, but this suggestion has been persistently ignored.[21] Indeed, by superimposing a completely undemocratic FDA on top of the already non-autonomous VFCs and by pumping all afforestation funds into the programme via the National Afforestation Plan 2000, the government has further reduced any possibility of bottom-up community-based forestry.

This then lends strong credence to the political economy argument. Notwithstanding all their rhetoric, the FDs have had a very ambivalent attitude towards JFM. From being hostile to the idea of decentralization, the FDs quickly realized the necessity and advantages of this particular brand of 'collaborative' management. On the one hand, adopting the JFM rhetoric has enabled them to attract large amounts of funds from international donors. Indeed, the first JFM order in Karnataka openly states that 'the "Process Plan" of the Western Ghats Forestry and Environmental Project funded by ODA [Official Development Assistance] of the UK depends mainly on the JFPM and the ODA authorities have insisted for a Government Order authorizing the principles of Joint Forest Planning and Joint Forest Planning and Management'! And the biggest donors, such as JBIC, appear to be the ones least concerned about the quality of community participation or jointness, as the Karnataka experience shows.[22] On the other hand, JFM can also be used simply as an instrument for co-opting communities, especially their elite, for the limited purpose of raising commercial plantations and for evicting those that the FD sees as encroachers.

In other words, the core tenets of the forest bureaucracy seem to remain what they were (Kumar and Kant 2006) and were crisply summarized by the topmost official of the Karnataka Forest Department when confronted with a critical assessment of JFM: 'Our fundamental goal is forest conservation. Our main job is to plant and protect the forest, and to catch and punish the offenders. If, after this, we have time to spare, we will take up JFPM' (PCCF, Karnataka, May 2008, quoted in 8).

Some analysts have suggested that JFM represents a battle between a central government keen on effecting major changes in forest property rights and state governments resisting such changes (Damodaran and Engel 2003). But, in fact, there is no evidence to support this formulation. Even after the central government started funding JFM substantially from its own funds (the National Afforestation Programme that started in 2000), the Ministry of Environment and Forests has not made any changes in the structure of JFM and has, in fact, introduced the more retrograde concept of FDAs. And it has strongly opposed the more radical changes being attempted under the FRA, as Sarin's chapter in this book shows in detail.

On the other hand, there is evidence to lend some support to the political economy explanation advanced by Sundar *et al.* (2001), namely, that the forces of capitalism, whose mantra of market penetration and privatization has dominated Indian economic policy during this same period, have also played a significant role in shaping the practice of JFM. Specifically, the persistent focus on commercially valuable species (sal, teak, acacia, and eucalyptus), in the guise of income-generation for local communities when the forest policy categorically de-emphasized the commercial objective of forestry, is a sign of this influence. In the overall context of environmental concerns taking a back seat, the general lack of political attention to JFM at the centre and in most states is not surprising.

At the same time, the other actors who have been involved with JFM are not entirely blameless. Given that the JFM orders were so deficient at the outset, JFM clearly did not constitute a meaningful window of opportunity for a radical change in forest governance in the country (see Lele 1995). But most of the NGOs, academics, and donors who engaged with JFM failed to scrutinize the JFM orders rigorously. NGOs assumed that the state had finally changed, and welcomed the flow (if not flood) of funds for implementing JFM in the field. They kept interpreting the obstacles they encountered as 'second-generation' issues (see, for example, Saigal 2000), when actually these were (to use the same terminology) 'first-generation issues', issues of tenure security and autonomy in day-to-day operations that should have been addressed at the outset. Many groups that had in the past rigorously critiqued state forest policy (for example, groups in Karnataka in 1990) got co-opted by the funds available for JFM, or the intransigent ones got sidelined as the less discerning ones grabbed the opportunities (Lele 2000).

Researchers fell into a similar trap: a cottage industry of JFM assessments sprang up, with few willing to examine the assumptions or even to apply rigorous frameworks. Donors jumped onto the bandwagon, with the best of them hoping against hope that they could use their funds to arm-twist the FD into changing itself, and others simply seeing this as another opportunity to extend loans to state governments. Civil society actors—researchers and NGOs alike—could not resist the temptation of engaging with donors and the department without asking whether adequate conditions (both at the top and at the bottom) had actually been created for ensuring significant and irreversible change, whether the apparent window of opportunity was a real one or a mirage. The engagement with political parties and individuals was minimal, except perhaps in West Bengal and to an extent in Andhra Pradesh, although the results have not been substantially different, at least in the latter state.

***

I began this chapter by asking through what framework one should assess JFM. It is clear that one must use the correct normative lenses.

For instance, criticizing JFM for (say) its failure to achieve poverty alleviation, as the World Bank did (Kumar *et al.* 1999), or asking poverty alleviation to be made the main goal, as the British DFID insisted when discussing second-phase funds for JFM in Karnataka (P.J.Dilipkumar pers.comm.) is unfair, as these are not the primary objectives of JFM.

But if the appropriate objective against which to measure JFM is the putting in place of a system of JPMF, it is equally clear from the empirical evidence provided by a variety of studies that JFM is today not achieving this objective by far. All official claims that 'about one-third of the land with the forest department is being managed today in partnership with the people of the country' (Bahuguna 2011) are patently unsupportable. The land is only officially under JFM; actual joint management that meets local needs and regenerates natural forests is hardly occurring in a fraction of this landscape.

Official reports attempting to explain these gross shortcomings typically gloss over the problem with statements such as 'the functioning of village level JFMCs has been akin to the functioning of any nascent organization' even as they acknowledge that the 'functioning of forest committees remain still beyond the grasp of a majority of forest committee members and forest department in majority of cases' (EIA Division 2008: ii). A large semi-academic literature provides a predictable laundry list of recommendations, ranging from more 'capacity-building of villagers' and more 'training of forest officials' to 'providing continuous budgetary support' and 'promoting federations' or 'promoting NTFP-based enterprises' (for example, Bhattacharya *et al.* 2008; Sudha *et al.* 2004b). Such interpretations and recommendations result from the lack of a clear theoretical framework to explain observed outcomes. Even the framework from CPR theory often adopted in academic analyses only explains some of the design-level shortcomings in the JFM setup. A bigger lacuna, however, is the theoretically untenable idea of 'jointness = operational control by the FD over community decisions' that the concept of JFM embodies. That day-to-day operational control of community-managed forests is both feasible and necessary is the primary conceptual flaw in JFM.

The persistence of these design flaws even 20 years after the first JFM programmes were launched and the huge shortcomings in just the process of implementation itself forces us to seek a deeper explanation, and only a political economy framework provides that. While the official rhetoric may be of decentralization and even democratization (EIA Division 2008, preface and p. ii), in fact, the vast majority of forest officials are opposed to any serious devolution of powers, and even to a pragmatic de-concentration of functions. They only appear to use JFM as a means for retaining control over local communities through the co-optation of the village elite and a means for garnering more resources from donor agencies through the cynical use of the rhetoric of participation. The de facto and de jure reduction in authority that the FD officials would face if any substantive and meaningful devolution of responsibilities creates a major conflict of interest when they set about implementing such devolution. The 'J' in JFM then becomes the means of resolving this conflict: retaining full operational control with the FD while appearing to collaborate with local communities.

In other words, if the forestry debate has been polarized since the 1980s by the two diametrically opposite formulations of exclusive state control and complete community control (Guha 2012), JFM as a proposed middle path has turned out to be a failure, both conceptually and practically for reasons outlined above. Conceptually, the way forward lies in re-thinking the role of the FD (Lele 2011), shifting from being owner, manager, protector, regulator, and policy-maker all rolled into one to primarily a regulatory role with significant improvements in transparency and accountability. Politically, this shift may have to be preceded by a shift in the political economy. Both processes have begun under the FRA 2006 and the social movement that has been associated with it. It is hoped that understanding the limitations of JFM will constructively inform the new arrangements that are emerging under this Act.

Notes

1. Other 'objectives' listed in the document can either be subsumed under ecological balance (such as reducing soil erosion) or are more in the nature of strategies to achieve these objectives (such as carrying out afforestation).

2. Different sources give somewhat different figures. Sudha and Ravindranath (2004) estimate total external funding at Rs 4,881 crore, which translates to approximately $1.5 billion, depending upon the exchange rate used.

3. See http://www.naeb.nic.in/documents/NAP%20statistics_Funds_Released.xls.

4. See, for example, the plan submitted by the Karnataka Forest Department for utilizing CAMPA funds: http://www.moef.nic.in/downloads/public-information/Karnataka-APO-STATE-CAMPA-2010-11.pdf.

5. For instance, when a farmer cooperates with the implementing agency in a watershed development programme to bring his farm plot under 'treatment'.

6. Note that the term 'devolution' has been used interchangeably with decentralization by some authors (Sundar 2001).

7. Note that participation and decentralization/devolution are not exactly interchangeable. Decentralization is a structural change, which may lead to better or greater participation of local communities in decision-making. The question is to what degree this decision-making is autonomous.

8. A fourth reason could be that decentralization will give greater voice or visibility to marginalized groups within the local community (Larson and Ribot 2005), but I have not seen this argument stated with any frequency, at least in the Indian context.

9. See, for example, Sundar *et al.*'s critique of the National Forest Policy and the 1990 JFM circular (p. 43).

10. As is made explicit in the phrase Joint Forest Planning and Management (JFPM) used in Karnataka state.

11. Rao *et al.* (2004b) indicate that they found 200 studies till about 2003, of which only 99 indicated their methods, most were donor supported, and most did not cover ecological aspects. They discuss 11 studies in detail, most of which were again commissioned by the FDs.

12. JFM committees go under various names and acronyms, including Van Samrakshan Samitis (VSSs), Forest Protection Committees (FPCs) and so on. We use the acronym VFCs to cover all these variations.

13. 'The functioning of forest committees remain still beyond the grasp of a majority of forest committee members and forest department in majority of cases (p. 21).... The concept of collective ownership and management of forest resources by village level institution for collective good for the members has not evolved in a majority of cases' (p. 22).

14. Not all encroachers are poor, but there is ample evidence that only poor encroachers get evicted.

15. I use the term 'theoretical framework' to refer to the core assumptions which enable one to build hypotheses to *explain* observed outcomes. It is distinct from the normative framework in that the latter enables one to characterize the outcomes themselves based upon one's underlying concerns.

16. This point is made by Behera and Engel (2006b), but then they only focus on how dominant the household feels the FD official has been, ignoring the structural factors such as the rules themselves.

17. A potential third line of explanation, namely, that other factors such as lack of resources, political support or skills may have impeded the implementing agency from proper implementation or making corrections in design does not hold water, since initial resources were made available and the programme itself emerged due to pressure from outside the FD.

18. This section draws heavily on Lele (2001) and Annexure I in Lele *et al.* (2005).

19. Surprisingly, academic studies still seem to take for granted that FDAs are federations (see, for example, Ghate and Mehra 2008).

20. Even in Gujarat, as elsewhere, the micro-plans have to be approved by the FD, and all major silvicultural operations involving trees, including thinning and cutback and, of course, felling, require FD permission. Transport of all forest products outside the village again requires FD permits.

21. Even in states such as Karnataka where the Village Forest chapter has been amended to include JFM, there has been no real progress. In Karnataka specifically, the rules under the amended Act have not been issued, and no new areas have been notified legally as JFM areas under the Act.

22. The JBIC read the critical report (8) of the first phase of the eastern plains project in Karnataka. Its team met with the author and assured that the issues raised in the report would be addressed in the second phase of funding (when in fact one of the core issues was the retrograde impact of funding itself). JBIC then went ahead with funding the second phase, with no substantive changes being made in the JFM orders.

References

Agrawal, A. 2001. 'Common Property Institutions and Sustainable Governance of Resources', *World Development*, 29 (10): 1649–72.

Agrawal, A. and K. Gupta. 2005. 'Decentralization and Participation: The Governance of Common Pool Resources in Nepal's Terai'. *World Development*, 33 (7): 1101–14.

Agarwal, B. 2010. 'Does Women's Proportional Strength Affect their Participation? Governing Local Forests in South Asia', *World Development*, 38 (1): 98–112.

Anonymous (ed.). 1999. 'Proceedings: National Workshop on Joint Forest Management', 24–26 February 1999, Ahmedabad. Aga Khan Foundation (India), Gujarat Forest Department, Society for Promotion of Wasteland Development and Vikram Sarabhai Centre for Development Interaction (VIKSAT), Ahmedabad.

Agriculture Operations Division (AOD). 1993. 'Project Completion Report: Karnataka Social Forestry Project (Credit No.1432-IN)'. Report no. 11929, AOD, Country Department II, South Asia Regional Office. Washington, DC: The World Bank.

Arnold, J.E.M., and W.C. Stewart. 1991. 'Common Property Resource Management in India', Tropical Forestry Paper no. 24, Oxford Forestry Institute, Oxford, UK.

Bahuguna, V.K. 2011. 'Foreword', in V.R.R. Singh, D. Mishra, and V.K. Dhawan (eds), *Status of Joint Forest Management in India (as on June 2011)*, p. 1. Dehra Dun: Forest Research Institute.

Banerjee, A. 2007. 'Joint Forest Management in West Bengal', in O. Springate-Baginski and P. Blaikie (eds), *Forests, People & Power: The Political Ecology of Reform in South Asia*, pp. 221–60. London: Earthscan.

Behera, B. 2009. 'Explaining the Performance of State-community Joint Forest Management in India', *Ecological Economics*, 69 (1): 177–85.

Behera, B. and S. Engel. 2006a. 'Institutional Analysis of Evolution of Joint Forest Management in India: A New Institutional Economics Approach', *Forest Policy and Economics*, 8 (4): 350–62.

———. 2006b. 'Who Forms Local Institutions? Levels of Household Participation in India's Joint Forest Management Program', Discussion Papers on Development Policy no. 103, Bonn: Zentrum für Entwicklungsforschung (Center for Development Research: ZEF).

Bhattacharya, P., K.N. Krishna Kumar, R. Prasad, K.C. Malhotra, D. Debnath, G. Yadav, B.K. Prasad, S. Roy, L. Pradhan, and A. Singh. 2008. 'JFM at Crossroads: Future Strategy and Action Programme for Institutionalizing Community Forestry', in P. Bhattacharya, A.K. Kandya, and K.N. Krishna Kumar (eds), *Joint Forest Management in India*, vol. 1, pp. 3–31. Jaipur: Aavishkar Publishers Distributors.

Chhattre, A. and A. Agrawal. 2009. 'Trade-offs and Synergies between Carbon Storage and Livelihood Benefits from Forest Commons', *Proceedings of the National Academy of Sciences of the United States of America*, 106 (42): 17667–70.

Cohen, J. and N. Uphoff. 1980. 'Participation's Place in Rural Development: Seeking to Clarify through Specificity', *World Development*, 8 (3): 213–35.

Corbridge, S. and S. Jewitt. 1997. 'From Forest Struggles to Forest Citizens? Joint Forest Management in the Unquiet Woods of India's Jharkhand', *Environment and Planning* A, 29 (12): 2145–64.

Correa, M. 1996. 'No Role for Women: Karnataka's Joint Forest Management Programmes', *Economic and Political Weekly*, 31 (23): 1382–3.

Campaign for Survival and Dignity (CSD). 2010. 'Forked Tongue of the Forest Bureaucracy'. New Delhi: CSD.

Damodaran, A. and S. Engel. 2003. 'Joint Forest Management in India: Assessment of Performance and Evaluation of Impacts. Discussion Papers on Development Policy no. 77, Bonn: Zentrum für Entwicklungsforschung (Center for Development Research: ZEF).

Dove, M.R. 1992. 'The Dialectical History of "Jungle" in Pakistan: An Examination of the Relationship between Nature and Culture', *Journal of Anthropological Research*, 48 (3): 231–53.

Dutta, M., S. Roy, D.S. Maiti, and S. Saha. 2005. 'Protecting India's Forests: The Effectiveness of Forest Protection Committees, the Case of Southern West Bengal', *The International Journal of Sustainable Development and World Ecology*, 12 (1): 68–77.

Dutta, M., S. Roy, S. Saha, and D.S. Maity. 2004. 'Forest Protection Policies and Local Benefits from NTFP: Lessons from West Bengal', *Economic and Political Weekly*, 39 (6): 587–91.

Environmental Impact Assessment (EIA) Division. 2008. 'Mid-term Evaluation of the National Afforestation Programme (NAP) Schemes Implemented through Forest Development Agencies (FDAs)'. New Delhi: National Afforestation and Ecodevelopment Board, Ministry of Environment and Forests, Government of India.

Gadgil, M., S.N. Prasad, and R. Ali. 1983. 'Forest Management and Forest Policy in India: A Critical Review', *Forest, Environment and People*, 33 (2): 127–55.

Ghate, R. and D. Mehra. 2008. 'Good in Intention, Bad in Practice: Forest Development Agency: Nesting of JFM Committees', in P. Bhattacharya, A.K. Kandya, and K.N. Krishna Kumar (eds), *Joint Forest Management in India*, vol. 2, pp. 519–33. Jaipur: Aavishkar Publishers Distributors.

Government of India (GoI). 1988. *National Forest Policy*. New Delhi: Ministry of Environment and Forests.

———. 1990. 'Involvement of Village Communities and Voluntary Agencies for Regeneration of Degraded Forest Lands'. Memorandum to Forest Secretaries of All States and Union Territories no. 6-21/89-F.P. dt. 1 June 1990. New Delhi: Ministry of Environment and Forests, Government of India.

———. 2000a. 'Guidelines for Strengthening of Joint Forest Management (JFM) Programme'. No. 22-8/2000-JFM (FPD), New Delhi: Ministry of Environment and Forests (Forest Protection Division), Government of India.

———. 2000b. 'Samanvit Gram Vanikaran Samirddhi Yojana (SGYSY) (Integrated Village Afforestation and Eco-Development Scheme) and creation of Forest Development Agency (FDA) Guidelines', no..22-8/2000-JFM(FPD), National Afforestation and Eco Development Board, New Delhi: Ministry of Environment and Forests, Government of India.

———. 2002. 'Strengthening of Joint Forest Management (JFM) Programme'. No. 22-8/2000-JFM (FPD), New Delhi: Ministry of Environment and Forests (JFM Cell), Government of India.

Government of United Provinces (GoUP). 1931. Kumaon Panchayat Forest Rules no. 441/XIV-366, Forest Department, Government of United Provinces.

Guha, R. 2012. 'The Past and Future of Indian Forestry', in Anonymous (ed.), *Deeper Roots of Historical Injustice, Rights and Resources Initiative*, pp. 1–13. Washington, DC: Rights and Resources Initiative.

Guha, R. and M. Gadgil. 1989. 'State Forestry and Social Conflict in British India'. *Past and Present*, 123: 141–77.

Joint Committee. 2010. 'Manthan: Report of the National Committee on Forest Rights Act'. New Delhi: Ministry of Environment and Forests and Ministry of Tribal Affairs, Government of India.

Joy, K.J., S. Paranjape, A.K. Kiran Kumar, R. Lele, and R. Adagale. 2004. 'Watershed Development Review: Issues and Prospects'. Centre for Interdisciplinary Studies in Environment and Development (CISED) Technical Report. Bangalore: CISED.

Kumar, N., N.C. Saxena, Y.K. Alagh, and K. Mitra. 1999. 'Alleviating Poverty Through Participatory Forestry Development: An Evaluation of India's Forest Development and World Bank Assistance'. Operations Evaluation Department, Washington, DC: World Bank.

Kumar, S. 2002. 'Does "Participation" in Common Pool Resource Management Help the Poor? A Social cost–benefit Analysis of Joint Forest Management in Jharkhand, India', *World Development* 30 (5): 763–82.

Kumar, S. and S. Kant. 2006. 'Organizational Resistance to Participatory Approaches in Public Agencies: An Analysis of Forest Department's Resistance to Community-based Forest Management', *International Public Management Journal*, 9 (2): 141–73.

Larson, A.M. and J.C. Ribot. 2005. 'Introduction', in J.C. Ribot and A.M. Larson (eds), *Democratic Decentralisation through a Natural Resource Lens*, pp. 1–25. Oxon, UK: Routledge.

Lele, S. 1994. 'Sustainable Use of Biomass Resources: A Note on Definitions, Criteria, and Practical Applications', *Energy for Sustainable Development*, 1 (4): 42–6.

———. 1995. 'Voices from the Past: An *Ex-ante* Analysis of Karnataka GO on JFPM'. Paper presented at Seminar on JFPM in Karnataka organized by the Institute for Social and Economic Change at Bangalore, on 9–10 April.

———. 1998. 'Why, Who, and How of Jointness in Joint Forest Management: Theoretical Considerations and Empirical Insights from the Western Ghats of Karnataka'. Paper presented at International Workshop on Shared Resource Management in South Asia organized by the Institute of Rural Management Anand (IRMA) at Anand on 17–19 February.

———. 1999. 'Institutional Issues in (J)FM(& R)', in Anonymous (ed.), *National Workshop on Joint Forest Management*, pp. 19–29. Ahmedabad: VIKSAT, Gujarat Forest Department and Aga Khan Foundation.

————. 2000. 'Godsend, Sleight of Hand, or Just Muddling Through: Joint Water and Forest Management in India'. Overseas Development Institute (ODI) Natural Resource Perspectives no. 53, London: ODI.

————. 2001, 'What is Wrong with JFPM?', Paper presented at State-level Convention on Campaign for Participatory Forest Management, organized by Jana Aranya Vedike at Bangalore on 4–5 December 2001.

————. 2004. 'Beyond State-Community Polarisations and Bogus "Joint"ness: Crafting Institutional Solutions for Resource Management', in M. Spoor (ed.) *Globalisation, Poverty and Conflict: A Critical 'Development' Reader*, pp. 283–303. Boston and London: Kluwer Academic Publishers, Dordrecht.

————. 2011. 'Rethinking Forest Governance: Towards a Perspective beyond JFM, the Godavarman Case and FRA', in Anonymous (ed.), *The Hindu Survey of the Environment 2011*, pp. 95–103. Chennai: *The Hindu*.

Lele, S., A.K. Kiran Kumar, and P. Shivashankar. 2005. 'Joint Forest Planning and Management in the Eastern Plains Region of Karnataka: A Rapid Assessment'. Centre for Interdisciplinary Studies in Environment and Development (CISED) Technical Report, Bangalore: CISED.

Lele, S., M. Pattanaik, and N.D. Rai. 2010. 'NTFPs in India: Rhetoric and Reality', in S.A. Laird, R. McLain, and R.P. Wynberg (eds), *Wild Product Governance: Finding Policies that Work for Non-Timber Forest Products*, pp. 85–112. London: Earthscan.

Lele, S., I. Patil, S. Badiger, A. Menon, and R. Kumar. 2011. 'Forests, Hydrological Services, and Agricultural Income: A Case Study from Mysore District of the Western Ghats of India', in A.K.E. Haque, M.N. Murty, and P. Shyamsundar (eds), *Environmental Valuation in South Asia*, pp. 141–69. Cambridge: Cambridge University Press.

Lele, S. and V. Srinivasan. 2013. 'Disaggregated Economic Impact Analysis Incorporating Ecological and Social Trade-offs and Techno-institutional Context: A Case from the Western Ghats of India', *Ecological Economics*, 91: 98–112.

Lise, W. 2000. 'Factors Influencing People's Participation in Forest Management in India', *Ecological Economics*, 34 (3): 379–92.

Midgley, J. 1986. 'Community Participation: History, Concepts, and Controversies', in J. Midgely, A. Hall, M. Hardiman, and D. Narine (eds), *Community Participation, Social Development and the State*, pp. 13–44. London: Metheun.

Ministry of Environment and Forests (MOEF). 2010. 'National Mission for a Green India', draft submitted to Prime Minister's Council on Climate Change, New Delhi: Ministry of Environment and Forests, Government of India.

Murphree, M.W. 2000. 'Boundaries and Borders: The Question of Scale in the Theory and Practice of Common Property Management. Paper presented at Constituting the Commons: Crafting Sustainable Commons in the Millennium, organized by International Association for the Study of Common Property (IASCP) at Bloomington on 31 May–4 June.

Naik, G. 1997. 'Joint Forest Management: Factors Influencing Household Participation', *Economic and Political Weekly*, 32 (48): 3084–9.

Ostrom, E. 1990. *Governing the Commons: The Evolution of Institutions for Collective Action*. New York: Cambridge University Press.

Pathak, S. 1997. 'State, Society and Natural Resources in Himalaya: Dynamics of Change in Colonial and Post-Colonial Uttarakhand', *Economic and Political Weekly*, 32 (17): 908–12.

Putz, F.E. and K.H. Redford. 2010. 'The Importance of Defining 'Forest': Tropical Forest Degradation, Deforestation, Long-term Phase Shifts, and Further Transitions', *Biotropica*, 42 (1): 10–20.

Ramakrishnan, R., M. Dubey, R.K. Raman, P. Baumann, and J. Farrington. 2002. 'Panchayati Raj and Natural Resources Management: How to Decentralise Management over Natural Resources'. National Synthesis Report, Overseas Development Institute, London (along with Taru Leading Edge (New Delhi and Hyderabad), Centre for Budget and Policy Studies (Bangalore), Centre for World Solidarity (Hyderabad), and Sanket (Bhopal).

Rangachari, C.S. and S.D. Mukherji. 2000. *Old Roots New Shoots: A Study of Joint Forest Management in Andhra Pradesh, India*. New Delhi: Winrock International and Ford Foundation.

Rangarajan, M. 1996. *Fencing the Forest: Conservation and Ecological Change in India's Central Provinces 1860–1914*. New Delhi: Oxford University Press.

Rao, K.K., P.V.V. Prasada Rao, K. Anil, J. Chourey, and S. Kanna Kumar. 2004a. 'Joint Forest Management in Andhra Pradesh: Its Spread, Performance and Impact', in N. H. Ravindranath and P. Sudha (eds), *Joint Forest Management in India: Spread, Performance and Impact*, pp. 41–65. Hyderabad: Universities Press.

Rao, R.J., K.S. Murali, and I.K. Murthy. 2004b. 'Joint Forest Management Studies in India: A Review of the Monitoring and Evaluation Methods', in N.H. Ravindranath and P. Sudha (eds), *Joint Forest Management in India: Spread, Performance and Impact*, pp. 26–40. Hyderabad: Universities Press.

Ravindranath, N.H., K.S. Murali, and K.C. Malhotra (eds). 2000. *Joint Forest Management and Community Forestry in India: An Ecological and Institutional Assessment*. New Delhi: Oxford and IBH.

Ravindranath, N.H. and P. Sudha (eds). 2004. *Joint Forest Management in India: Spread, Performance and Impact*. Hyderabad: Universities Press.

Ravindranath, N.H., P. Sudha, K.C. Malhotra, and S. Palit. 2004. 'Sustaining Joint Forest Management in India', in N.H. Ravindranath and P. Sudha (eds), *Joint Forest Management in India: Spread, Performance and Impact*. Hyderabad: Universities Press, pp. 317–40.

Reddy, V.R., M.G. Reddy, M. Bandi, V.M. Ravi Kumar, M.S. Reddy, and O. Springate-Baginski. 2007. 'Participatory Forest Management in Andhra Pradesh: Implementation, Outcomes and Livelihood Impacts', in O. Springate-Baginski and P. Blaikie (eds), *Forests, People & Power: The Political Ecology of Reform in South Asia*, pp. 302–34. London: Earthscan.

Ribot, J. 2002. *Democratic Decentralisation of Natural Resources: Institutionalizing Popular Participation*. Washington, DC: World Resources Institute.

Rodríguez, J.P., T.D. Beard Jr, E.M. Bennett, G.S. Cumming, S. Cork, J. Agard, A.P. Dobson, and G.D. Peterson. 2006. 'Trade-offs across Space, Time, and Ecosystem Services', *Ecology and Society*, 11 (1): 28.

Roy, S.B. and S. Roy. 2008. 'Institutional Sustainability of JFM Programme in India', in P. Bhattacharya, A.K. Kandya, and K.N. Krishna Kumar (eds), *Joint Forest Management in India*, vol. 2, pp. 341–9. Jaipur: Aavishkar Publishers Distributors.

Saigal, S. 2000. 'Beyond Experimentation: Emerging Issues in the Institutionalization of Joint Forest Management in India', *Environmental Management*, 26 (3): 269–81.

Sarin, M. 1997. 'Who is Gaining? Who is Losing? Gender and Equity Concerns in Joint Forest Management', in Proceedings of the Regional Workshop on Community-Based Conservation. New Delhi: Indian Institute of Public Administration.

Sarin, M., N.M. Singh, N. Sundar, and R.K. Bhogal. 2003. 'Devolution as a Threat to Democratic Decision-making in Forestry? Findings from Three States in India'. Overseas Development Institute (ODI), Working Paper no. 197. London: ODI.

Saxena, N.C., M. Sarin, R.V. Singh, and T. Shah. 1997. 'Independent Study of Implementation Experience in Kanara Circle'. Review committee report, Bangalore: Karnataka Forest Department.

Singh, V.R.R., D. Mishra, and V.K. Dhawan (eds). 2011. *Status of Joint Forest Management in India (as on June 2011)*, Dehra Dun: Forest Research Institute.

Springate-Baginski, O. and P. Blaikie (eds). 2007. *Forests, People & Power: The Political Ecology of Reform in South Asia*. London: Earthscan.

Sudha, P., P.R. Bhat, R. Jagannatha Rao, B.C. Nagaraja, G.T. Hegde, C.M. Shastri, G.N. Hegde, D.M. Shetty, K.S. Murali, D.M. Bhat, and N.H. Ravindranath. 2004a. 'Joint Forest Planning and Management in Karnataka: Its Spread, Performance and Impact', in N.H. Ravindranath and P. Sudha (eds), *Joint Forest Management in India: Spread, Performance and Impact*, pp. 86–121. Hyderabad: Universities Press.

Sudha, P., K.C. Malhotra, S. Palit, K. Kameswara Rao, M. Srinivas, N.K. Negi, B.K. Tiwari, R. Jagannatha Rao, P.R. Bhat, I.K. Murthy, and N.H. Ravindranath. 2004b. 'Joint Forest Management: Synthesis of its Spread,

Performance and Impact in Andhra Pradesh, Gujarat, Karnataka, Rajasthan, Tripura and West Bengal', in N.H. Ravindranath and P. Sudha (eds), *Joint Forest Management in India: Spread, Performance and Impact*, pp. 196–219. Hyderabad: Universities Press.

Sudha, P. and N.H. Ravindranath. 2004. 'Evolution of Policies and the Spread of Joint Forest Management in India', in N.H. Ravindranath and P. Sudha (eds), *Joint Forest Management in India: Spread, Performance and Impact*, pp.1–25. Hyderabad: Universities Press.

Sundar, N. 2001. 'Is Devolution Democratisation?', *World Development*, 29 (12): 2007–24.

Sundar, N., R. Jeffery, and N. Thin. 2001. *Branching Out: Joint Forest Management in India*. New Delhi: Oxford University Press.

Verma, D.P.S. 2008. 'Policy, Legal Issues and Property Rights in JFM', in P. Bhattacharya, A.K. Kandya, and K.N. Krishna Kumar (eds), *Joint Forest Management in India*, vol. 1, pp. 120–7. Jaipur: Aavishkar Publishers Distributors.

2

SAGARI R. RAMDAS

Adivasis, Pastoralists, and Forest Governance

Challenges and Opportunities

The history of conflict between the state and adivasis, peasants and pastoralists who graze their animals in forests and other non-forest common property resources dates back to the second half of the nineteenth century, when the British colonial state extended its laws and models of private property and state monopoly over forests and other natural resources in India. This conflict continued after independence because the Indian state[1] persisted with an identical set of laws and ideologies, pertaining to the governance of forests and other non-forest common property resources. People were 'encroachers' and needed to be kept out. Systems of grazing, and shifting cultivation, have in particular and consistently been singled out by both the precolonial and postcolonial state for being ecologically destructive and economically inefficient. The state has persisted in taking aggressive steps to wipe out these 'pernicious practices' and through legislations, policies, and development programmes attempted to delegitimize and criminalize these livelihood practices. It is a testimony to the resilience of people, and their 'traditional' livelihoods that, despite 200 years of the state's attempt to stamp them out, they survive. Grazing on common property resources (forest, non-forest) and on agriculture fallows, continues to be the single most important means by which livestock (particularly small ruminants) in India, obtain their feed and nutrition (Foundation for Ecological Security 2010).

Two hundred years of resistance has finally resulted in the state legally recognizing 'grazing' and seasonal nomadic pastoralism and transhumance, under the aegis of the Scheduled Tribes and Other Traditional Forest Dwellers (Recognition of Forest Rights) Act 2006 (FRA). This legislation recognizes communities as integral to the survival and sustainability of the forest eco-system, and aims to undo a historic injustice. It provides a legal framework for the state to record and recognize age-old collective and individual rights of adivasi communities and other traditional forest-dwellers to their ancestral homes and habitats in the forests, including the path-breaking recognition of grazing as a legal right. This innocuous-sounding clause in the legislation, we hypothesize, is going to pave the way for profound changes in governance of forests, and will also force the state to wake up and respond to the demands and development plans of communities for their livestock. The provisions in the FRA reaffirm the powers of the *gram sabha* as envisaged in the Panchayats (Extension to the Scheduled Areas) Act, 1996 (Act No. 40 of 1996) (PESA) that pertain to Schedule V regions, and the powers of panchayati raj institutions as listed in Schedule II (Article 243)[2] that pertain to all other rural villages, with respect to the governance of village resources. The community rights of the FRA coupled with the empowering legislations of local governance (PESA and the Panchyati Raj Act), add strength to people's movements in their attempt to challenge fossilized mindsets of the state and those of global capital interested in industrializing people's livestock livelihoods.

This chapter, located in the state of Andhra Pradesh, is an account of how the state has systematically delegitimized people's customary livestock-rearing practices that are anchored on grazing regimes embedded within forests and other non-forest common property resources. It also highlights how people have resisted the enclosure of their commons, and the implications of the FRA in recognizing customary practices and knowledge systems, hence strengthening traditional livelihoods.

People, Land, and Livestock: Understanding the Relationship and People's Customary Practices

Historically in India, livestock (large and small ruminants) have obtained their nutritional and water requirements by grazing on common lands, forests, and harvested agriculture fields. Large-scale legal restrictions and imposition of private property by the colonial state were responsible for alienating communities from their natural resource base (Bhattacharya 1995; Murali 1995). This drastically transformed the complex, mutually sustaining relationship that had evolved hitherto between agriculture, forests, and the non-forest commons.

Despite legal obstacles, livestock-rearers in India, cutting across different ecological terrains, landholding categories, castes and genders, and livestock type continue to depend on the non-forest commons, forests, and private agriculture lands to meet their fodder and water requirements. A continuum exists of households that primarily depend on common property resources (forest and/or non-forest) and private lands within their own village to those who are mostly dependent on these same categories of resources beyond the village boundaries, sometimes extending to distances of over 400 km from the 'homebase'. The 54th National Sample Survey Organisation (NSSO) report defines common property resources (CPRs) to include village pasture lands and grazing grounds, village forests and woodlots, protected and unclassed government forests, wasteland, common threshing grounds, watershed drainage, ponds and tanks, rivers, rivulets, water reservoirs, canals and irrigation channels (NSSO 1999). CPRs constitute 15 per cent of India's total geographic area at 0.31 hectares (ha) per household and rural India still depends significantly on CPRs to rear livestock. At the all-India level, 20 per cent of households depended on CPRs for grazing livestock, 13 per cent collected fodder from CPRs, and only a small percentage (2 per cent) reported cultivation of fodder on CPRs.

Livestock-owning communities respond in different ways when there is a decline or decrease in access to and/or availability of fodder/water. There could be a shift in species reared. Erstwhile pastoralists who reared cattle have switched to rearing sheep and goats (FES 2010; Kavoori 1999) while settled farmers who reared cattle are today rearing buffaloes (Brara 2006) and sheep (FES 2010). They respond by adjusting the size of their flock such as increasing or decreasing the numbers of animals in their flock/herd. There may be spatial movements of migration to 'greener pastures' as it were, in search of new areas to access fodder and water from CPRs beyond the village or from harvested private fields on mutually beneficial terms, which are negotiated with farmers. The landowners allow the animals to be penned on their fields and graze on the stubble of harvested crop-residue in exchange for manure and urine that is valuable for enriching the soils. Pastoralists with their animals may alternately migrate to forest areas for extended periods of time. The small peasants, pastoralists, or adivasis will relinquish their animals and become 'non-livestock owners' only when there appear to be no other avenues to care for their livestock.

The relationship between livestock, land, and people in Andhra Pradesh has been particularly strong, as a large fraction of the state is semi-arid and forested. Livestock is seen as wealth by families in rural Andhra. Animals provide dung and urine to keep the soil healthy and fertile, energy for agricultural operations and transport, food (milk, meat, and eggs) for human consumption, and fibre for clothing. Livestock is the dependable bank on hooves. This circle of connectedness and inter-dependency between animals, crops, land, forests, and people is captured in a rhyme commonly cited by farmers in Chittoor:

> *Kasu leka Pashu ledu, Pashu leka, Penta ledu, Penta leka, Panta ledu, Panta leka Pashu ledu.*

> Without fodder we have no animals, without animals we have no dung, without dung we have no crops, without crops we have no fodder.

Local names for animals, proverbs, and songs from different parts of the state, reinforce and reflect the intrinsic value and irreplaceable role of livestock and the sense of loss when a farmer loses any stock. In the northern tracts of Andhra Pradesh, livestock are referred to as *sommulu*, literally meaning 'wealth' whereas the Kuruma shepherds of Telangana worship their Deccani sheep as Lakshmi. Koya adivasis of East Godavari have a saying: '*kodi valla koti labhalu*', meaning a crore and more benefits from one hen. The Chenchu proverb '*Bhari chechi poyi, muntha pagili poyee*' means the death of a buffalo leads to a broken, empty life, like a broken pot.

Livestock, both large and small ruminants, have historically been raised under grazing-based production systems where animals are seasonally managed and herded to different parts of the village so as to obtain their fodder and water. Alternatively, livestock are herded to other villages and forests both within and beyond the state. Transhumant livestock livelihoods have existed in the region for thousands of years (Bhangya 2010; Murali 1995; Sontheimer and Murty 2004). Natural vegetation including grasses, trees, shrubs, creepers, and climbers is used as fodder and medicines. This vegetation was available on community grazing/pasture land (known as *charayi zameen*), *gautan* land (village settlement), *poromboke* land (land communally used by the village but unusable for agriculture), *banjar* land (fallow lands), *shikam* land (areas bordering village tanks), and lands situated along canals and forests. Villages used to have clear-cut mechanisms and customary grazing practices in different locations within and beyond the village. The village also had a practice of appointing a person (or two people), known as *Jangali Kasewan* in Telangana, who was responsible for grazing the village cattle and buffaloes. Dalit families frequently provided this service, and each owner paid a fixed monthly amount to them.[3] In addition to animals being grazed on CPRs, animals grazed on harvested agriculture fallows. What is well recorded is that even 'privately' owned lands, become commonly grazed lands, post-harvest, and are extremely important as a fodder source for animals, particularly during summer months, and especially in the semi-arid regions (Anthra 2008; FES 2010). In Telangana, a portion of private land is kept for grazing animals and is known as *woralu*. Grazing on the stubble of assorted crop residues (millets, pulses and oilseeds), or being stall-fed the crop-residue during the summer months, supplemented with lopped tree fodder, completes the feeding regime for animals. During the monsoons, farmers collect naturally available grasses and other herbage from common and private land, to feed their animals.

Small ruminant owners (shepherds) additionally pen or fold their sheep/goats on farmers' fields, and in return for the urine and manure, the farmer pays the shepherds/ pastoralists in kind, mostly grain. Villages also developed special arrangements amongst themselves to accommodate animals from one village grazing in another village. For example, in Gummadidalla, Medak district, Telangana region, shepherds have traditionally grazed their sheep in the forests of Mambapur. In exchange, the shepherds of Gummadidalla contributed one sheep from their flock to the Mambapur villagers at the time of the 'Peerla

Panduga', or Moharram, a Muslim festival celebrated by all (Anthra 2009). Many small farmers and shepherds enter into lease and rental arrangements with landowners within their village or in neighbouring villages, where they pay the owner a rent for a period of four to six months in return for exclusive grazing rights. Some may actually give their animals to other owners to rear on a 'sharing' basis, where the recipient grazes and takes care of the animals and both parties share the offspring on a 50–50 basis.

Forest grazing systems are also well demarcated with distinct parts of the forest grazed at specific periods through the year and different livestock preferring areas with specific fodder types. There are defined periods of rest when the forest as a whole, or parts of the forest, are left undisturbed and allowed to rejuvenate. Animals contribute to forest wealth and diversity, enriching the soil with manure, controlling the undergrowth and grass and thereby minimizing the chances of summer fires and propagating different tree and shrub species. The interaction of animal and forest is one of reciprocity, each one nurturing the other. Each region and each community has over the years developed a vast repertoire of indigenous livestock management, feeding, shelter, breeding, and healing practices appropriate to their area. The survival and practice of traditional knowledge is intrinsically linked to the availability and access to these local genetic resources (Ghotge *et al.* 2002; Ramdas *et al.* 2004).

Traditional institutions of governance,[4] such as the gram sabha/gram panchayat, played a critical role in the governance of the commons—forest, non-forest and waterbodies, and decision-making around land-use for grazing. While some traditional practices and institutions of governance have collapsed and broken down, others continue to survive, as illustrated in the case of the village Gummadidalla.

Laying Waste to the 'Non-forest' Commons: The Role of the State

The Government of India (GoI) has its own lexicon of land governance. Lands are categorized as gautan (habitation area), poromboke (public land accessible to all), banjar (wasteland), but all these are considered a part of 'government revenue wastelands', administered by the Revenue

Department. Forests are administered by the Forest Department (FD). The state takes its own unilateral decisions vis-à-vis most of these lands, which have seriously restricted and constrained people's customary usage of the land, weakened institutions of local governance, and in turn negatively affected their livelihoods.

A government report titled 'The Report of the Committee on State Agrarian Relations and the Unfinished Task of Land Reforms' (GoI 2009a) highlights that the state is unable to distinguish which land-use categories actually are commons. This, the report suggests, has resulted in a long history of improper public interventions in CPRs. The report goes on to say that various development projects have resulted in the over-exploitation of the commons and that different departments work at cross-purposes. Other committees such as the Task Force on Grasslands and Deserts, appointed by the Ministry of Environment and Forests, talks about the 'orphan' status of grassland, a valuable 'common' resource, but not the responsibility of the FD, the Agriculture Department, nor the Veterinary Department. The latter are concerned with livestock, but not the grass on which the livestock are dependent (GoI 2006). Finally, the report on land reforms, points out that the term 'wasteland'[5] continues to be used as in the colonial period and hence are deemed to be of 'no use' and available to the government to be utilized for 'public purpose'. The problem for rural and adivasi people, is that so-called 'government wastelands', is part of customary grazing land.

In this section, we illustrate how key legislations and public policies cutting across multiple sectors, have served to systematically undermine and disrupt grazing-based livestock production systems in Andhra Pradesh 'forest' and 'non-forest' CPRs. Land reforms that involved CPR distribution, agriculture, and animal husbandry policies that have promoted intensification and industrial production, expanding irrigation coupled with privatization of water resources, the spread of monoculture forestry, horticulture and biofuel plantations, and the concentration of land in the hands of 'neo-zamindars' and the real-estate mafia, enabled by amendments to existing land laws, have profoundly impacted land-use and livestock in Andhra Pradesh. These land-use transformations and accompanying loss of local resource governance have triggered new pressures on forests, 83 per cent of which are located in the Schedule V regions of the state.

Faulty Land Reforms: Village Pastures and Grazing Grounds Distributed as Private Pattas

Land reforms and land distribution were key thrust areas of state public policy and economic planning in the 1950s. The state's inability to enforce a just process of land distribution stands out as one of the earliest and most crucial factors post-independence that resulted in mass-level conversion of common land/grazing land into private land in Andhra Pradesh. According to the Andhra Pradesh 'Human Development Report, 2007' (Government of Andhra Pradesh [GoAP] and Centre for Economic and Social Sciences [CESS] 2007), land reforms carried out in the period 1950–74, had three major components: abolition of intermediaries, tenancy reforms, and the redistribution of land acquired through the land ceiling. While the abolition of intermediaries was relatively successful, tenancy reforms and land ceilings were a dismal failure. Under the Ceiling Acts (1961 and 1973), big landlords did not declare their excess land, and where they did, they handed over fallow land or the most infertile land, all of which was permitted under the law. Land distribution became a process whereby the state compensated landlords for surplus land and large scale benami transactions. Very little surplus land was obtained. Moreover, much of this land was uncultivable, and in many cases the distribution of land only took place on paper (GoAP and CESS 2007). In the early 1970s (1973–4), agitations for distribution of banjar lands, rights for small and marginal peasants on temple land, and distribution of forest land took place as expressions of dissent against the tardy implementation of land reforms by the state (GoAP and CESS 2007). The state thus began to assign government land (essentially banjar, poromboke, and other common grazing land) to the landless poor—12.5 per cent of the total net sown area comprised such land.

These changes are evident if one examines land-use records: in 1955–6, permanent pastures comprised 1.17 million ha or 4.3 per cent of the entire geographic landmass of Andhra Pradesh. This decreased to 0.84 million ha or approxmiately 2 per cent of the landmass by 1990–1 (Table 2.1). The 1.2 million ha of pastures, grazing land and culturable wastes 'lost' between 1955–6 and 1990–1 (−28 per cent) is approximately equivalent to the net sown area that was distributed to landless communities. The figures also point to a drastic decline in public grazing

TABLE 2.1 Land-use in Andhra Pradesh, 1960–1 to 2009–10 (in million ha)

Land-use	1955–6	1960–1	1970–1	1980–1	1990–1	% change (30 years)	2000–1	2009–10	% change (18 years)
Total geographic area	26.98	27.42	27.44	27.44	27.44		27.44	27.5	
1. Forest area	5.66	5.97	6.34	6.18	6.2		6.2	6.21	
2. Non-agricultural uses	1.35	1.83	2.12	2.17	2.30	+70%	2.51	2.67	16%
a. Barren and uncultivable land	2.93	2.35	2.1	2.34	2.09		2.11	2.04	
3. Permanent pastures and grazing land	1.17	1.2	1.07	0.92	0.84	–28%	0.67	0.56	–36%
Miscellaneous a) Trees	0.25	0.29	0.29	0.27	0.26		0.27	0.29	
b. Culturable waste	1.66	1.62	1.11	0.87	0.78	–53%	0.72	0.64	–18%
4. Total fallows	2.64	3.09	3	3.65	3.86	46%	3.86	4.98	29%
a. Other fallows	0.69	0.89	0.88	1.35	1.37		1.41	1.62	
b. Current fallow	1.95	2.45	1.77	2.56	2.48		2.31	3.36	
5. Net sown area	11.29	10.78	11.73	10.73	11.02		11.22	10.08	–10%
6. Gross cropped area	11.82	12.77	12.77	12.7	13.63	15%	13.63	13.56	

Source: Compendium of Area and Land Use Statistics of Andhra Pradesh 1955–56 to 2004–5 and An Outline of Agricultural Situation in Andhra Pradesh 2007–8, both from the Directorate of Economics and Statistics, Hyderabad.

lands with an increase in primary grazing on private land. Secondary grazing areas with public access (forests, barren and uncultivable lands, non-agriculture lands) increased slightly whereas the net cultivated area decreased (Table 2.1). The decline in public grazing areas has been accompanied by a huge loss of traditional fodder varieties. The 54th NSSO report on CPRs points out that Andhra Pradesh has a mere 9 per cent of its total geographic area classified as CPRs, with an availability of 0.17 ha per household, which is lower than the national average. This data from Andhra Pradesh, supports N.S. Jodha's findings pertaining to Andhra Pradesh (Jodha 1986, 1995a, 1995b).

In short, land reforms which were meant to bring justice to the landless, ended up as an exercise of distributing the most uncultivable marginal land, which typified the grassland ecology of the region, to communities on the margins. At the same time land reforms dispossessed the entire village (including the new landowners) and nomadic pastoralists, of their customary grazing land. The landless/near landless poor and agro-pastoral farming communities have been hardest hit by the declines in public and common grazing land, because of their greater dependence on common property resources, in the absence of private grazing land, for meeting the fodder and water needs of their livestock.

An unfortunate fallout of this faulty land reform has been heightened conflict between two marginalized communities, namely, the shepherds and the Dalit cultivators. This is an aspect not brought out in Jodha's studies but observed in Andhra Pradesh (Ramdas 2011) and other parts of the country (Vikas Adhyayan Kendra 2006). The traditional shepherd communities (Gollas and Kurumas, who are classified as backward caste communities in the Telangana and Andhra regions of the state) and the Lambadas, the erstwhile cattle-rearing pastoralists of the Deccan, used to graze their animals on CPRs, which were subsequently distributed or 'assigned' to the landless, many of whom were Dalits (Srinivasalu 2002). Shepherds in particular, challenged by shrinking grazing spaces, hotly contested the conversion of CPRs into private land titles. The government, in fact, recognized these traditional claims in Government Order (Ms) No. 559, Revenue Department, GoAP. Some shepherds utilized this GO to secure their access. However, this provision clashed with the government's strategy of re-distributing common land as private assigned land to landless communities and illustrates the absence of an overall vision and plan, which encompasses the need for land reforms along with the collective need for common land. As an alternative to the commons, many shepherds began to lease in fallow land from landlords for fixed seasons of the year, to graze their flocks.

It is perhaps the greatest 'tragedy of the commons' that distributing a village common property resource as 'private land' to the most discriminated community in the village, namely, the Dalits, has served to heighten and deepen caste tensions and divisions, pit the poor against the poor, a slightly less oppressed group against the most oppressed group, and ultimately allow powerful classes/castes to continue to control land. Moreover, the land distributed to Dalits is nominal and hence it is not possible for them to graze animals on this land. Today, Dalits own negligible livestock (small and large ruminants). Lack of resources (financial and material) has often meant that Dalits have not been able to cultivate their assigned land, leaving them fallow. Shepherds and other communities in the village have in such situations continued to graze their animals on such land. Whenever state-supported land, water, and soil conservation programmes[6] for developing lands assigned to Dalits are operationalized, it triggers fresh tensions and conflicts between the communities, as the land once again becomes 'out of bounds' to shepherds. Tensions between agriculturists and shepherds peak in the agriculture season, when almost all land gets cropped, and there is virtually no fallow land left where livestock can graze. Many Dalits, marginal peasants, and shepherds, who are unable to migrate nor can afford to lease in land, are forced to sell their animals.

Post 1990s: The New Wave of Enclosures and the Emergence of a New Class of Zamindars

Since 1990, in tune with the Indian government's move to liberalize the economy, the Andhra Pradesh government has taken several major steps to open its economy, cut back state expenditure, and implement neoliberal reforms and legislations. These measures have facilitated the transfer of land from agriculture to industrial, real-estate, and other non-agriculture uses such as plantations. Several amendments to existing land legislations have been executed and new laws enacted to facilitate the process. This has led to further shrinking of the remaining CPRs, the alienation of 'assigned' lands, which were erstwhile CPRs, and the rapid conversion of privately owned agriculture lands to non-agriculture uses.

Loss of Commons Facilitated by Amendments to Land Laws

Changing the laws of the land, to facilitate unprecedented transfer and concentration of resources in the hands of corporations and industrialists, enabled the mushrooming of real-estate businesses and land markets. In 2006, the GoAP enacted The Andhra Pradesh Agricultural Land (Conversion for Non-Agricultural Purposes) Act (Act No. 3 of 2006) to regulate the conversion of agricultural land to non-agricultural purposes. Simultaneously, the The Andhra Pradesh Non-Agricultural Land Assessment Act, 1963 (NALA) was repealed and abolished. The 2006 Act defines 'non-agricultural land' in extremely broad terms, including virtually all land except agricultural land. Hence, the Act has enabled the quick conversion/diversion of agricultural land for non-agricultural purposes, including most notably real-estate. In 2007, the GoAP amended the Andhra Pradesh Assigned Lands (Prohibition of Transfers) Act, 1977 that was enacted to protect the interests of the original assignees of the land. The amendment to the Act, No. 8 of 2007, empowered the government to use the resumed assigned land in the notified areas for infrastructural development and promotion of industries, without assigning the same to the original assignee. This amendment empowered the state to steamroll the landless poor who had been assigned land from the commons in favour of the wealthy in the name of public purpose. For example, large areas of land have been handed over to private parties by the government to set up Special Economic Zones (SEZs) (Seethalakshmi 2010).

The AP Land Reforms (Ceiling on Agriculture Holdings) Act of 1973 was amended twice in 2009 to facilitate land transfer for public purposes like industries and infrastructure projects. The amendments were brought in to enable the government to sell ceiling surplus land to industries, instead of distributing them to the landless poor in accordance with the objectives of land reforms (Seethalakshmi 2010).

Land put to non-agricultural purposes for the period 1990–1 to 2009–10 increased by 16 per cent from 23.06 lakh ha to 26.37 lakh ha. This, too, may be an underestimation since many land transactions, especially real-estate ones, are not officially recorded (Seethalakshmi

2010). Permanent pastures declined by 36 per cent from 8.42 lakh ha in 1990–1 to 5.6 lakh ha in 2009–10, a much faster decline than in the decades preceding liberalization.

These changes, coupled with the centre's Special Economic Zone Act, 2005 and the Andhra Pradesh Special Economic Zones Policy, facilitated a mushrooming of SEZs across the state.[7] There are no clear-cut norms related to land acquisition, especially agricultural lands, for setting up SEZs. The state has facilitated the easy transfer of land at desirable locations and attractive prices to attract SEZ developers. The liberalization of land laws, has paved the way for real-estate companies to lure vulnerable debt-ridden farmers to sell their land (Seethalakshmi 2010).

The visible evidence of these 'land banks' is the acres upon acres of 'fenced' lands cordoned off with white pillars and barbed wires. If one drives south from Hyderabad towards the Tungabhadra, or north towards Warangal, or east towards Nalgonda or west towards Mumbai, the barbed wire is an eyesore. It has disrupted free movement of live-stock owned by small peasants and pastoralists, and traditions of har-vested agriculture fallow land have become commonly grazed areas. Ironically, the land monopoly by a handful of wealthy Indians has facili-tated the emergence of a new class of livestock entrepreneurs, whose land wealth affords them the opportunities to invest in and establish goat, sheep, emu, or dairy farms, managed under intensive stall-fed or extensive grazing systems!

Diversion of Land for Non-food Cash Crops and Horticulture Crops

Farmland is rapidly being converted into horticulture plantations. 1.9 million ha of land in 2009–10 was cropped with fresh fruits, plantation crops (cashew, coconut, palm oil), vegetables, spices (chillies, turmeric, tamarind, ginger, garlic), flowers and medicinal plants (GoAP 2010). India's National Biofuel Policy passed in December 2009 (GoI 2009b) has also encouraged the mass cultivation of non-edible oil seeds to pro-duce biodiesel. Together, these two developments constitute the 'sec-ond wave of enclosures of the commons'. Although the biofuel policy claims to be a policy that avoids a conflict of fuel versus food, proposed targets of biodiesel are to be met through raising biodiesel plantations on 'wasteland'. The mission document claims that 55 million ha of land in India is wasteland, of which 72 per cent is non-forest degraded land, and 28 per cent is forest degraded land. Like previous interventions, the promotion of biofuel has ignored the fact that wasteland is not, in fact, wasteland at all.

These trends within agriculture are the outcome of policies originating in the 1960s. Green Revolution models of farming, comprising high external input technologies such as intensive irrigation, fertilizers and pesticides, chemicals, high-yielding varieties, mechanization, and monocrops, entered Andhra Pradesh in the coastal 'rice-bowl' regions around the Krishna and Godavari deltas (Subramaniam and Shekhar 2003), and expanded to the rainfed dryland areas of Telangana, Rayalseema, and coastal Andhra Pradesh in the 1980s. It triggered a shift from food crops, primarily millets, to supposed high-value crops such as oilseeds, cotton, and pulses (Subramaniam and Shekhar 2003). The share of millets in gross cropped area declined from 27 per cent during 1978–9 to less than 10 per cent in 1998–9. Today, in all districts of Andhra Pradesh, it is fair to say that food-based systems of mixed crop-livestock farming have been converted to commercial cash crop systems. In other words, the rich diversity of different pulse-millet/cereal-legume mixed crops, which were a nutritionally rich source of crop-residue as fodder for animals and feed for poultry, have been replaced by monocrops such as high-yielding varieties of paddy, sugarcane, and cotton, all of which yield poorer residue for fodder. Several such farms today are devoid of cows and bullocks, and a gradual separation of livestock from crops is evident at the level of the farm. Harvested cotton plants have become the main source of fodder for sheep for two to three months in a year in intensive cotton cultivated regions, which have virtually transformed to GMO Bt cotton in the last decade. However, Bt cotton is under the scanner for possible adverse impacts on animals that graze/feed on the byproducts, namely crop-residues (GoI Standing Parliamentary Committee, 2012). Moreover, the past decade has seen the explosion of commercially cultivated maize and soya to be used as poultry and cattle feed, suggesting the absence of other sources of natural fodder.

Two new threats to crop-residue availability are the emergence of mechanized harvesters, which are rapidly replacing human labour in harvesting paddy, and the demand for crop-residues as a raw material for biomass-based energy industries (Kandhari and Pallavi 2010). The harvesters cut the crop about a foot and a half high, resulting in huge wastage of paddy crop-residue. The standing crop-residue is too high for cattle to graze on and hence results in large wastage. Farmers end up burning the standing crop-residue in preparation for the next agriculture season. In short, a landscape of growing scarcity of crop-residues is emerging.

These transformations in agriculture have been accompanied by government livestock development programmes that have focused on replacing so-called low-producing nondescript livestock with 'high-yielding/producing' superior breeds, and aim to replace the multi-functional roles of livestock in food-farming systems, with animals that have specialized production traits (for example, high milk production) (Ghotge 2004).

Scarcity of Drinking Water-led Migration of People and Animals

Livestock is primarily dependent on public water sources such as rivers, streams, canals, and tanks that are available along their grazing route. Between 1955–6 and 2004–5, the contribution of tanks to total irrigation in Andhra Pradesh declined from 39 per cent to about 12 per cent. In Telangana, tank irrigation declined by 58 per cent from 1956–7 to 2005–9 whereas in Rayalseema and coastal Andhra Pradesh, the decline was 70 per cent and 25 per cent respectively (Pingle 2011). This has been accompanied by a sharp increase in irrigation from tube wells. In 2007–8, groundwater contributed 77 per cent, 75 per cent, and 25 per cent of the total net area irrigated in Telangana, Rayalseema, and coastal Andhra (Directorate of Economics and Statistics, Government of Andhra Pradesh). The area sowed more than once increased by 29 per cent between 1990–1 and 2009–10 at the state level, indicating expansion in irrigation infrastructure and private investments in bore wells in particular (Pingle 2011). Several regions have been declared over-exploited, critical and semi-critical, indicating a near exhaustion of groundwater resources.

Privatization of water has aggravated the collapse of traditional water harvesting and water management systems in the dry lands. Village tanks and smaller watering ponds were perhaps the only public/common source of drinking water for villagers and migratory livestock. The privatization of water and in particular, the expansion of borewells has meant that livestock-rearers are dependent upon the 'largess' and goodwill of borewell-owners to water their animals. Sometimes they are able to negotiate with the borewell-owner through payment in cash or kind. Particularly in drought periods, however, borewell-owners are reluctant to allow the use of their 'private water'. In Andhra Pradesh, efforts were made in the recent past by the Animal Husbandry Department to provide funds to panchayats for the construction of water-troughs; however, these were designed for large ruminants and hence did not benefit small ruminants.

Other research on livestock and water in the larger context of water, equity, and development has come up with similar findings (Dhas 2006; Kher 2006; Phansalkar 2007). This research has, in particular, highlighted the vulnerability of livestock-rearers and migrant-rearers in terms of their access to water. It is pointed out that neither decentralized governance at the village level nor local institutions necessarily provide for the watering needs of livestock owned by migrant herders or for that matter, settled farmers. The poor use public watering sources by default rather than as a legally recognized right in addition to remaining water ponds. And the state only steps in when there is acute scarcity throughout entire regions and that, too, by banning irrigation, a solution that does not help the poor in terms of their grazing needs. In fact, the poor in periods of scarcity are left at the mercy of richer farmers who often charge for water.

Forests and Livestock: Policies to Restrict and Eliminate Grazing

Andhra Pradesh is the third largest state in terms of forest cover in the country. 23.2 per cent or 6.2 million hectares (mha) of the total area of the state is classified as forests, of which 79 per cent are categorized as reserved forests, 19.4 per cent protected forests, and the balance are unclassified forests (Government of Andhra Pradesh, Andhra Pradesh Forest Department). Adivasis, who comprise 6.3 per cent of the state population, live in these forests; 83 per cent or 5.3 mha of forests in Andhra Pradesh are located in the scheduled areas; 60 per cent of the total areas of Schedule V areas are forests.

Colonial forest policy was very restrictive towards grazing. This was true for the forests of the Eastern Ghats that were located in the erstwhile Madras State under direct British governance, and later on, in the forests located in Hyderabad State, which was ruled by the Nizam. Under the Nizam's rule, Hyderabad State did not regulate agriculture practices, nor restrict the movement of cattle and nomadic communities. This was to change in the latter part of the nineteenth century due to British colonial pressure. Laws such as the Cattle Trespass Act, 1871 (Act 1 of 1871) and the Indian Forest Act, 1927, were enacted that regulated cattle movements, cattle markets, and cattle grazing. Fines could be levied under Section 12 of the latter act. Scientific forestry led to the shrinking of grazing land, the reduction in fodder and fuelwood and consequently, the decline in cattle populations that impacted upon livelihoods (Bhangya 2010: 73–116; Satya 2004).

Unfortunately, these policies continued after independence. The National Forest Policy of 1952 stated that in general all grazing in forests is incompatible with scientific forestry. The policy recommended restrictions on sheep grazing in forests and the total exclusion of goats from forests, as their presence was deemed contradictory to the aims of 'scientific forestry management'. 'Illegality' of grazing remained unchanged in Chapter VIII of the Andhra Pradesh Forest Act, 1967. However, in view of the recurrent drought in many parts of the state in 1968, and in view of the consequent inconvenience caused to farmers, the Andhra Pradesh Forest Department did issue a Government Order (Ms) No. 387, directing that from 1 April 1968, free grazing of all livestock (save goats) would be allowed in the reserved forests, excepting in plantation areas. This was watered down in 1975 in another government order which directed that grazing fees be collected from livestock-owners of cattle and other livestock from other states that graze in the reserved forests of Andhra Pradesh.

The 1988 Forest Policy, while praised for its stance that apparently supported forest-dependent communities and participatory management, continued to be anti-grazing. Grazing was still seen as a major cause of deforestation that needed to be restricted and regulated with grazing fees. The policy focused on the need to restrict 'unproductive animals' and replace them with high-producing stall-fed breeds.

Joint Forest Management (JFM),[8] often hailed for its increasing emphasis on people's participation in the spirit of the 1988 National Forest Policy, was also aimed at reducing and completely eliminating the presence of livestock within forests. Livestock-rearers for the first time were faced with a situation where their fellow villagers, as office bearers or members of newly formed village forest protection committees (FPCs) or Vana Samrakshan Samitis (VSSs), enforced rules that prevented them from grazing in the neighbouring forests. While preventing grazing and tree lopping, FPCs established plantations of eucalyptus and pongamia in natural forests. JFM, as part of the Andhra Pradesh Community Forest Management Project (APCFMP), aimed at keeping livestock out of forests, by imposing grazing fees, which would be collected by the VSS. Payment for access to forests, would only serve to favour a rich class within the village, with the capacity to 'pay', and dispossess the poor, who were unable to pay, thus forcing them to sell their stock. This was a key factor that triggered state-wise opposition to the proposed grazing policy (see Box 2.1).

Policing the Graziers: The Participatory JFM Way

Adivasis, farmers, and shepherds stretching from Chittoor in the south to Srikakulam in the north, Medak in the west to Vishakapatnam in the east, have been at the receiving end of JFM's 'policing through people's participation'. The state, through the aegis of the VSS, has been particularly virulent against goat-rearers, threatening them, imposing fines. and confiscating goats. Mukherjee *et al.* (2004) highlight how grazing restrictions in 10 JFM villages in the Telangana region of Andhra Pradesh resulted in significant conflict. The study points out that in five of the 10 villages studied (Nizampet VSS, Seetharamapuram VSS, Kodur Thanda VSS, Kondapur VSS, and Boddugunda VSS,) the VSS imposed severe restrictions on grazing forcing shepherds to sell their goats. Similar findings emerge from Carter and Gronow's (2005) study. They illustrated how shifting cultivators, head-loaders, and goat-herds have been denied access to the forests in the name of forest protection. The FD, ignoring the community, took most of the management decisions. Goats were made 'scapegoats' and goat-rearers were forced to sell their animals.

Reddy *et al.* (2007) illustrate, based on data from three districts of the state, how JFM led to landless households reducing their livestock numbers. The decline in goat-rearing, an important livelihood of the poor, was most acute in Kadapa district, due to the department's strict enforcement of its policy. Springate-Baginski and Blaikie (2007: 15) point out how banning grazing forced wage-laboureres to hand-harvest

Box 2.1   Responses to the proposed grazing policy

Reasons Farmers, Pastoralists, and Adivasis Reject the Policy

The policy is aimed at keeping livestock out of the forests, by imposing grazing fees. This will soon make it impossible for poor farmers to keep livestock, thus pushing them out of agriculture, their only means of livelihood.

1. The farmers firmly reject the argument that livestock has destroyed forests. Farmers highlight the fact that the disappearance of forests and grazing lands over the past 50 years has forced them to reduce their flock sizes drastically over the years. This has been a response to the declining availability of grazing resources, and not vice-versa. Commercial felling of trees for timber, industrialization, building dams have destroyed forests. Village grazing lands have also been systematically diverted for other uses.
2. Livestock is critical for the health and fertility of forests, as their dung enriches the soil. Livestock is also critical for reseeding and thus regeneration of forest species. (Many varieties of seeds pass through the gut of the animal and get treated; thus are able to germinate.)
3. Goat-rearing is a vital means of livelihood for the poor. What will be the fate of thousands of goats and goat-rearers when only a few are permitted into forests for grazing?
4. Lopping fodder for feeding goats, has never harmed the trees, but has enhanced growth. Goat-rearers have developed very advanced techniques for lopping that do not endanger the survival of the tree. All trees need some pruning, for their growth and the FD, too. carries out activities such as coppicing and singling, which are similar to lopping. Lopping is not carried out at random but is restricted to only a couple of key species and in selective months.

Specific Responses from Semi-arid and Arid Areas (Telangana, Rayalseema)

1. Drought is a chronic problem that confronts farmers of these districts. Earlier there used to be good crops of millets, cereals, and pulses, thus there was plentiful availability of crop-residues, which were fed to the livestock during the summer months. With the continuous drought and failure of crops for the past three to four years, the only coping strategy to save the few heads of cattle per household, is the use of forests and migration. Drought has also severely impinged on watering sources for

(Cont'd)

animals, and often the only remaining waterholes in summer months, are located in the forest areas. Thus farmers are forced to migrate with their cattle in search of water.

Specific Responses from Adivasi Areas (hilly, forested regions of the state)

The grazing policy proposed is in contradiction of PESA which gives the powers of managing local resources to panchayats under the aegis of the gram sabha.

1. In adivasi areas, JFM has already placed severe restrictions on the grazing rights of farmers. This policy will only serve to further compromise peoples' livelihoods.
2. Grazing practices in adivasi areas are based on six months of controlled grazing (between June and December), when the crops are being grown. The animals graze on bunds, open land, areas bordering rivers and border areas of the forests. In the dry months (January–May), animals are left loose to find fodder wherever they can (including forests). This system has to be taken into account.
3. Sheep flocks from plain areas migrate into forests between September and January, as fodder is scarce in the plain areas during those times. They return to the plain areas in January when the crops are harvested and agriculture lands are once more free to be grazed upon.

Source: 'Summary of Key Event of Grazing Policy', unpublished report, 2003, Anthra, Hyderabad.

fodder, which resulted in a loss of daily wages. Saito-Jensen and Jensen (2010) narrate how the creation of new boundaries through JFM, and the strict regulations and severe fines that accompanied it, adversely impacted many goat-herders of Kanchanpally village in Medak district. Between 1998 and 2008, the goat population declined from 2,500 to 1,000. Not surprisingly a majority (81 per cent) of the 26 respondents perceived JFM in negative terms. Ramdas and Ghotge (2007: 52) focus on the ways in which the FD and VSS committees restrict graziers, especially women, from grazing their cattle, sheep, and goats. Amongst the measures imposed are restrictions on lopping fodder with axes, the digging of trenches and the planting of thorny bushes to prevent entry into the forest, and the demand for an animal in exchange for permission to enter the forests. Given the fact that graziers are usually not members of FPCs, also suggests that their interests are neither represented nor included whilst forest management plans are evolved.

Thus, except for a handful of studies that report a positive impact of JFM on animals and grazing (such as TERI 1999), the verdict on JFM from a grazing perspective has generally been negative. Customary arrangements of sharing and using the forests for the purposes of grazing appear to have been broken. The political ideology of JFM is that forests can be most effectively managed by privileging a select sub-group within the larger community, and 'endowing' them with rights to forest resources. In effect, private forests for the exclusive use of some, have been created at the expense of others. The establishment of VSS/FPC committees and the effective privatization of forest areas by these groups is in complete contrast to the earlier experience where a forest resource specifically with respect to grazing, was shared by several communities often cutting across many villages and involving intricate and complex patterns of use. Moreover, JFM 'micro-plans', have to be in tune with the larger 'working plans' of the FD, resulting in the neglect of community needs and fodder needs in particular. JFM has unleashed a huge number of conflicts within and between villages, centred on grazing, where there had been none. Forest work carried out through JFM, such as 'coppicing' and 'bush clearing', and digging trenches ostensibly for stemming soil erosion and plantations, have completely destroyed the natural fodder base of many of these forests and promoted species which have absolutely minimum or no fodder value.

The enclosure of the commons to forcibly stop grazing regimes has been accompanied by development efforts to persuade and often, to coerce communities to replace their local breeds with so-called high-yielding crossbreds or goats with sheep. The supposed advantage of high-yielding crossbreds is that they can be stall-fed or reared on zero-grazing regimes.[9,10,11] Such measures would in the eyes of the FD result in less pressure on forest resources albeit at the expense of marginal farmers who depend upon the indigenous/local animal breeds.

When a new grazing policy was being formulated in the early 2000s, sheep- and goat-rearers' associations and civil society allies mobilized. This led to the creation of the Andhra Pradesh Forestry Committee (Turner 2004). People's organizations were able to successfully pre-empt this new policy from coming into force through a state-wide protest and campaign. The position taken was that the new policy would severely undermine the ability of poor farmers to keep livestock and thus compromise their means of livelihood (see Box 2.1). Moreover, adivasi communities pointed out that the policy designed by the FD, infringed on adivasi rights as spelt out in PESA. Despite these protest, the FD in April 2002 announced a new policy through G.O. (M) 34. An entire section (Section 3.1.2) was devoted to the issue of restricting *free grazing in forests. The solution proposed was upgrading cattle- and stall-feeding.*

Despite various restrictions placed on grazing, small peasants, adivasis, and shepherds have continued to oppose new laws. In addition, they have managed to sustain extremely intricate and complex ways of grazing their animals in forests, eluding forest guards where possible or paying fees and fines allowing them to graze their animals.

In Chittoor, shepherds organized themselves into *sanghams*/collectives in 2004–5, and began passing resolutions asserting that they would continue to graze their animals in forests and resist the pressure of the VSS and forest guards (Anthra 2008). Women shepherds countered threats from forest guards by getting sanghams to stop FD harassment (Ramdas and Ghotge 2007). In Medak, shepherds expressed their anger by lopping trees, in ways that actually harmed the tree, as a sign of their non-cooperation with the department (Ramdas and Ghotge 2007). In some places, shepherds stopped paying grazing fees to the VSS or the FD (Anthra 2009). In several villages located in scheduled areas, adivasis challenged the authority of the FD and the VSS chairman, and resumed grazing their animals in the VSS enclosures. People protested against the restrictions by wilfully uprooting plantation saplings and grazing their animals directly on the saplings. People also protested by refusing to accept new breeds, and stall-fed regimes, holding on to to their own breeds of goats, cattle, and local buffaloes (Ramdas and Ghotge 2007).

Along with opposing laws, the shepherd sanghams (collectives) negotiated and persuaded VSS committees of the legitimacy of their customary rights to graze in forests. Sanghams in Chittoor and Medak, engaged with the Animal Husbandry Department in order to convince them to provide veterinary services to goats. After initial resistance, the department gradually relented. In order to avoid rebuke from higher officials, local veterinary doctors recorded 'goats' as 'sheep' in their books (Anthra 2008).

Impact of Land-use Shifts on Livestock Ownership, Demography and Practices in 'Plain' and Schedule V Regions

Despite islands of protest, mostly in Scheduled V regions or by shepherds in the plains, the overall trend of shrinking grazing spaces has had a negative impact on the livestock economy in the Telangana, Andhra, and Rayalseema regions, including Schedule V areas. According to livestock census data, between 1983 and 1993, the cattle population declined by 17 per cent. In the following decade (1993–2003), it declined by another 14 per cent. This decline in cattle population was matched by a 200 per cent increase in sheep, 17 per cent increase in goat and 24 per cent increase in buffalo populations. The most recent census of 2007 points towards a five-year rapid increase in cattle population bringing it on par with population levels found in 1983. Micro-level studies carried out in the same period, however, highlight depleting stocks of all species, particularly cattle, at the village level (Akter *et al.* 2007; Deshingkar *et al.* 2008; Rao and Charyulu 2007). NSSO data[12] supports micro-level studies and reveals significant declines of livestock amongst small, marginal, and landless households who are joining the ranks of 'non-livestock owners'.

To some extent, peasants and pastoralists in the plains appear to have responded to declining grazing resources and new market forces by changing the species/ breed reared.[13] There is a clear trend of selling/ reducing cattle herds, and shifting to rearing sheep, goats, and buffaloes which can survive in harsher and less productive environments (FES 2010). However, adivasis have held on to their indigenous cattle and goats albeit in reduced numbers, and continue to reject sheep, buffaloes, and cross-bred cows, despite huge efforts by government departments to finance the latter. Their decisions are based on deep knowledge of their hilly-forested homelands that are conducive to indigenous cattle and goats. Deteriorating and fewer pastures have not only resulted in changes in the type of livestock held but have also triggered new conflicts amongst livestock owners (Ramdas *et al.* 2013). This is because communities in the plains have had to leave their villages in search of fodder and water, finding new routes and staying in forests for extended periods of time. This has sparked conflicts between permanent forest-dwelling adivasis and seasonal migratory pastoralists and pastoralists and settled peasants, despite the long history of inter-dependence between agriculture, pastoral, and forest livelihoods.

A New Opportunity for Democratic Governance of Forests in Support of Grazing-based Livestock Production Systems: FRA 2006

The local level struggles discussed above merged into larger state and national level mobilization of adivasi and other forest dependent/dwelling communities against evictions from forests. Part of this struggle against eviction was the struggle for recognizing people's rights (individual and collective) to forests. The struggles brought pressure on the government to enact the FRA.

The FRA has provided a new lease of life to the idea that conservation of forests and people's livelihoods are not antagonistic but complementary. This historic legislation in addition to recognizing the collective rights of adivasis and other traditional forest-dwelling communities to remain in their homelands, protect and govern their community forest resources, and practise traditional forest-dependent livelihoods, also recognizes the *grazing rights (settled or transhumant) and traditional seasonal resource access of nomadic or pastoralist communities to graze their animals in forests* (FRA 2006, Chapter II, Section 3 (1) d). While the FRA does not address the seasonal usage by pastoralists of 'non-forest' land, it gives legality to grazing-based livestock systems that have been branded as 'destructive and harmful' all these years.

Pastoralists, moreover, can negotiate grazing terms directly with forest-dwelling adivasis in a manner that would empower both sets of communities to collectively exercise their rights to protect and conserve forests (Chapter III, Section 5 [a, b, c, d] of FRA 2006).[14] No longer do shepherds have to pay 'arbitrarily decided grazing fees' and bribe officials to graze animals in the forests (Rao 2011). Monetary payments to forest officials have broken down age-old customary governance practices where seasonal pastoralists discussed and obtained consent from the relevant adivasi gram sabhas, in whose forests they sought seasonal grazing rights. Chenchu adivasis of the Nallamalla forests, Gonds and Kollams in Adilabad, and Konda Reddis and Koyas in East Godavari, bemoan the fact that shepherds and communities such as the Lambadas[15] have stopped dialoguing with them. Adivasis believe that only dialogue with shepherds will result in decisions about grazing that are more nuanced and regulated so as to be sensitive to the changing fodder cycles in the forests.

People-centred decentralization needs be backed with pro-active action to prevent further loss and destruction of the non-forest commons and grassland, in particular. Transhumant communities' migratory cycles have been disrupted resulting in greater pressure on forests. What the government can do is operationalize the Supreme Court[16] directive to the states on restoring CPRs to gram sabhas/panchayats (GS/GP) and using the provisions of PESA[17] and the Panchayati Raj Act to strengthen decentralized governance of village natural resources. Given the fact that 83 per cent of forests in Andhra Pradesh are located in Schedule V regions, the notification of PESA rules in March 2011, although 16 years after the law was enacted paves the way along with the FRA for people to govern their villages and resources including forests in ways amenable to sustainable livelihoods and social justice.

Adivasi groups in Andhra Pradesh[18] began to pro-actively distribute Telugu copies of the FRA, 2006,[19] long before rules were notified. Community activists in different regions informed villagers, sanghams, and other organizations, of the provisions of the Act, including the right of grazing animals in forests. In Chittoor district, for example, activists of shepherd sanghams in KVB Puram Mandal, armed each family with a copy of the act. They organized meetings with VSS committee members in the surrounding villages and explained the provisions of the Act and how it recognizes the grazing rights of forest-dependent communities and other pastoralists. Forests guards and the VSS chairman were warned that they could be prosecuted if they did not allow animals to graze in the forests (Anthra 2008).

Once the rules were notified, the Andhra Pradesh Tribal Welfare Department (APTWD) actively encouraged people to file individual claims. However, from 2008 up to mid-2009, little was done by government staff in the way of encouraging gram sabhas to file community claims. Community forest rights (CFR) claims, including rights to graze, nonetheless were extensively filed in regions where movements were strong (Ramdas 2009).

Even as village communities made claims to community forests, the APTWD, advised by the FD, attempted to recognize VSS-managed forests as 'community forests'. In some districts, VSSs were identified as the 'holders' of community rights, completely bypassing the authority of the gram sabha. CFR titles were at times given in the name of the VSS , not gram sabhas (Reddy *et al.* 2011). Adivasi struggle organizations

immediately protested, claiming the Act was being subverted and
JFM institutions empowered. Several gram sabhas passed resolutions
to reject the false CFR titles (Venkatesh Dora, Adivasi Aikya Vedika,
personal communication, October 2011). Adivasi Aikya Vedika's cam-
paigned for customary boundaries to be the basis for recognition of
community rights. Its efforts resulted in the APTWD issuing Memo
No. 355 in January 2011, directing all their officials to ensure that community rights are prepared according to the customary boundaries of
a village.

Shepherds have been much more conservative in using the FRA.[20]
This is largely because the state has abdicated its responsibility of
spreading awareness to the shepherd community. Those shepherds in
Narasapur division of Medak district and KVB Puram and Madanapalle
divisions of Chittoor district, who have used the FRA, have done so
because of advocacy work carried out by shepherd sanghams and civil
society groups such as Anthra. First, shepherds and other livestock-rearers mapped their traditional forest grazing locations, waterbodies, and
other forest uses. Then they filed community claims through the gram
sabhas (Anthra 2011; Nazareth 2011). Shepherds from 12 villages that lie
adjacent to Horsley Hills and the Thettu forests of Thettu panchayat,
Chittoor district were the first community of shepherds in June 2011
to file for CFR rights to graze in the forests (Anthra 2011). In Medak
district, the submission of the first claim was publicized through local
media (July 2012), which resulted in a spill-on effect to other villages.
This said, the majority of shepherds in Andhra Pradesh continue to
be unaware of the Act and their rights to graze in and govern forests
through this Act. Dialogue with adivasi communities has, however,
resulted in greater awareness amongst shepherds in the Eastern Ghats
and Nallamala, suggesting the potential for the act to empower shepherds remains (Madavi *et al.* 2011).

It would be fair to say that community systems of governance and
decision-making, which are central to effective forest governance, are
much stronger in Schedule V areas as compared to the plains. A recent
study carried out by Anthra and CESS (Ramdas *et al.* 2013), comparing
forests and livestock in Scheduled V and the plains, highlights how traditional institutions such as the Panch in Adilabad and the Gotti in East
Godavari and Vishakapatnam remain robust. The study also argues that
since all families are in one way or another dependent on forests, there

is unity within the village with respect to forest governance. Moreover, the knowledge of how to graze, where to graze, and where not to graze is imbibed by even the youngest child in the village. This is not to say that institutions or practices are static. Institutions modify or update traditional norms or are flexible enough to adapt to existing situations. For example, the decision about whether and on what terms migratory pastoralists can graze their animals in the forests located within the customary boundaries of an adivasi village is debated each year within the Panch or Gotti. Permission is granted or denied after considering and debating factors such as the state of the forests and the health of the visitor's flocks.

In contrast, villages at the forest-interface in the plains only have small numbers who are dependent on forests on a day-to-day basis. Moreover, the gram sabha in many of these villages has become defunct. Elected representatives of the gram sabhas, have allowed the FD by default, to exercise and consolidate their powers. Even though shepherds, Dalits, erstwhile nomadic pastoral communities such as the Lambadas, traditional healers, the few remaining families who collect forest produce, and women from small peasant families who collect firewood, wild vegetables, herbs, and medicinal plants continue to depend on the forests, agriculture remains the mainstay for most of these groups. The forest connection is, therefore, more tenuous than in the case of adivasis.

The FRA, nonetheless, has the potential to open up the 'legal space' for communities to rear their animals without fear while managing the forests. The latter is possible because the FRA gives communities the legal right to help conserve. Although villages even in Schedule V areas are yet to formally discuss conservation plans, legally that possibility exists.

Regrettably, government officials are extremely ill-informed about the provisions and scope of the FRA. Furthermore, they often misinterpret the provisions of the law and continue to allow the diktat of the FD to prevail. Finally, the state apparatus in general is far from ready to respect the powers that are bestowed on gram sabhas with respect to the governance of forests. A number of examples illustrate this. First, while the FRA empowers adivasis through their gram sabhas to decide how to use and manage their community resources, the Integrated Tribal Development Authorities (ITDAs) and the FD, utilizing funds from the National Rural Employment Guarantee Scheme (NREGS), have been forcing adivasi families to grow rubber, fruit trees, and biofuels, on their private and community lands (Council of Social Development 2010;

Ramdas 2009). Such crops not only undermine food sovereignty and security but disrupt the regeneration of natural vegetation on which animals graze. Moreover, as suggested above, the MoEF persists in trying to prop up what many consider redundant institutions under JFM (GoI 2010).This completely dilutes the self-rule and decision-making powers of the gram sabha as spelt out in the FRA and more comprehensively in PESA.

Finally, the Machiavellian state along with its corporate allies has identified forests as a major means by which India can demonstrate its commitment to carbon emission reduction. India has endorsed the International Agreement on REDD+ (Reduced Emissions from Deforestation and Forest Degradation), which globally in the climate change debate, is considered as the most cost-effective way of reducing emissions. India considers REDD+ a bargaining chip for its increasing emissions in global negotiations. The huge network of JFM FPCs are seen to be the means to facilitate REDD+, as part of the National Green India Mission, one of the eight national programmes for climate change adaptation. REDD+ works on the assumption that the forest will be allowed to regenerateby stopping grazing, forest fires, shifting cultivators, and all other unnecessary human interference. JFM committees will police people and livestock (TERI 2009). Initial analysis of the Clean Development Mechanism (CDM) (Ramdas 2009) highlights how it disrupts the complex ways in which communities interact with forest space, in the process displacing livelihoods. Perhaps more importantly, how can the assumption be made that forest people should live their lives by 'watching the trees grow', in exchange for money, while being forced to stop all their forest-based production activities such as grazing goats, as was seen in the case of Adilabad (Ramdas 2009), and appears to be unfolding in the case of India's first REDD+ projects underway in the Northeast (Community Forestry International 2011; Ghosh 2011). Resistance to and critique of these new market-based solutions to climate change is already evident in Andhra Pradesh (Ramdas 2010; Venkatesh Dora, personal communication).

Looking Ahead

The FRA, along with recent rulings on the commons, offers the opportunity to revitalize and democratize the governance of the commons—

forest and non-forest. It is still, however, early days, with respect to implementation of the legislation. The experience from the ground suggests that adivasis in Schedule V regions, and pastoralists in plain areas, cannot fearlessly deepen their engagement with the sustainable use and conservation aspects of forest governance, so long as they have to contend with a tyrannical FD and a non-motivated Tribal Welfare Department/District Level Committee (DLC)/Sub-Divisional Level Committee (SDLC) whose combined goal appears to be to obstruct the process of the law, and prevent legal recognition of people's forest governance. The seeds of conservation are embedded in the complex set of traditions, knowledge, and practice of people, and will blossom if only administrators would abdicate their power and step aside. As of now, people's energies are spent in contesting the everyday hurdles set up in their path by the bureaucracy whose approach to the law and justice seems to be based on the premise that the claimant (read community), is 'illegal and guilty' until proven innocent. The onus of proof lies with the community, if they are to 'be eligible for the right'. In reality what we have seen is that no proof appears to be powerful enough for the bureaucracy who has pre-decided to reject most claims, and stop people from exercising their powers of decentralized forest governance.

What is evident is that strong and democratic institutions of local governance and self-rule, with a clear political perspective and vision of the kind of livestock livelihoods that are desirable, is paramount for utilizing the FRA in empowering ways. The law can be used either to invite market-based systems of forest management or conservation and livestock production, or be used to strengthen forest governance that is founded on deliberative democratic principles. The latter is more likely to enhance bio-diverse low-input, mixed-crop-livestock, food-farming systems, that nurture local autonomy, sovereignty, justice, and resilience. Such a conclusion is based on the fact that traditional adivasi institutions in Schedule V areas of the state continue to be dynamic and vibrant, and capable of promoting sustainable and resilient livelihoods rooted in customary laws and worldviews.

Communities including pastoralists and others in 'plain areas' who continue to be dependent on the non-forest commons and forests need to exercise their citizenship to protect the remaining commons that exist in their villages. This will complement the task of decentralized forest governance being led by adivasis. The bottom line is: will the state be a willing ally in the process, or will it have to be a continued battle for justice?

Notes

1. I use the term 'state' to refer to the set of governing and administrative institutions that have sovereignty over a definite territory and population.

2. Henceforth, I refer to these areas as 'plain areas' in this chapter.

3. In 2011, for instance, the rates for grazing and herding animals was Rs 100 per cow or adult buffalo.

4. Traditional land-use systems and their governance were certainly not ideal, and as pointed out, were particularly unequal with respect to private landownership. However, what we can glean from the literature is that the erstwhile Hyderabad State ruled by the Nizam, which included significant portions of today's Andhra Pradesh, had clearly demarcated lands for livestock to graze that were accessible to all (Bhangya 2010). Dalits in several parts of India report how they continue to be excluded from freely accessing common grazing lands (Shah 2006; Vikas Adhyayan Kendra 2006).

5. Wasteland is a term used in a legal sense to refer to land that was unoccupied, undeveloped, or uncultivated. As a result, the land could not be the source of any tax or other revenue to its owner. The current formal definition includes land with or without scrub, waterlogged and marshy land, land affected by salinity/alkalinity, coastal/inland, shifting cultivation area, degraded pastures/grazing land, degraded land under plantation crops, sands/inland coastal, mining/industrial wasteland.

6. Such as the Watershed Development, Drought-Prone Area Development, Desert-Area Development, Wasteland Development, and Comprehensive Land Development Programmes initiated in 2004, and the NREGS.

7. Andhra Pradesh had 106 SEZs as of March 2010, the second highest in the country after Maharashtra. Of the 106, 73 SEZs have been notified, which is the highest in the country. Notified SEZs have already acquired the land. In the remaining SEZs, land acquisition is still in progress (Seethalakshmi 2010).

8. The JFM programme in Andhra Pradesh, implemented by the Andhra Pradesh Forest Department under the aegis of the Andhra Pradesh Forestry Project, was started in 1992, with the financial backing of the World Bank worth Rs 365 crore. The second phase of the project, known as Andhra Pradesh Community Forestry Management Project (APCFMP), was financed by the World Bank from 2001 to 2007, with a loan of Rs 650 crore. See http://forest.ap.nic.in/JFM%20CFM/CFM/World%20Bank/PAD.htm#e.

9. The World Bank-funded APCFMP, with its intended objective of 'reducing poverty' alongside improving forests, begins by stating how grazing

is the major cause of degradation of forests. The solutions offered are replacing animals with superior ones that will be stall-fed, and augmenting grasses through developing separate fodder plots (http://forest.ap.nic.in/JFM%20 CFM/CFM/World%20Bank/PID.htm).

10. See http://archive.is/www.envfor.nic.in/divisions/forprt/terijfm.html.

11. The Pashu Kranti programme of the Government of Andhra Pradesh from 2008 to 2010 financed loans for 'crossbred' improved varieties of cows and buffaloes. Women were forced to purchase these breeds and the programme did not recognize or support the purchase of local breeds. This massive programme, which saw crores being invested, ended up burdening communities who could barely afford to feed themselves, leave alone their buffaloes which were given to them through the Pashu Kranti (Ramdas 2007).

12. Trend analysis of NSSO data for Andhra Pradesh, with respect to the numbers of in-milk bovines (cows and buffaloes), male cattle, young stock (male and female cattle), and ovines per 100 households, indicates a drastic and significant drop in the numbers of 'in-milk' cattle and buffaloes, per 100 households from 1971–2 to 2002–3.

13. Reforms in livestock markets have been towards promoting dairy production and meat production, especially encouraging the export of meat.

14. a) Protect wildlife, forest, and biodiversity; b) ensure that the adjoining catchments area, water sources, and other ecological sensitive areas are adequately protected; c) ensure that the habitat of forest dwelling Scheduled Tribes and other traditional forest-dwellers is preserved from any form of destructive practices affecting their cultural and natural heritage; and d) ensure that the decisions taken in the gram sabha to regulate access to community and forest resources and stop any activity which adversely affects wild animals, forest, and the biodiversity are complied with.

15. They were recognized as Scheduled tribes in Andhra Pradesh in 1977, and are the original transhumant cattle pastoralists of the Deccan.

16. In a landmark decision given by the Supreme Court on 28 January 2011, in the case No. 1132 /2011 @ SLP(C) No.3109/2011 concerning encroachment of common land in a village in Punjab, the court gave instructions that the land be restored to the gram panchayat. The court directed all state governments in the country to prepare schemes for eviction of illegal/unauthorized occupants of gram sabha land, and to comply.

17. The Andhra Pradesh Panchayat Raj Act, 1994 amended in 2006, which pertains to local governance of all rural villages, has specified roles and responsibilities of the GP vis-a-vis decisions to be taken on managing land, animal wealth, agriculture, as also forests. The gram sabhas at the smallest hamlet level in Scheduled V regions have been given sweeping powers to govern their land, common land, forests, minor waterbodies, and other resources, as stated in the

Andhra Pradesh Panchayat Raj (Amendment) Act, 1998 (Act No. 7 of 1998), 1998.

18. Adivasi Aikya Vedika, a coalition of adivasi people's organizations and struggle groups, is in the forefront in this respect assisting communities across Andhra Pradesh.

19. Yakshi, an organization that works with Adivasi Communities and People's Organizations in Andhra Pradesh translated the Act soon after its enactment, and printed multiple copies of the same, as also trained several activists across AP about the Act and its contents.

20. Several shepherd and pastoralist communities across India are still exploring how best to use FRA to assert their rights to graze in forests. Their general awareness about the Act, however, has been low.

References

Akter, S., J. Farrington, P. Deshingkar, L. Rao, and A. Freeman. 2007. 'Species Diversification, Livestock Production and Income of the Poor in the Indian State of Andhra Pradesh', *Livestock Research for Rural Development*, vol. 19, Article #175; http://www.lrrd.org/lrrd19/11/akte19175.htm, accessed on 2 August 2013.

Andhra Pradesh Community Forest Management Programme; http://forest.ap.nic.in/JFM%20CFM/CFM/World%20Bank/PID.htm, accessed on 2 August 2013.

Anthra. 2008. *Bridging the Knowledge Divide: Livestock and Livelihoods in the Emerging Context*. Hyderabad: Anthra Publications.

———. 2009. 'Anthra Annual Report', April 2008–March 2009. Hyderabad, India: Anthra.

———. 2011. 'Anthra Annual Report', April 2012–March 2011. Hyderabad, India: Anthra.

Bhangya, B. 2010. *Subjugated Nomads: The Lambadas Under the Rule of the Nizam*. Hyderabad: Orient BlackSwan.

Bhattacharya, N. 1995. 'Pastoralists in a Colonial World', in D. Arnold and R. Guha (eds), *Nature, Culture and Imperialism: Essays on the Environmental History of South Asia*. New Delhi: Oxford University Press.

Brara, R. 2006. *Shifting Landscapes. The Making and Remaking of Village Commons in India*. New Delhi: Oxford University Press.

Carter, J. and J. Gronow. 2005. 'Recent Experiences in Collaborative Forest Management'. CIFOR Occasional Papers No. 43. Bogor: Center for International Forestry Research.

Community Forestry International. 2011. 'Payment For Environmental Servces. A Case Study from Meghalaya, North East India'; www.communityforestry. org, accessed on 1 February 2014.

Council of Social Development. 2010. 'Summary Report on Implementation of the Forest Rights Act'; http://www.forestrightsact.com/component/ k2/item/15, accessed on 1 February 2014.

Deshingkar, P., J. Farrington, L. Rao, S. Akter, P. Sharma, A. Freeman, and J. Reddy. 2008. 'Livestock and Poverty Reduction in India: Findings from the ODI Livelihood Options Project', Discussion Paper No. 8. Targeting and Innovation Series. Nairobi, Kenya: International Livestock Research Institute (ILRI).

Dhas, M. 2006. 'How the Migrant Sheep and Goat Rearers of Maharashtra Manage Water for Their Livestock'. Paper presented at Annual Partners' Meeting organized by International Water Management Institute—Tata Water Policy Research Programme, Anand, India.

Foundation for Ecological Security (FES). 2010. *A Commons Story: In the Rain Shadow of Green Revolution*. Ahmedabad, Gujarat: FES.

Ghosh, S. 2011. 'REDD+ in India, and India's First REDD+ project: A Critical Examination', *Mausam*, 3 (1): 32–50.

Ghotge, N.S. 2004. *Livestock and Livelihoods: The Indian Context*. Foundation Books. New Delhi: Cambridge University Press India.

Ghotge N.S., S. Ramdas, S. Ashalatha, N.P. Mathur, V.G. Broome, and M.L. Sanyasi Rao. 2002. 'A Social Approach to the Validation of Traditional Veterinary Remedies—The Anthra Project', *Tropical Animal Health and Production*, 34 (2): 121–43.

Government of Andhra Pradesh (GoAP). 2010. 'Socio-Economic Survey of Andhra Pradesh 2009–2010'. Available at http://www.aponline.gov.in/ Apportal/AP%20Govt%20Information/APSES.html

GoAP and Centre for Economic and Social Sciences (CESS). 2007. 'Human Development Report, 2007. Andhra Pradesh'. Hyderabad: CESS. Available at http://www.in.undp.org/content/dam/india/docs/human_revelop_ report_andhra_pradesh_2007_full_report.pdf.

Government of India (GoI). 2003. Ministry of Agriculture, Department of Animal Husbandry and Dairying, 17th Indian Livestock Census, All India Summary Report. New Delhi.

———. 2006. 'Report of the Task Force on Grasslands and Deserts'. New Delhi: Ministry of Environment and Forests.

———. 2009a. 'Report of the Committee on State Agrarian Relations and the Unfinished Task in Land Reforms'. New Delhi: Department of Land Resources and Ministry of Rural Development.

Government of India (GoI). 2009b. 'National Policy on Biofuels'. New Delhi: Ministry of New and Renewable Energy. Available at http://mnre.gov.in/file-manager/UserFiles/biofuel_policy.pdf.

———. 2010. New Initiative on Panchayats and Forests'. New Delhi: Ministry of Environment and Forests. Available at http://www.moef.nic.in/downloads/public-information/Panchayats%20and%20Forests,.pdf.

———. 2012. 'Cultivation of Genetically Modified Food Crops—Prospects and Effects'. Thirty seventh Report. New Delhi: Parliamentary Standing Committee on Agriculture, Lok Sabha. Available at http://164.100.47.134/lsscommittee/Agriculture/GM_Report.pdf

Jodha, N.S. 1986. 'Common Property Resources and Rural Poor in Dry Regions of India', *Economic and Political Weekly*, 21 (27): 1169–81.

———. 1995a. 'Studying Common Property Resources: Biography of a Research Project', *Economic and Political Weekly*, 30 (11): 556–60.

———. 1995b. 'Common Property Resources and the Environmental Context', *Economic and Political Weekly*, 30 (51): 3278–84.

Kandhari, R. and A. Pallavi. 2010. 'Biomass Market in a Flux'. *Down To Earth*, 31 March 2010.

Kavoori, P.S. 1999. *Pastoralism in Expansion: The Transhuming Herders of Western Rajasthan*. University of Michigan: Oxford University Press.

Kher, V. 2006. 'How the Migrant Livestock Herd Owners of Gujarat and Rajasthan Manage the Water Requirement of Their Herds'. Presented at Annual Partners' Meeting, organized by (IWMI—Tata Water Policy Research Programme, Annual Partners Meeting), Anand, India.

Madavi, S., N.H. Swami, S.A. Reddy, and S.R. Ramdas. 2011. 'Adivasis, Pastoralists and Forest Governance'. Presented at the 13th Biennial Conference of the International Association for the Study of the Commons. Hyderabad, India, January 2011.

Mukherjee, S.D., S. Galab, and B. Sundar. 2004. 'From Policy to Practice: A Study on Joint Forest Management in Andhra Pradesh (Telangana Region)'. ISNRMPA, SDC-IC NGO Programme, Hyderabad, Andhra Pradesh, India.

Murali, A. 1995. 'Whose Trees? Forest Practices and Local Communities in Andhra, 1600–1922', in D. Arnold and R. Guha (eds), *Nature, Culture and Imperialism, Essays on the Environmental History of South Asia*, pp. 86–122. New Delhi: Oxford University Press.

National Forest Policy. 1988. No. 3-1/86-FP. MOEF, Department of Environment, Forests and Wildlife.

National Sample Survey Organisation (NSSO). 1999. *Common Property Resources in India*. Report no.452 (54/31/4). New Delhi: NSSO, Department of Statistics and Programme Implementation, Government of India.

Nazareth, M. 2011. *Understanding the Concept of Commons: A Shared Responsibility* (cited 4 February 2011); Countercurrents.org, accessed on 1 February 2014.

Phansalkar, S. 2007. 'How the Poor Manage the Water Requirement of their Livestock'. *International Journal of Rural Management*, 3 (1): 1–25.

Pingle, G. 2011. 'Irrigation in Telangana: The Rise and Fall of Tanks', *Economic and Political Weekly Supplement*, June 25, 46 (26 and 27): 123–30.

Ramdas, S., N.S. Ghotge, S. Ashalatha, N. Mathur, M.L. Sanyasi Rao, N. Madhusudhan, S. Seethalakshmi, N. Pandu Dora, N. Kantham, E. Venkatesh, and J. Savithri. 2004. 'Overcoming Gender Barriers: Local Knowledge Systems and Animal Health Healing in Andhra Pradesh and Maharashtra', in S. Krishna (ed.), *Livelihood and Gender: Equity in Community Resource Management*, pp. 67–91. New Delhi, Thousand Oaks, and London: Sage Publications.

Ramdas, S. and N.S. Ghotge. 2007. 'Whose Rights? Women in Pastoralist and Shifting Cultivation Communities: A Continuing Struggle for Recognition and Rights to Livelihood Resources', in S. Krishna (ed.) *Women's Livelihood Rights: Recasting Citizenship for Development*, pp. 41–61. New Delhi: Sage Publications.

Ramdas, S., S. Ashalatha, and M.L. Sanyasi Rao. 2013. 'Livestock Dependent Livelihoods at the Forest Interface in Schedule V and Plain Areas of Telangana and Andhra Regions of Andhra Pradesh', RULNR Working Paper No. 16. Hyderabad: Centre for Economic and Social Studies.

Ramdas, S. 2007. 'Yet Another Revolution... Pashu Kranti', *Andhra Jyoti*, 11 October.

———. 2009. 'Women, Forest Spaces and the Law: Transgressing the Boundarie', *Economic and Political Weekly*, 31 October, 44 (44): 65–73.

———. 2010. '*Vaatavarana Maarpu, Vyapara Vannare*—Part I and II, Climate Change up for Global Trade', *Andha Jyothi*, 22 and 24 December 2010.

———. 2011. 'Pushed to the Brink: Livestock Dependent Livelihoods at the Forest Interface. The Case of Andhra Pradesh', Working Paper No. 103. RULNR Working Paper No. 11. September 2011. Hyderabad: Centre for Economic and Social Studies.

Rao, K.P.C. and D. Kumara Charyulu. 2007. 'Changes in Agriculture and Village Economies'. *Research Bulletin* no. 21. Patancheru, Andhra Pradesh, India: International Crops Research Institute for the Semi-Arid Tropics.

Rao, S.C. 2011. 'Drought Prone Regions and Survival Strategies: A Study of Semi-arid Andhra Pradesh. PhD, University of Hyderabad, Hyderabad, India'; http://hdl.handle.net/10603/1597, or http://shodhganga.inflibnet.ac.in/handle/10603/1597?mode=ful, accessed on 1 February 2014.

Reddy R.V., M. Gopinath Reddy, M. Bandi, V.M. Ravi Kumar, M. Srinivas Reddy, and O. Springate-Baginiski. 2007. 'Participatory Forest Management in Andhra Pradesh: Implementation, Outcomes and Livelihood Impacts', in O. Springate-Baginiski and P. Laikie (eds), *Forests, People and Power: The Political Ecology of Reform In South Asia*. London: Earthscan.

Reddy, M. Gopinath, K. Anil Kumar, P. Trinadha Rao, and O. Springate-Baginski. 2011. 'Issues Related to the Implementation of the Forest Rights Act in Andhra Pradesh', *Economic and Political Weekly*, 30 April 30–6 May, 46 (18): 73–81.

Saito-Jensen, M. and C.B. Jensen. 2010. 'Rearranging Social Space: Boundary-making and Boundary-work in a Joint Forest Management Project, Andhra Pradesh, India'. *Conservation Society*, 8: 196–208.

Satya, L.D. 2004. *Ecology, Colonialism and Cattle: Central India in the Nineteenth Century*. New Delhi: Oxford University Press.

Seethalakshmi, S. 2010. 'Shifting Land Use Patterns in Andhra Pradesh: Implications for Agriculture and Food Security', Study report. Hyderabad: Centre for Sustainable Agriculture.

Shah, G. 2006. *Untouchability in Rural India*. New Delhi: Sage Publications.

Sontheimer, G.D. and M.L. Murty. 2004. 'Pre-historic Background to Pastoralism in the Southern Deccan in the Light of Oral Traditions and Cults of Some Pastoral Communities', in H. Bruckner, A. Feldhaus, and A. Malik (eds), *Essays on Religion, Literature and Law*, p. 161. Delhi: Manohar Publishers.

Srinivasalu, K. 2002. 'Caste, Class and Social Articulation in Andhra Pradesh: Mapping Differential Regional Trajectories', Working Paper 179. London: Oversees Development Institute.

Subramaniam, S. and P. Satya Sekhar. 2003. 'Agriculture Growth: Pattern and Prospects', in C.H. Hanumantha Rao and S. Mahendra Dev (eds), *Andhra Pradesh Development. Economic Reforms and Challenges Ahead*, pp. 221–45. New Delhi: Manohar Publishers and Distributors.

Tata Energy Research Institute (TERI). (n.d.) 'Study on Joint Forest Management'. Ministry of Environment and Forests; http://envfor.nic.in/divisions/forprt/terijfm.html, or http://archive.is/www.envfor.nic.in/divisions/forprt/terijfm.html.

———. 1999. 'Study on Joint Forest Management', final report prepared for the Ministry of Environment and Forests. TERI project report no. 98 SF 64. New Delhi: Tata Energy Research Institute.

———. 2009. 'Is India Ready to Implement REDD? A Preliminary Assessment'; http://www.teriin.org/events/CoP15/Forests.pdf, accessed on 1 February 2014.

Turner, R.L. 2004. 'Livestock Production and the Rural Poor in Andhra Pradesh and Orissa States, India', PPLPI Working Paper 9; http://www.fao.org/ag/againfo/programmes/en/pplpi/docarc/pb_wp9.pdf.

Vikas Adhyayan Kendra. 2006. 'Dalit and Adivasi Land Ownership', Fact Sheet. Quarterly Journal of Dalit Resource Centre', April. Mumbai: Vikas Adhyayan Kendra.

3

MADHU SARIN

Undoing Historical Injustice

Reclaiming Citizenship Rights and Democratic Forest Governance through the Forest Rights Act

India's forest lands are an arena of intense conflicts today. These conflicts are rooted in the historical–political processes by which huge swathes of ecologically diverse lands, inhabited by culturally diverse communities managing them for multiple uses and values, have been legally (mis)classified or recorded as 'forests' and brought under unifunctional and centralized forest management. Even in areas where customary tenures and community resource management systems are constitutionally or legally protected, forest laws, aided by Supreme Court orders that transcend judicial boundaries into the legislative and executive domains, are continuing to overrule them. The official conceptualization of 'forests' as unifunctional land-use systems primarily for sustained timber production was inherited from colonial rule. With post-independent India failing to review this colonial forest policy in light of the new mandate of an independent, democratic nation, and instead continuing with the policies of state appropriation of the commons for commercial exploitation with even greater vigour, state–community conflicts related to forest land intensified manifold. The policy of exclusionary conservation initiated in the 1970s compounded the survival crisis of forest-dwelling communities, and the recent attention to environmental services, including carbon markets threatens to further aggravate the problem.

The government's flagship Joint Forest Management (JFM) programme emerged in the 1990s in response to civil society and grassroots pressures to restore community stakes in the country's forests. JFM, however, has reached an impasse due to its skirting around critical issues of tenure, the livelihood functions of land classified or claimed to be 'forest', and the resource rights and customary management institutions of indigenous forest-dwelling communities (see Chapter 1 in this book). Instead of addressing these issues, JFM has been used as an instrument to extend forest boundaries to additional settled and shifting cultivation, and grazing/pasture land by claiming such land to be state 'forests' and indeed, evicting forest-dwellers.

Finally, the multiple orders issued by the Supreme Court under the ongoing Godavarman public interest litigation (PIL) case[1] have further narrowed the ecological focus to protecting *trees* and forest *land* (rather than forest ecosystems) through centralized administrative control. The cumulative impact of these processes on the citizenship and survival rights of many already marginalized scheduled tribes (STs) and other forest-dwelling communities have been devastating. Matters reached a flashpoint in 2002 with the Ministry of Environment and Forests (MoEF) ordering large-scale evictions of forest-dwellers (MoEF 2002), in many cases from their ancestral lands, by treating them as illegal 'encroachers' on state forests.

The Scheduled Tribes and Other Traditional Forest Dwellers (Recognition of Forest Rights) Act, 2006 (FRA for short) emerged as a legislative means to centre-stage the critical issue of forest-dwelling communities' citizenship and resource rights in the arena of forest land classification and its exclusionary management. This chapter traces the origins and basis of this historic Act, and its potential and constraints. Finally, the experience with implementation of the Act from 2008 till 2011 and the challenges ahead are summarized.

The next section of the chapter delineates the legal construction of state forests during the colonial and post-independence periods, highlighting state appropriation of the commons and the entrenchment of private property rights. This is followed by an analysis of post-independence forest and wildlife conservation policy and law, including the emergence of JFM, culminating in the Godavarman PIL orders effectively re-writing forest law. Included in this analysis is an examination of the large-scale declaration of constitutionally protected tribal land as state forests under a colonial forest law and how due legal process even under that law for recognizing rights was bypassed. Attention is then given to the FRA's potential to create a new rights-based framework for democratic forest governance and early trends in its implementation. The concluding section examines the potential and limitations of the new law in the context of globalization and the continuing extension of the state forest boundary through legally ambiguous means facilitated by Supreme Court orders.

Legal Construction of 'National' Forests

Myths and Assumptions

One of the most vociferous criticisms of the FRA while it was being framed was that it would lead to destruction and privatization of state forests in the country through their distribution among the country's tribal minority. The ahistorical assumptions underlying this criticism are that the national forests are clearly defined, are legally constituted, are really 'national' in nature, and all consist of real forests. All these are questionable assumptions. Due to the poor condition of the country's land records, for instance, forest and revenue land records do not tally. In 2003, whereas the recorded forest area (RFA) according to the MoEF records was 77 million hectares (ha), according to the land-use records maintained by the Ministry of Agriculture, it was only 67.87 million ha (FSI 2005: 1). This implies that as much as 9.13 million ha of land is disputed between the revenue and forest departments. As discussed later in this section, a lot of this land is under cultivation or other uses with little forest cover on it but it continues to be treated as legal forest by the MoEF.

Similarly, day-to-day discourses on 'forest' management seldom question the legality or rationality of the premises and processes by which the 77.47 million ha of RFA, representing 23.57 per cent of the country's geographic area, has been assembled. As much as 17.6 per cent of this area consists of unclassified forests under diverse owners and tenures which are not even legally notified as forest (FSI 2005: 5). A large part of this consists of shifting cultivation lands in the Northeast governed by customary tenures protected by Schedule VI or other provisions of the Constitution. Even in the case of the 51.6 per cent of the RFA stated to be Reserved Forests (RFs) and 30.8 per cent that is Protected Forests (PFs), the required legal process of settling the rights of existing users has yet to be completed in most tribal areas in central India despite these being governed by Schedule V of the Constitution. Thus, a major contradiction in the current approach to forest conservation is its focus on protecting *land* that has often arbitrarily been (mis) classified as (legal) forest, instead of focusing on real *forests*.

A brief look at how the national forest estate has been assembled will help understand the three major roots of the problem: a) classification of the country's notified or 'recorded' forest land has been done without following due process of law, often in violation of constitutional provisions for safeguarding tribal cultures and rights; b) significant areas of this land never had, or should not have (from the point of view of biodiversity conservation),[2] or are ecologically *incapable* of supporting forests, and c) centralized and unifunctional tree-focused forest management has been superimposed on these lands, irrespective of their pre-existing multifunctional uses, customary tenures and rights, thereby disenfranchising their residents of their basic citizenship and livelihood rights.

Legal Construction of Indian Forest Boundaries during the Colonial Period

Historically, cultivated lands and the uncultivated commons in many of the country's forested landscapes were managed as an integrated resource base by diverse communal resource management traditions and institutions[3] under notional control of different rulers. Most of these systems rested on customary boundaries defining communal property rights regulated and enforced by traditional community institutions (Agarwal 1996; Guha 1989 and 2001; Sundar 2001). The system of revenue administration introduced during colonial rule delegitimized many of these systems by instituting private property rights for land under settled cultivation to facilitate revenue collection. This was combined with the declaration of large areas of the non-privatized commons as state property in areas under direct British rule. Community or common property rights in such areas were divested of legal protection under the new statutory regime unless specifically included in the record of rights prepared during revenue or forest settlements. State appropriated commons, termed 'the wastes'[4] by the colonial government due to their not yielding land revenue, were either categorized as 'forests' or revenue 'wastelands'. While good forests were selectively reserved for commercial exploitation, large areas of other common land were arbitrarily declared state forests through blanket notifications without any vegetational or socio-economic surveys. Rather than identifying forests, the objective was to assert state ownership over non-private lands. All uncultivated lands, including those under permanent snow and alpine pastures in British Kumaon, for example, were declared state-owned 'District Protected Forests' through a blanket notification in 1893. Under the Assam Forest Regulation, 1891, vast areas of uncultivated land considered to be 'at the disposal of the government' were categorized as 'unsettled forests', and recorded as 'unclassed state forests' despite these having little woody growth even at that time (Upadhyay and Jain 2004). The legal designation of such land as state forests has not been reviewed till today.

In areas such as Himachal Pradesh and British Kumaon, early forest reservation was accompanied by extensive forest 'settlements' involving the recording of customary rights of users. Many tribal forested areas, however, were reserved without a forest settlement process. In part, this was due to the high costs and difficulties of surveying them, their low potential of generating land revenue, and the regularity of tribal rebellions against external interference in their relatively autonomous governance systems. Concerted efforts were also made to stop the widespread practice of shifting cultivation by the use of force and declaring it illegal.[5] Settlement of rights only occurred sometimes with those who had permanently occupied land, resulting in the exclusion of the majority of shifting cultivators. Many tribal areas where shifting cultivation was practised lay in the territories of the nominally sovereign princely states where land administration lagged behind the British-ruled areas (Kumar 2008).

Colonial efforts to impose private-property-based land revenue administration and restrict customary forest access were met with over 150 tribal rebellions. While many were suppressed brutally, in other areas, the colonial government was compelled to either recognize diverse community tenures or keep them outside the ambit of regular land and revenue administration. Many tribal areas were declared excluded (in the Northeast) or partially excluded and brought under direct administration through the Governor or an agent of the Crown to enable continuation of their customary resource use and governance systems.

Due to historical discussions on forestry being primarily focused on forest reservation, a relatively less discussed aspect of the colonial period is that substantial, generally less valuable, forest areas were set aside for community use in the revenue settlements of villages. The record of rights of these had extensive recording of community rights. Revenue settlements in the Central Provinces and Berar (undivided Madhya Pradesh after independence) for example, included extensive forests for *nistar* (Garg 2005).[6] Settlements in Odisha provided Gramya jungles, Khesra forests, and so on for community use. In the Damin-i-Koh hill tract in the Santhal Parganas, the Santhal Pargana Protected Forest Rules even recognize shifting cultivation rights of the Paharias (Rao 2005).

In other cases, the tenants of *zamindars*, princely states, and private forest-owners not only enjoyed both customary and legal rights to use of common land and forests but local forest management was also left in the hands of traditional community institutions (Sarin *et al.* 2003). The most famous example is, of course, that of Van Panchayats in British Kumaon. Here, in response to violent protests against forest reservation, provision for Van Panchayats was made under the Scheduled Districts Act of 1894, till today almost the only example of legally demarcated and notified community forests managed by communities.[7]

A less well known, but perhaps more dramatic, example is that of the erstwhile Chhota Nagpur region (present-day Jharkhand). Here, in response to the repeated tribal rebellions throughout the nineteenth century, a series of legislations was enacted, culminating in the Chhota Nagpur Tenancy Act (CNTA) of 1908. Besides providing for the creation and maintenance of land records, the CNTA also created a special tenure category of 'Mundari khuntkattidars' (considered to be the original settlers of the land among Mundas) and restricted the transfer of tribal land to non-tribals. Most significantly, the CNTA provides for the recording of various customary community rights in land and 'jungle or wasteland', such as the right to take produce and to graze cattle, as well as the right to reclaim 'wastes'. The colonial land revenue laws in Chhota Nagpur were perhaps unique in India in the extent to which community rights in common land and other resources were recognized (Upadhyay 2005). The Santhal Parganas Tenancy Act, 1949 has similar provisions and, between them, the two Acts still cover the whole of present-day Jharkhand.

In the Western Ghats of Karnataka (which include parts from the erstwhile Bombay Presidency, Mysore princely state, Coorg princely state, and Madras Presidency), a combination of different physical and social geographies resulted in the settlement of rights taking a different turn. Significant areas of forest were assigned to individuals or groups of households within the landowning class or a subset of this class, although with much local variation in names (*soppinabetta, kumki, haadi, bane*), in extent of area and rights granted, in the nature of survey and demarcation, and in administrative arrangements (Srinidhi and Lele 2001). With the exception of Uttara Kannada, where the soppinabettas were demarcated within PFs (because virtually all the non-agricultural land had been taken over in the British forest settlement process), these forest tenures were recorded in the revenue settlement and administered by the revenue departments.

At the end of colonial rule, thus, a complex and wide diversity of tenurial regimes, both customary and legally recognized, still prevailed in British Indian provinces, Indian princely states, and other territories in British India. Within each of these land tenure systems, there were a large number of categories of tenancies and land and forest rights that were recognized by the owners or intermediaries. The nature and extent of these rights, as well as the extent to which these were recorded varied widely. In areas where revenue and/or forest settlements had been done, records of rights had been prepared. Most revenue settlements provided for substantial areas of common land and forests for local use. Most hilly and forested tribal areas left partially or fully excluded from normal revenue administration, however, had not been surveyed and, despite the colonial protection provided to tribal governance systems in them, had poor or no records of customary rights of their inhabitants.

Post-Independence Expansion of the Forest Estate

Instead of enabling indigenous forest-dwelling communities to claim restitution of their lands forcibly appropriated during colonial rule,[8] post-independent India did the opposite. Contrary to the general impression of massive diversion of forest land to other uses since independence, the *net* area of state forest land increased by 26 million ha between 1951 and 1988 (from 41 million ha to 67 million ha), largely as RFs in which there are limited or no rights (Saxena 1995 and 1999).[9] This was done by 'vesting' in the state diverse categories of non-private land of the ex-princely states and zamindars by a stroke of the pen without surveying their vegetation/ecological status, and declaring them RFs, PFs, or 'deemed' state forests irrespective of their existing users or uses.

As this expansion of the national forest estate was mostly done using the Indian Forest Act (IFA), 1927, it is useful to look at its key provisions. Chapter II to V of IFA clearly provide that no forest or land should be so notified unless the existing rights of individuals and communities have been fully enquired into and taken into account. Sections 3 and 29 allow only land that is government property or where government has some proprietary rights to be declared PF or RF. Sections 7 and 29 require an inquiry into pre-existing rights of villagers before such declaration. Sections 6, 21, and 31 specify that a vernacular notification of intent is essential. All these sections were violated in the creation of new RFs and PFs in most tribal areas after independence.

In Odisha and Madhya Pradesh, the addition of Section 20A through an amendment to the IFA was used to circumvent the requirement of settling pre-existing rights by declaring the vested forest land as 'deemed' RFs or PFs. However, in each case it was mentioned that such declaration (as RF or PF) shall be subject to recognizing the existing land rights and usage customs of individuals and communities. Unfortunately, this was never done in either of the two states. Thus, village forests and common land, with extensive recorded rights, were simply 'vested' in the state and handed over to either the revenue or forest department fairly arbitrarily.

Due to difficulties in dealing with the different types of land tenure records at the time of the vesting of private forests in the state, and the fact that in many cases there were no proper land survey records (especially in the erstwhile princely states), the process of forest settlement in them is still far from being complete. In many tribal areas, forest settlement is yet to be undertaken even 60 years after independence.[10] The rights of *podu* (shifting) cultivators as well as settled cultivators on land with slopes above 10 degrees were simply ignored in both Odisha and Andhra Pradesh.[11] Despite this, over time, state forest departments have de facto extinguished the pre-existing rights of forest-dwellers and established their exclusive legal jurisdiction over such 'forest' land.

While zamindari abolition freed tenant cultivators in the plains from landlord oppression, declaration of zamindari forests as state forests often illegally deprived them of their forest rights.[12] In poorly surveyed hilly forested landscapes, it threw millions of predominantly tribal forest-dwellers in the clutches of a far more oppressive zamindar—the Forest Department (FD), which declared even their unsurveyed cultivated lands and settlements as state forests. In the process, large numbers of the most vulnerable STs and other forest-dwellers were disenfranchised of their customary resource rights without even their knowledge and labelled 'encroachers' and thieves on their ancestral lands. Even in areas with good records of rights, there was a near wholesale reclassification of legally recognized community land and forests into 'national' forests, a fact which has escaped serious questioning to date.

The cases of Madhya Pradesh and Jharkhand illustrate the post-independence processes by which state forest boundaries were expanded across the country with drastic impacts on people's land and forest rights.[13]

Conversion of Nistari Land and Forests into 'National' Forests in Madhya Pradesh[14]

Revenue settlements in the princely states constituting undivided Madhya Pradesh were carried out in 1910. These included preparing a record of (private property) rights and a *nistar patrak*, which recorded usufruct rights in common land. Common land areas were classified as *nistari van*, *malguzari/zamindari van*, revenue *van*, *bade jhad ke jungle*, *chhote jhad ke jungle*, *ghas*, *charnoi*, *charagah*, among others (all representing different types of common forests or grazing land) and settled on a similar basis in all villages. Provision was also made for common *gothan*, *khalihan*, *kabristan*, *shamshan*, land for skinning hides, playgrounds, *padav*, bazaar, etc. In the malguzari and zamindari villages, these common land areas were controlled by the malguzars and zamindars while in the *ryotwari* villages their control was with revenue officials (Garg 2005).

After independence, Madhya Pradesh's Abolition of Malguzari and Zamindari Act vested all proprietary rights in such estates with the state free from all encumbrances. Under this, 94,78,000 ha of common nistar land controlled by malguzars and zamindars was handed over to the revenue department. In 1958, these same lands were notified as Undemarcated Protected Forests and transferred to the FD. The said notification clarified that, pending the settlement of rights under Section 29 of the IFA, existing rights of individuals or communities in such land shall not be abridged or affected in any manner except in so far as they may be modified by the state government from time to time. The revenue department, however, made no changes in its records and, in 1959, declared the same land as *dakhal rahit bhoomi* (land free from all encumbrances) and set it aside for nistar rights of the people, similar to those in the earlier nistar patrak, as per the new Madhya Pradesh Land Revenue Code, 1959 (Garg 2005).

While the revenue department recorded community uses of these lands in great detail, a notification entrusted the FD to manage the same land as state PFs. During demarcation of the PFs, 12,37,000 ha of scattered or poorer quality forests was left undemarcated and marked orange on the maps (known as 'Orange Areas'). With the focus on increasing food production at that time, under instructions from the state government and without changing the status of the land to revenue land, the revenue department granted pattas and leases to an estimated 10,00,000 landless Scheduled Caste (SC) and ST families for cultivation, under the Government of India's Grow More Food Scheme.

Several surveys and settlement procedures for settlement of rights in the Orange Areas have taken place over the years. In 1966 and 1988, the FD undertook surveys. Simultaneously, a survey and settlement process under the Land Revenue Code was undertaken by the revenue department around 1968. Joint surveys by both departments were also conducted with further surveys in 1990 and 1994. This has resulted in total confusion regarding the exact legal status as well as the total area of lands that were once malguzari and zamindari forest land originally set aside for meeting the bona fide requirements of the local population (Garg 2005).

This confusion about the legal status of 12,37,400 ha of Orange Areas that continue to be recorded as both revenue and forest land has been the cause of immense conflict between the two departments. It has been particularly disastrous for the people distributed land under various government schemes by the revenue department but whom the FD now treats as 'encroachers' on forest land.[15] The situation got further complicated after the enactment of the Forest Conservation Act (FCA), 1980, and even more so after the Supreme Court order of December 1996.

In 2003, Ekta Parishad, a non-governmental organization (NGO) in Madhya Pradesh, filed an intervention application (IA) with the Central Empowered Committee (CEC) pleading early resolution of the matter:

> ... as the fate of about ten lakh families who are predominantly tribal people is hanging in uncertainty due to the negligence on the part of the states to resolve the contentions on the Orange Areas and its boundaries and jurisdiction. That the difference and erroneous interpretation of the forest boundaries especially in the light of the Supreme Court order dated December 12th, 1996 in C.W.P. No. 202 of 1995 relating to the definition of forest, is resulting in totally exploitative steps to evict tribals who are validly staying in their lands. This is affecting both the sustainability of forests and people who are dependent on them.

The matter has continued being shunted between the CEC and the Supreme Court without any resolution. In response to an inquiry under the Right to Information Act, the otherwise proactive CEC informed Anil Garg on 8 February 2012 that it did not have copies of the notices it had issued to the state governments of Madhya Pradesh and Chhattisgarh and that it had not made any recommendations in Ekta Parishad's application no. 196 of 2003 filed before it. In the meantime, the FD is undertaking demarcation of the Orange Areas and people are discovering new 'forest' boundaries cutting through their fields and homes (Khanna 2008).[16]

Madhya Pradesh's case indicates the totally arbitrary basis on which almost 9.5 million ha of common lands and nistari forests with extensive recorded rights were reclassified as 'national' state forests by a stroke of the pen after independence. While the legally recorded community rights in the land seem to have evaporated into thin air, the new rights granted to cultivators by one government department continue to be ignored by another. Massive destruction of the forests on these lands took place during the period of their transfer from private owners to the state and much of what remained was destroyed either through converting them into open access forests with the two departments locked in a jurisdictional dispute, or through distributing the land to the landless.

Dilution of Legal Community Rights in Jharkhand[17]

The extensive rights recognized under the CNTA and the Santhal Parganas Tenancy Act (SPTA) in Jharkhand have similarly been progressively diluted with the various changes in the law and land revenue system, creating conflicts over access to, and control over, common land and forests.

Zamindari abolition in undivided Bihar was effected by the Bihar Land Reforms Act, 1950, which provided for the 'vesting' in the state of all land, estates and interests (other than *raiyati* lands), abolishing all intermediate tenures, and the transfer of all land recorded in the names of zamindars and other tenure-holders to the state.[18] After the abolition of zamindari, there remained basically two categories of non-raiyati land in this region—'Mundari khuntkattidars' and 'vested'. This significantly altered the land tenure system in Jharkhand (Upadhyay 2005).

The CNTA and the original land records recognized both community rights and settler rights on *gair mazrua* (gm) or *parti* land (traditional common land in pre-independence revenue records of present-day Jharkhand), subject to certain restrictions. However, legal ambiguities in the process of the vesting of such land in the state have eroded these rights and effectively dispossessed many cultivators who have been occupying such land for years, but who have no legal proof of their possession. For instance, families that had been settled on gm land by a zamindar or Munda, but who had only *hukumnamas* or rent receipts as proof of possession, could not get their names recorded during revision surveys. The 'vesting' of gm lands has also led to the erosion of community rights in such land, which has given rise to conflicts between the state, which now claims the sole right to settle or use such land, and local communities, who regard these lands as their traditional common land for their own use—rights that were confirmed in the original land records throughout the region. The revision land revenue settlements, in the few districts where these have been completed, have surreptitiously reclassified '*gair mazrua khas*' land as '*anabad Bihar sarkar*' (uninhabited Bihar government land) without recording or upholding existing user and settler rights recognized under the CNTA despite it still being in force. The government can now legally dispose of gm land without seeking permission from the local community and without a land acquisition process. Because of this, people settled on gm land cannot claim compensation for such acquisition, nor can the local community claim compensation for loss of access to such common land (Upadhyay 2005).

In the same fashion, over 20,000 sq. km of land in undivided Bihar, most of it in Jharkhand, has been notified as PF after independence through nationalization of a category called Private Protected Forests (PPF), land which was zamindari land on which tenants had extensive rights under the CNTA. Several records indicate that acquisition of these lands and their classification as PF violated legal procedures laid out in the IFA (Vasan 2005).

To summarize, through the 'vesting' of the non-private lands of princes, zamindars, and intermediary tenure-holders with the state after independence, land with a complex diversity of customary and legal common property tenures and land-uses were converted either into revenue 'wasteland' or state forest land and brought under centralized management by large bureaucracies. Forest boundaries were arbitrarily defined in poor correlation with the ecological characteristics of the land; even legally recognized rights were eroded, diluted, or extinguished, often without following due legal process, and community resources reclassified as 'national' forests. The requirement under Section 4 of the IFA is that while declaring the state's intention to reserve an area as forest, a settlement officer should be appointed to settle the claims of its pre-existing occupants and users. This requirement was often dispensed with. Many of these lands have still not been surveyed, and the land and forest rights of their pre-existing occupants and users remain unrecognized. Large numbers of their predominantly tribal inhabitants were converted into 'encroachers' and thieves on their ancestral lands, with even their unsurveyed villages notified as state 'forests'. In many cases, these lands are yet to be finally notified as RFs under Section 20 of the IFA. Hence, their legal status as state 'forests' is highly problematic (Sarin 2005a).

Post-independence Forest Policy and Law

1952 Forest Policy and Commercial Forest Exploitation

The 1952 national forest policy reflected a continued contempt for local rights and livelihoods by stating that 'the accident of a village being situated close to a forest does not prejudice the right of the country as a whole to receive benefits of a national asset.'[19] And while the policy talked of the benefits derived from this asset in environmental terms, in practice, the focus of state forestry in the first three decades after independence was on commercial exploitation of the forest resource for industry and urban markets.

This form of forestry changed the nature of the forest itself through replacement of multi-species forests by commercial plantations. This further deprived forest-based communities of their livelihoods while simultaneously destroying rich biodiversity under the rubric of 'scientific' forest management. During the 1970s, even important NTFPs were nationalized. In 1976, by when most multi-species forests had been exhausted, the National Commission on Agriculture (NCA) announced that: 'Production of industrial wood would have to be the raison d'être for the existence of forests.' As pointed out by Saxena (1999: 12), 'the entire thrust of forestry during the first four decades after Independence was towards the production of a uniform industrial cropping system, created after clear felling and ruthless cutting back of all growth, except of the species chosen for dominance.' Forest Development Corporations set up for raising commercial plantations turned themselves (in the words of Dr Salim Ali and Mrs Indira Gandhi) into Forest Destruction Corporations and clear felled huge tracts of rich natural forests without ensuring their replacement. Forest-based industries were given bamboo or trees for pulpwood at throwaway prices and they promptly exhausted these resources. FDs did not spare even the sacred groves protected by communities for generations. The plywood industry was provided access to giant wild mango trees, which yielded fruits famous for pickles worth hundreds of rupees every year for local communities, for as little as 60 rupees (Gadgil 2008).

This transformation of the forest landscape brought local communities in perpetual conflict with forest departments. A wave of protests against commercial fellings and replacement of natural forests by commercial monocultural plantations swept the country during the 1970s in Uttarakhand (the Chipko movement), Bastar, Jharkhand, and other areas. Unfortunately, the changes in the legal regime that occurred during the 1970s and 1980s did not quite address these concerns.

Centralization of Control with Growing Environmental Concerns

While local communities were thus demanding forest and environmental management focused on supporting local livelihoods, elite environmental concerns found expression in two new central laws in the 1970s.

The Wildlife (Protection) Act (WLPA), 1972, adopted the western exclusionary approach to wildlife conservation, namely, the setting aside of large tracts of land where little or no human presence is to be permitted. For this, the Act requires all legal and customary rights in national parks to be extinguished while severely restricting them in wildlife sanctuaries. More remarkably, it vests unfettered authority with the state to declare any area a protected area (PA) without any process of public consultation or the right for affected people to file their objections. PA managers are empowered to stop the exercise of rights from the day of the preliminary notification by providing alternatives till rights are settled. With little awareness among forest-dwellers about provisions of the law, the inaccessibility of judicial recompense for the average non-literate villager living in such areas, and the immense powers and authority enjoyed by forest officials, de facto extinguishment of even legally recorded rights in most PAs has been the order of the day. In practice, even notional alternatives have not been provided for the loss of rights despite the fact that the final settlement of rights yet to be completed in over 60 per cent of the PAs. In any case, the WLPA provides for settling only 'recorded' rights despite the fact that in most tribal areas, where most PAs are concentrated, few customary rights are recorded. It is next to impossible for the affected people to seek any legal remedy as all decisions related to PAs must now be approved by the Supreme Court and the National Board of Wildlife.

The Forest Conservation Act (FCA), 1980, enacted after forests had been moved from the State to the Concurrent List in 1976, made central government permission mandatory for diverting even small parcels of forest land to non-forest uses irrespective of the diversity of contexts across the country. The FCA froze legal land use for land declared 'state forests' through the highly deficient processes described above. Initially considered applicable only to finally notified RFs and PFs, over time its mandate was extended even to lands with preliminary notifications where rights are yet to be settled, in addition to 'any area recorded as forest in the government records' despite the notoriously poor quality of government records. The word 'forest' has been used generically for recording even community grazing and other common land and customary community land.

Although the FCA has nothing to do with the settlement of rights, it brought even the ongoing slow and inefficient forest survey and settlement processes for forest land vested in the state after independence in different states to a near halt. The Odisha government, for example, had identified 276,000 acres of forest land being cultivated by STs and other landless people for settling in favour of the cultivators in 1972. Enactment of the FCA before the plan could be implemented left the forest land cultivators in the lurch (Kumar *et al.* 2005). Even the recognition of existing rights started being treated as diversion of forest land to non-forest uses requiring central clearance and compensatory afforestation (CA). In so doing, the FCA prevented any remedy to the several million forest-dwellers who had been cast as illegal occupants of their ancestral lands due to the faulty settlement process described earlier. The Forest Advisory Committee (FAC) constituted by MoEF for advising on diversion of forest land under the FCA has no accountability to the local people whose land and forests it is empowered to permit for diversion and is not required to take legally recognized rights into consideration.

Impact of the Godavarman PIL

Matters were further complicated by the Supreme Court order of December 1996 under the Godavarman PIL, which extended application of the FCA even to all lands conforming to the dictionary definition of forest, irrespective of ownership. All such 'forest' land now has to be managed in accordance with working plans/schemes prepared by FDs and approved by the MoEF.

State FDs have been identifying such 'forest-like land' to bring them under their management control with little discussion about the legal processes to be followed, the livelihood impacts on people dependent on such land or how their legal rights under other existing laws or constitutional provisions are to be dealt with. Under the Santhal Pargana Tenancy Act (SPTA), for example, traditional village heads are legally empowered to settle scrub village forest land in the name of ryots (Rao

2005). The interim court order has effectively overruled this without the state legislature amending the law.[20] The situation is even more contradictory in the Northeastern states where community rights and customary tenures enjoy constitutional protection.

Proactive Interlocutory Applications (IAs) filed by the amicus curiae in the case have led to further interim court orders with drastic impacts on the rights and livelihoods of impoverished tribal and other forest-dwellers. Besides staying regularization of even eligible pre-1980 'encroachments' (Order dated 23 November 2001) and de-reservation of forest land or PAs, irrespective of whether these have been finally notified after due settlement of rights (Order dated 13 November 2000 in WP(C) 337/95), the Court has also stayed the 'removal of dead, diseased, dying or wind fallen trees, drift wood and grasses, etc.' from all National Parks (NP) and Wildlife Sanctuaries (WLS) (Order dated 14 February 2000). Although the last order was directed at FDs to prevent them from using the removal of dead and dying trees as a cover for unauthorized felling from PAs, the MoEF and the CEC interpreted it to mean that 'no rights can now be exercised' in PAs and have banned the collection and sale of all NTFP from them. This is despite people having legally admitted rights in many finally notified PAs.

In one stroke, between three to four million of the poorest forest-dwellers, who were living inside PAs long before their notification as forests or PAs, were deprived of their citizenship rights and access to livelihood resources. There was no due legal process or any scientific studies substantiating that all collection of forest produce is harmful to wildlife habitats or biodiversity. In Odisha's infamous 'starvation deaths' tribal region, some PA managers have been refusing permission for gram sabha meetings for information dissemination, entry of health workers and in one case, even the delivery of public distribution system (PDS) rations to villages inside PAs.[21] Impoverished tribals have been driven to giving their children in bondage and resorting to large-scale distress migration (Anonymous 2004; Rao 2004). While the Court's focus on holding the executive accountable for protecting forests and wildlife may be laudable, its orders have totally overlooked, and in fact reinforced, the even more grave failures of the executive in enforcing the constitutional protection to tribal rights and governance systems in the same areas. With state FDs being the principal respondents in the PIL, the affected PA-dwellers have had little representation in the ongoing court proceedings. As mentioned earlier, in 2003, Ekta Parishad had filed an intervention application (IA) with the CEC, pleading for early resolution of the Orange Areas and their boundaries and jurisdiction affecting the tenurial rights of about ten lakh predominantly tribal families. In February 2012, the CEC informed the petitioner that it had made no recommendation on the matter. Dominated by strong supporters of an exclusionary approach to conservation, the CEC assisting the Court has no representation of either the constitutional authority or the ministry responsible for tribal affairs (Sarin 2005b).

The bringing of community land with diverse tenures and livelihood functions under the FCA's purview has confused their management objectives, diluted or erased legal and constitutionally protected community rights, created jurisdictional conflicts between forest and revenue departments and panchayats and traditional community institutions, while being difficult to enforce. As pointed out by the CEC itself in its recommendations to the Court on dealing with 'encroachments' on forest land, 'In respect of deemed forest area, unclassed forest and areas recorded as forest in government records, which are not legally constituted forests, the provisions under which an offence can be booked are not clear' (CEC 2002).

Court orders under this PIL have reshaped forest management in the country, effectively re-written the law, and also significantly changed centre–state–local relations concerning land-use. Despite land being a state subject, due to land under diverse owners, tenures, and uses being brought within the extended forest boundary by Supreme Court orders, the MoEF is now responsible for enforcing their management in accordance with forest 'working plans' without having any legal jurisdiction over them. While court proceedings in this PIL have tended to negatively equate all references to the rights of forest-dwellers with 'encroachment', both the MoEF and the court have been increasingly liberal in permitting the destruction of rich forests and tribal and wildlife habitats for mining, industry, and hydro projects (CSE 2011; Khanna 2008) (see also Chapters 6 and 7 in this book). The biggest beneficiary of the Court's orders has been the forest bureaucracy, as their powers to control land and forest use have been extended to areas hitherto outside their domain. This is particularly ironic, given that Godavarman filed the original PIL against the mismanagement of forests by the department.

Dissonance Between Tribal and Conservation Laws

The Indian Constitution continued protection for the partially excluded and excluded tribal areas through Schedules V and VI of the Constitution under Article 244. Any government interventions in the scheduled areas need to be in harmony with the constitutional provisions and other policy directives for safeguarding the culture, resource rights, and livelihoods of tribal communities. Schedule V of the Constitution empowers the state governor to withhold the application of any laws considered detrimental to tribal interests from Scheduled Areas.[22] Article 338 (9) of the Constitution requires that the National Commission for Scheduled Castes and Scheduled Tribes (now bifurcated into separate commissions for SCs and STs) must be consulted by the Union and state governments on all major policy matters affecting SCs and STs.

Yet, massive legal expansion of the national forest (and revenue 'wasteland') estate in Schedule V areas after independence has violated all the above constitutional provisions. Due to the poor recording of tribals' customary rights and tenures, Schedule V areas bore the brunt of the post-independence spree of state takeover. The state has been the biggest violator of the spirit of the Constitution by 'vesting' huge areas of customary tribal land in itself as state forests, PAs, or 'wasteland', without recognizing ancestral rights of the tribals and extending all its coercive laws to them. At best, rights only over land under settled cultivation were recognized, largely leaving out shifting cultivators and nomadic and extremely vulnerable pre-agricultural hunting-gathering communities. The poor recognition of communal tenures in Indian statutory law[23] has decimated their economies and cultures. Instead of withholding or adapting the Land Acquisition Act (LAA), IFA, FCA, and WLPA to accommodate the tribals' customary tenures and governance systems, their indiscriminate use even in Schedule V areas as if these are the only laws applicable to them, has progressively negated even the hard-fought-for rights tribals gained during colonial rule.

Deprivation of their customary resource rights, holistic land-use systems without rigid forest-non-forest boundaries and a rich diversity of resource management institutions has been accompanied by tribals being labelled 'encroachers' on their ancestral lands. Millions have been displaced without any compensation or rehabilitation due to not having legally recorded rights.

Seventy-six per cent of Odisha's Schedule V areas have been declared state property—50 per cent as forests and 26 per cent as revenue wasteland, while the vast majority of the tribals have been left legally landless. Land with over 10-degree slopes were left unsurveyed simply because the cost of surveying them was too high or because shifting cultivators were considered ineligible for a grant of land titles as they did not occupy the same piece of land continuously for 12 years (Kumar 2008). Hundreds of tribal villages on land declared to be state forests have never been surveyed, depriving them of access to basic development facilities and citizenship rights.[24] A similar situation prevails in Andhra Pradesh where over 60 per cent of Schedule V areas have been declared RFs without following the due legal process.

Implementation of the Provisions of the Panchayats (Extension to the Scheduled Areas) Act, 1996 (PESA) has met the same fate. PESA makes the gram sabha (the body of all adult voters of a self-defined community) 'competent to safeguard and preserve the traditions and customs of the people, their cultural identity, community resources and the customary mode of dispute resolution' (Clause 4d). Every gram sabha is also empowered to approve the plans, programmes, and projects for its social and economic development before their implementation, besides being endowed with ownership of minor forest produce (MFP). PESA effectively mandates gram sabhas to undertake community-based management of their customary forests. Yet, despite PESA being applicable to entire Schedule V areas, the MoEF claims exclusive jurisdiction over forest land within them and has continued to enforce its unilateral interpretations of PESA in the absence of any other agency forcefully protecting tribal interests. Similarly, ownership of MFPs has not been granted to gram sabhas on the ground that the Act has not 'defined' MFPs.

A Frame for Resolving Tribal–Forest Conflicts and MoEF's 1990 Circulars

In his 29th report (1987–9) to the President of India, the commissioner for SCs and STs brought the disquiet prevalent in tribal-forest areas to the government's notice and recommended a framework for resolving disputes related to forest land between tribal people and the state. This was discussed and approved by a committee of secretaries and in a conference of state forest ministers. Based on the commissioner's recommendations, the MoEF issued a set of six circulars on 18 September 1990.[25]

Only the first of these related to regularizing pre-1980 'encroachments' on forest land. The second circular required resolution of disputed claims over forest land arising out of incomplete or faulty forest settlements. Instead of penalizing villagers for the government's own failures, the third circular required recognition of *pattas*, leases, or grants issued under due legal authority by the revenue department for land also recorded as forest land. The fifth circular required conversion of an estimated 2,500 to 3,000 'forest villages',[26] created by FDs themselves in the past for ensuring availability of bonded labour for forestry operations, and old habitations to revenue villages. On paper, the land is recorded as 'forest'; on the ground, these are legally constituted villages. However, their residents have no titles to their land, cannot obtain domicile certificates or benefit from development programmes as other departments cannot work on forest land. Their residents remain at the FD's exclusive mercy for most of their basic needs.

No state government took any meaningful action on these circulars.[27] The MoEF only pursued enforcement of the circular related to 'encroachments' without emphasizing the distinction between 'encroachers' and those with disputed claims and pattas. Indeed, the ministry admitted the same in an affidavit filed in the Supreme Court in July 2004, where it stated that 'the state/UT governments could not maintain a distinction between the guidelines for regularization of encroachments and the settlement of disputed claims of tribals over forest lands... the state/UT governments have mixed up the whole issue.' Not surprisingly, all forest-dwellers with long-pending disputed claims have become equated with 'encroachers' on forest land in the public mind reflected in the vitriolic attack on the FRA by elite wildlifers, conservationists, and the MoEF itself. Unfortunately, the Supreme Court displayed the same bias in the *Godavarman* case hearings with its orders effectively negating the MoEF's 1990 circulars. Interestingly, although the 1990 circulars highlighted irregularities in the notification of forests, they did not question the fundamental anomaly in the declaration of constitutionally protected tribal areas as state forests.

Joint Forest Management (JFM): De-concentration of Administration Rather Than Devolution of Control

The JFM programme in the country was the result of a herculean effort by a handful of proactive bureaucrats within the MoEF and outside to bring about the issuance of the 1 June 1990 circular asking states to implement JFM. It was initially welcomed by many activists (including this author) as a progressive shift from centralized forest management to a collaborative approach, which recognized the importance of satisfying local livelihood needs and making villagers partners in sustainable forest management. In hindsight, the narrow focus of the 1990 JFM guidelines is evident from the fact that at that very time, a committee of secretaries was deliberating the 18 September 1990 circulars dealing with more fundamental issues of non-recognition of rights and disputed claims over forest land. Only historians may be able to shed light on how five of the six September 1990 circulars, other than the one on regularizing pre-1980 'encroachments', got ignored by the MoEF while JFM became its flagship programme, attracting huge donor funding.

In the early years, several innovative initiatives were taken in Haryana, West Bengal, and Gujarat to evolve state-specific JFM frameworks through establishing field-based learning loops, process documentation, and multi-stakeholder state-level working groups. Due to being based on a more holistic community-based natural resource management approach developed in Sukhomajri, the work in Haryana, in particular, generated rich learnings (Sarin 1996a and 1996b). These attempted to tailor JFM to recognizing socio-economic and gender-differentiated forest dependence within communities combined with developing autonomous, inclusive, and democratic community institutions to function as the FD's JFM partners. Haryana's benefit-sharing model included access to water for irrigation, fodder grasses, bamboo, and bhabbar grass for strengthening existing livelihoods instead of the mechanical timber-sharing formula used in West Bengal. Analysis of inter- and intra-community conflicts arising out of exclusion of existing users, pre-existing rights, and usage patterns of forest land and gender and equity concerns were integrated in the development of JFM agreements in Haryana (Sarin 1996b).

Early FD responses were reticent and most forest officers had to be cajoled into giving collaborative management a try. This, however, changed dramatically once donor agencies started funding large JFM projects. The role of NGOs in evolving JFM frameworks started getting marginalized with the forestry establishment reasserting its supremacy in defining JFM parameters. The December 1996 Supreme Court order requiring management of all forests in accordance with MoEF-approved working plans came in handy for countering demands for greater devolution of management authority to JFM groups. The centrally sponsored National Afforestation Programme (NAP) introduced in 2002 (see NAEB 2002) negated most of the diversity of JFM frameworks developed by different states. The NAP created a standard two-tier structure of Forest Development Agencies (FDAs) and JFM Committees (JFMCs) for implementing JFM across the entire country, including in the Northeastern states where FDs own little land. Although pretending to be federations of JFMCs, the FDAs are structured such that all office-bearers are forest officials. The member-secretary-cum-joint account holder of *all* JFMCs under the NAP had to be the forest guard who *nominated* most JFMC office-bearers (NAEB 2002) despite many state JFM orders providing for more democratic frameworks.[28] In one stroke, the NAP restored almost total FD control over JFM, even in states which had provided for self-governing local institutions like the Hill Resource Management Societies of Haryana or the Tree Growers Co-operatives in Gujarat (Shah 2003).

The imbalance in power between communities and the department, lack of genuine participation in decision-making or preparation of JFM micro-plans, the lack of departmental accountability to villagers— essentially of JFM using 'community participation' as an instrument for achieving predetermined FD objectives, have been identified as major limitations of the official JFM approach by several studies (Sarin *et al.* 2003; see also Chapter 1 in this book). That JFM's 'success' at any given time is dependent on the availability of external funds has also been noted. Sustainability of improvements in forest conditions observed in many areas after the end of external funding remains doubtful as JFMCs enjoy no clear authority or tenurial security over their JFM forests. In Gujarat and Odisha, the FD permitted paper companies to harvest bamboo from JFM forests while most states have failed to honour their benefit-sharing commitments to the villagers (AKRSP (I) 2004; Bera *et al.* 2011; Sarin *et al.* 2003; Vasundhara 1998).

Slowly, however, grassroots protests against JFM started bringing far more fundamental issues to the fore. Enclosure of even community grazing and panchayat land for JFM plantations drew protests from livestock-herders. Conflicts resulting from land under shifting or settled cultivation being forcibly brought under JFM started being reported from tribal areas (Samata and CRYNet 2001). The FD supporting more powerful members of JFM groups to evict poorer so-called 'encroachers' on forest land, or getting JFM members to attack villagers challenging corruption by FD staff, started opening up the Pandora's box of questions related to unrecognized pre-existing rights and JFM being used as an instrument for converting disputed lands under multiple uses into state forest land (Diwan *et al.* 2001; PUDR 2001; Sarin *et al.* 2003).

In Andhra Pradesh, for example, a 1987 government memo required regularizing adivasi rights over 77,661 acres of RF land under cultivation since before 1980. After initiation of the Andhra Pradesh World Bank-funded Forestry Project, in 1995, a new memo of the FD overruled the 1987 memo and directed that the adivasis' lands be brought under joint 'forest' management. Among the Bank project's phase-I achievements, the FD proudly claimed it had retrieved 37,000 ha of forest land from 'encroachments'. Field investigations by adivasi youth of the Andhra Pradesh Adivasi Aikya Vedike revealed that several impoverished podu cultivators were cheated of their cultivated lands by the FD first encouraging them to plant trees on their land and then claiming that it was forest land to be brought under JFM. Instead of removing 'encroachments' on forest land, this was a cynical use of a 'participatory' programme to illegally convert adivasis' land into state-owned forest land.

A strong recommendation by the Steering Committee on Environment and Forests for the Eleventh Five-Year Plan to democratize JFM by launching a 'Mission Village Forest' under Section 28 of the IFA (GoI 2007: 80–3) was totally ignored by the MoEF. The idea of developing a different framework for working with traditional village institutions in the Northeast instead of extending the standard JFM framework to them has met the same fate. Instead, Section 28 of the IFA was abused by surreptitiously bringing even Uttarakhand's autonomously managed historic Van Panchayat forests created under the Scheduled Districts Act of 1874 under the IFA's purview by revising the Van Panchayat rules under a World Bank-funded forestry project (Sarin *et al.* 2003).

Brutal Evictions as the Last Straw

The last straw came with MoEF's circular of 3 May 2002. This circular cited the Supreme Court's concern over growing forest encroachments in its 23 November 2001 order and asked all states and union territories (UTs) to summarily evict all forest 'encroachers' within five months. The Court order only directed the Union and state governments to report the steps they had taken to prevent encroachments and removal of post-1980 ones; the Court had refrained from ordering the removal of encroachments prayed for by the amicus as the petitioner. But the misguided circular led to a spate of brutal evictions across the country, including the use of elephants to destroy the huts and crops of impoverished tribals during a drought year, which in turn led to an uproar of protests (CSD 2003). As the constitutional authority for STs under Article 338 (9), the chairman of the ST Commission wrote to the prime minister objecting to not even being informed, leave aside being consulted by the MoEF and the CEC (which had made draconian recommendations for eviction of forest 'encroachers' to the Court), in a matter drastically impacting an estimated 10 million tribals. The MoEF was compelled to issue a clarification order in October 2002 that the 1990 circulars remained valid and that not all forest-dwellers were 'encroachers'. Despite this, by the MoEF's own admission in Parliament on 16 Augist 2004, between May 2002 and August 2004, evictions were carried out from 1,52,000 ha of forest land. The Court itself has remained silent on the issue of an early resolution of disputed claims over forest land and the non-settlement of rights over vast areas classified as forest, while staying the regularization of even pre-1980 occupations without its permission.

In February 2004, just before the parliamentary elections, the MoEF issued two new circulars: one titled 'Regularisation of the rights of the tribals on the forest lands' that extended the cut-off date for regularization for tribals to December 1993 (instead of October 1980 under the first 1990 circular) and the other titled 'Stepping up of process for conversion of forest villages into revenue villages'. These were promptly stayed by the Supreme Court in response to an IA filed by the amicus that these were in violation of court orders staying both regularization and de-reservation of forest land. The Court's stay on the conversion of forest villages into revenue villages was particularly ironic as this has been the Government of India's stated policy since the 1970s. In its July 2004 affidavit to get the Court's stay vacated, the MoEF admitted that during the consolidation of state forests 'the rural people, especially tribals who have been living in the forests since time immemorial, were deprived of their traditional rights and livelihood and consequently, these tribals have become encroachers in the eyes of law' and that 'It should be understood clearly that the lands occupied by the tribals in forest areas do not have any forest vegetation'. It further asserted that its February 2004 circulars 'do not relate to encroachers, but to remedy a serious historical injustice' and that '(this) will also significantly lead to better forest conservation'. The Court has still not vacated the stay on the MoEF's 2004 February circulars.

With forest rights becoming a major national political issue due to the evictions, an informal alliance of grassroots movements, rights activists, and academics came together under the umbrella of the Campaign for Survival and Dignity (CSD). Together with left-wing political parties and other rights movements, the CSD undertook nationwide political mobilization with mass protests, rallies, public hearings, and conventions aimed at members of Parliament, state legislatures, and political parties. With a new central government in 2004 having made a commitment to stop forest evictions, the initial demand of the CSD and other mass movements was for the implementation of the 1990 circulars of the Ministry of Environment and Forests. However, this soon transformed into a demand for a comprehensive law for the statutory recognition of pre-existing rights, not only over cultivated lands but also over customary forest resources, and for the empowerment of village assemblies to protect, conserve, and manage such resources. The Scheduled Tribes and Other Traditional Forest Dwellers (Recognition of Forest Rights) Act, 2006, (referred to hereafter as the FRA), was an outcome of this prolonged struggle by grassroots movements for remedying a historical wrong through forest tenure reform.

The Forest Rights Act, 2006: A Game-changer

AND WHEREAS the forest rights on ancestral lands and their habitat were not adequately recognized in the consolidation of State forests during the colonial period as well as in independent India resulting in historical injustice to the forest dwelling Scheduled Tribes and other traditional forest dwellers *who are integral to the very survival and sustainability of the forest ecosystems...*

Preamble of the FRA (emphasis added)

The FRA represents a milestone in Indian legislative history with Parliament acknowledging the historical injustice done to India's tribal and other traditional forest-dwelling communities during the consolidation of state forests. By declaring that rights recognized under the Act include 'responsibilities and *authority* for sustainable use, conservation of biodiversity and maintenance of ecological balance and thereby strengthening the conservation regime... while ensuring livelihood and food security', the FRA questions the very basis of current state-controlled, exclusionary forest management. In doing so, it also lays the foundation for democratization of forest governance. For the millions treated as 'encroachers' on their forested ancestral lands, it implies restitution of their citizenship rights and a right to live with dignity.

Key Provisions of the Act

The Act has three major provisions: (a) the rights that may be claimed in all categories of forest land, including PAs; (b) the authorities and procedures for receiving and verifying the claims; and (c) the empowerment of right-holders and/or gram sabhas for conservation of forests, wildlife, and biodiversity and their natural and cultural heritage.

Rights to be Recognized Under the Act

The Act specifies 13 claimable rights providing individual and/or community tenure. Claimable rights over forest land include land under occupation, land disputed between forest-dwellers and the FD due to faulty forest settlements, land for which other government departments have issued pattas not recognized by the FD[29] and through the conversion of forest/unsurveyed villages into revenue villages. Claimable community forest rights include rights to nistar (usufructs); NTFPs; waterbodies; community tenure over customary habitat in the case of pre-agricultural communities; seasonal resource access for nomadic and pastoral communities; other traditional rights; and, most importantly, the 'right to protect, regenerate or conserve or manage any community forest resource which they have been traditionally protecting and conserving for sustainable use'. Simultaneously, the FRA protects all existing customary rights and rights recognized by state laws or Autonomous District Councils in Schedule VI areas in the Northeast.

Authorities and Procedures for Vesting of Forest Rights

The hamlet or village gram sabha (the assembly of all resident adults) is the authority for initiating the process to determine the nature and extent of forest rights claimed under the Act. This is a major departure from typical bureaucracy-controlled procedures and is designed to ensure transparency and accountability in the claim-making process. The claims verified and approved by the gram sabha are to be consolidated, examined, considered, and approved by committees at the subdivision and district levels consisting of representatives of the revenue, tribal, and forest departments and three elected local-government representatives at those levels. A state-level monitoring committee chaired by the chief secretary and with similar multi-departmental and political representation is to monitor implementation of the Act in each state. Although represented in the higher-level committees, forest officials must share decision-making authority with the elected representatives and officials of other departments.

The nodal agency for the law is the Ministry of Tribal Affairs (MoTA). This has ended the MoEF and FDs' exclusive hegemony on forest land, potentially liberating forest-dwelling communities from the unfettered control over their lives and livelihoods by an oppressive forestry establishment. Further, due to the FRA being an outcome of a prolonged struggle and mobilization by an informal alliance of grassroots movements, the organized sections of its intended beneficiaries are better informed about it, demanding its implementation compared to other top-down laws framed by benevolent lawmakers.[30]

Empowerment to Protect and Conserve

The rights recognized by the Act over community forest resources, combined with the power to protect adjoining forests, wildlife, and biodiversity and to prevent destruction of cultural and natural heritage, reinforce the empowerment of gram sabhas to manage community resources also mandated by the Panchayats (Extension to the Scheduled

Areas) Act, 1996. By creating space for statutory community forest governance (in contrast to FD-controlled JFM), the FRA challenges the forest bureaucracy's hegemonic control over the country's forested landscapes.

The FRA is also unique in making wildlife authorities accountable for decisions related to the relocation of communities from PAs. The modification of recognized rights in PAs is permitted only in 'critical wildlife habitats' identified within them through a transparent and consultative process. Relocation from such habitats can take place only after all rights have been recognized; it has been established that co-existence could lead to irreversible damage to threatened species or their habitats, and with the free and informed consent of the concerned gram sabhas. A resettlement package must ensure a secure livelihood and be acceptable to the concerned communities, and land allocation and the development of facilities in the new location must be complete before relocation.

Limitations of the FRA

A major limitation of the FRA lies in the ambiguities and shortcomings in the wording of the final Act due to intense contestation by wildlife conservationists and the forest bureaucracy during its formulation, leaving room for contrary interpretations. The most problematic among these is the differentiated eligibility of ST and Other Traditional Forest Dweller (OTFD) claimants. The requirement that OTFDs must prove continued residence in the area for three generations equated with 75 years has disadvantaged them severely. Seventy-five years ago, many of the concerned areas were under princely states or zamindars where no survey or demarcation of land had been undertaken because of which even the government has no records of that period. Equally forest-dependent OTFD claimants are largely being left out of the FRA's purview due to their inability to produce documentary evidence being demanded in support of 75 years of residence. Simultaneously, they are also not being able to claim regularization of pre-1980 occupation of forest land permitted under the MoEF's 1990 circular as the FRA is assumed to have superseded them.[31]

The second major limitation of the FRA is that it is applicable only to forest land. Because of the arbitrary manner in which customary community land has been categorized as forest or revenue land, there are large areas with similar ambiguity about people's rights which are categorized as revenue wasteland. In the absence of a law like the FCA, such lands are even more vulnerable to allocation to other users without respect for existing rights than forest land. Yet the affected people cannot claim recognition of their rights over such land under the FRA. In Odisha, for example, many of the land claims filed are turning out to be for revenue land despite their contiguity with similar land classified as forest.

Thirdly, the FRA only deals with the recognition of rights, saying little about post-recognition governance arrangements and the application of existing forestry/conservation laws or the role of the forest bureaucracy in them. While many consider this a weakness, this actually leaves room for evolving culture and region-specific plural governance systems through dialogue and negotiations with relevant authorities by rights-holding communities post the recognition of rights.

Rules for Implementing the Act

The rules finally notified by MoTA on 1 January 2008, a year after the Act was passed by Parliament, list the procedures, powers, and responsibilities of the authorities specified under the Act, the types of evidence acceptable in support of the claims and a clear procedure for dealing with petitions against rejection of claims. However, the rules failed to provide clear procedures for recognizing the rights of special groups such as transhumant pastoralists and pre-agricultural communities for whom the gram sabha-based claim-making process is unsuitable or to emphasize the distinction between claims for land under occupation and claims based on disputes arising from faulty forest settlements or for which pattas or leases had been issued by other government agencies (for which there is no upper limit for the area which may be claimed). The biggest shortcoming in the rules was a lack of clarification on the procedure for activating hamlet-level gram sabhas as defined in PESA in Schedule V areas or that in other areas with panchayats, the gram sabha had to be of the village. Instead, by requiring the gram panchayats to call initial gram sabha meetings (when the FRA does not even mention gram panchayats), the rules created the impression that the gram sabha for initiating the process of recognition of rights had to be that of the panchayat that has multiple villages in many states.

Dynamics of FRA Implementation Till Date

After over four years of implementation, the FRA has a long way to go before being able to undo the historical injustice to forest-dwelling communities. The intensity of bureaucratic and political resistance to the radical mandate of the FRA became evident in the concerted efforts made to stall the implementation of the Act even after it was passed unanimously by both houses of Parliament in December 2006. Finally brought into force on 1 January 2008, many state and non-state actors have attempted to undermine the FRA by ignoring it or subverting it through wilful misinterpretation. The Prime Minister's Office (PMO) exhorted the states to start time-bound implementation as soon as the FRA came into force, without ensuring that potential claimants and the officials responsible for implementation were fully informed about the law's provisions. Instead of clearly requiring hamlet or village gram sabhas to receive claims, at the PMO's behest, the Panchayati Raj minister wrote to all chief ministers on 15 February 2008 to get panchayats to call gram sabha meetings on the 28th of the same month to elect forest rights committees (FRCs). The states were expected to constitute the sub-division-, district-, and state-level committees and to make villagers and government officials aware of the Act's provisions during the few intervening days.[32]

Not surprisingly, the FRA's implementation had a disastrous start. Neither the villagers nor the officials who participated in the initial gram sabha meetings were informed about the Act's provisions or the role of the FRCs elected during those meetings. In one village in Odisha, only elderly people had shown up for the gram sabha meeting as they thought the meeting was about old age pensions. The school teacher deputed to conduct the meeting constituted an FRC from among the assembled elders without ensuring the required two-thirds quorum.

Since neither the rules nor the Panchayati Raj minister's letter clarified which gram sabha had to be called, practically all states that initiated implementation on the central directive ignored the Act's requirement of making hamlet-level gram sabhas in scheduled areas the initiating authority, using the much larger panchayat gram sabhas instead. Odisha was one of the few states to clarify early that revenue village *palli sabhas*, instead of multi-village panchayat gram sabhas, would be the initiating authorities under the Act. In areas where there are strong grassroots movements, such as Odisha, and some parts of Gujarat, Madhya Pradesh, and Maharashtra, they were successful in persuading the state governments to permit hamlet-level gram sabhas in scheduled areas where demanded by the villagers. Despite its large scheduled area and tribal population, Andhra Pradesh refused to permit hamlet gram sabhas during its first phase of implementation, thereby considerably disadvantaging adivasis living in remote hamlets of large multi-village panchayats. FRCs of large and heterogeneous panchayat gram sabhas are dominated by non-tribals who have often rejected the claims of their adivasi residents.

Similarly, statements from the PMO, the Ministry of Tribal Affairs, and the office of the President of India till 2009 referred only to the individual land rights claimable under the FRA, ignoring the diverse community forest rights and gram sabha empowerment also provided for. Statements of the central government also conveyed the impression that the FRA was meant only to recognize the rights of STs and not those of other traditional forest-dwellers, despite the law providing otherwise. Till recently, the nodal Ministry of Tribal Affairs provided poor leadership. The ministry has been monitoring only the number of claims received, rejected, or accepted, without any attention to compliance with gram sabha-centred procedures or to the diversity of rights claimed.

Taking a cue from the central government, the states started implementation with a primary focus on recognizing only individual land rights of STs, with an almost blanket rejection of claims by other traditional forest-dwellers. While some states are yet to begin or have been slow starters, others, which had upcoming state elections, initiated rapid, time-bound, and haphazard implementation with the intent of gaining electoral benefit. The tight time schedules for completing implementation initially prepared by Chhattisgarh and Andhra Pradesh (both had upcoming assembly elections) made it impossible to follow the democratic procedures laid down in the FRA. Chhattisgarh, with one of the largest forest-dwelling tribal populations, refused to distribute or accept claim forms for community forest rights and permitted lower-rung field officials to accept, reject, modify, or reduce individual claims as they liked, throwing all specified procedures to the wind. West Bengal similarly vested effective control over receiving and verifying claims in government officials instead of elected FRCs (see Joint Committee, 2010, and their Chhattisgarh trip report at http://fracommittee.icfre.org/

TripReports/Chhattisgarh/Chhattisgarh%20visit%20note,%20NC%20
Saxena%20_FINAL_.pdf; also Council for Social Development, 2010).

Odisha took the lead in issuing several circulars explaining the correct procedures to be followed, although the majority of the implementing field officials as well as villagers remain unaware of them till today. The rules permit rejected claimants to appeal, but this has been denied in most states. In almost all states there has been little generation of awareness or dissemination of information about the law's provisions among potential claimants or implementing officials. Members of most FRCs remain unaware of their role and most government officials and elected panchayati raj representatives involved with implementation are poorly informed about the law. Chhattisgarh's government spent millions on publicizing rice distribution at Rs 3 per kg but did not spare even a penny for publicizing the FRA.

Many states have not ensured one-third women's representation on FRCs and have issued titles in only men's names instead of in the names of both spouses. Most states are rejecting oral evidence permitted by the rules while illegally demanding other forms of evidence.

Almost across the board, the focus of implementation has remained only on one of the 13 claimable rights—that of rights over forest land under occupation with all other rights, even over land, being ignored. Community forest rights continue largely being ignored with some recent achievements. Diversion of small areas of forest land for community facilities, requiring a different procedure, has been reported as recognition of community rights creating a false impression of such rights being recognized.

State FDs in practically all the states continue attempting to obstruct proper implementation of the law. They have unilaterally been rejecting claims without authority, or demanding evidence not required by the rules. Even before the FRA had come into force, forcible evictions and/or plantations on cultivated forest land were undertaken in many states to prevent cultivators from claiming rights over the land. This continues to date. Grassroots movements continue reporting widespread use of JFM committees for forcible evictions for plantations. In many areas, JFM committees have been made into FRCs with the forest department controlling the claim-making process through them. FDs have also been spreading misinformation about the FRA—that it is not applicable to Reserve Forests (RFs) and PAs and that only

STs are eligible for claiming rights. By delaying the FRA's coming into force, over 30,000 sq. kms of forests under Project Tiger were hurriedly notified as Critical Tiger Habitats without following the specified legal process on the unfounded assumption that these could be kept out of the FRA's purview. The villagers living within these critical tiger habitats are now being pressured—illegally—to relocate without recognizing their rights, proving that co-existence will lead to irreversible damage, or obtaining their free and informed consent, all required under the FRA. Central and state governments even attempted to declare implementation complete by the end of 2009, despite the law having no such provision.

Grassroots Assertion of Rights and Its Impacts

Grassroots movements have maintained sustained pressure for the proper implementation of the FRA and have been challenging and protesting the illegalities being committed. In states and regions where such movements have a strong presence, mobilization and awareness among communities has been spreading, with the widespread filing of both individual and community claims. Through networking, experience-sharing, media coverage, and political mobilization, this awareness is beginning to reach new and distant areas and has had important impacts.

The arbitrary time-limit imposed on the implementation of the FRA has been withdrawn, and some states that had stopped accepting claims have re-opened their processes. In some states, inquiries into reasons for the excessive rejection of claims have been ordered, and attention is finally also being given to the recognition of community forest rights. Using the FRA as a weapon, sustained protests are being organized against the Ministry of Environment and Forests for continuing to divert forest land to non-forest uses, such as mining and dams, without recognizing forest rights. A significant milestone resulting from this was the denial of forest clearance in Odisha by the MoEF to mine bauxite in the sacred hill of the Dongria Kondh community for the benefit of Vedanta Alumina Ltd. The Supreme Court's judgement in the case challenging the MoEF's denial of final forest clearance has held that the gram sabha 'functioning under the Forest Rights Act read with Section 4(d) of PESA Act has an obligation to safeguard and preserve the traditions and customs of the STs and other forest dwellers, their cultural identity, community resources etc., which they have to discharge following the guidelines issued by the Ministry of Tribal Affairs vide its letter dated 12.7.2012'. The Court has directed the MoEF to take the final decision regarding forest clearance for bauxite mining in the Niyamgiri hills based on the gram sabhas'decision whether the mining would affect their right to worship their deity, Niyam Raja, which is a right which has to be preserved and protected. Also in Odisha, villagers resisting the allocation of their forest land to POSCO for a major steel plant have rejected the forest clearance granted to the company by the MoEF on the grounds that their forest rights have not been recognized.

The growing political importance of the FRA finally compelled the MoEF to undertake the first important complementary reform in the procedure for diverting forest land under the Forest Conservation Act, 1980. On 30 July 2009, the ministry issued an order to all state governments that no permission for forest diversion would be granted without evidence that the process of recognition of rights in the concerned forest area had been completed. Gram sabha resolutions certifying the recognition of the habitat rights of pre-agricultural communities and CFR rights, and giving informed community consent for use of the forest for a non-forest purpose, must now be attached with each application for forest diversion. The MoEF, however, has blatantly been violating its own order in permitting the large-scale diversion of forest land. Accepting most of the substantial recommendations lobbied for through the National Advisory Council (which advises the central government on policy matters), the Ministry of Tribal Affairs has issued new guidelines in July 2012 and amendment rules in September 2012 to ensure that implementation conforms with the law's provisions. Key changes include withdrawal of the arbitrary time-limit imposed on implementation of the FRA, several other procedural improvements, a specific focus on recognition of community forest rights and clarifying the gram sabhas' powers to issue transit permits for NTFPs and preparing their own conservation and management plans for CFRs.

Many developments have taken place in different states also. In Gujarat, with close to 90 per cent of claims being rejected initially without informing the claimants of the reasons for rejection or providing them with an opportunity to appeal, two NGO members of the Gujarat

Adivasi Mahasabha, a network of grassroots organizations working with forest-dwellers, had filed a PIL case in the state's High Court against the illegalities committed in the recognition of rights. The High Court judgement of 3 May 2013 has directed the state government to withdraw all earlier instructions which violated the amended FRA rules and to effectively treat all rejected claims as pending and review them strictly in compliance with the amended rules of September, 2012. In other states, rejected claimants have also mounted challenges, and rectifications are beginning to be made.

In response to grassroots protests and the observations of a couple of high-level committees, authorities in Andhra Pradesh found in January 2011, that as many as 71 per cent of the villages/habitations with a forest interface had been left out of the first, hurried phase of implementation of the FRA; fresh claims have been invited from both new and old villages for a second phase of implementation. The state has also permitted the reconstitution of FRCs at the hamlet-level in place of those formed for large, multi-village panchayats.

The most significant change has been in the recognition and assertion of community forest resource (CFR) rights. With MoTA now exhorting states to prioritize the recognition of community forest rights, the first breakthrough came in Maharashtra. Mendha and Marda villages in Gadchiroli district of Maharashtra were among the first in the country to receive CFR titles. Now, an estimated 1322 villages in the Gadchiroli and three other districts have received CFR titles for a forest area of over 625,890 acres. Despite the FRA defining bamboo as a non-timber forest produce over which the villagers have ownership and disposal rights, the Maharashtra FD insisted that under the Indian Forest Act, 1927, bamboo was classified as timber, which the gram sabha could not harvest and transport outside the village. The central Minister of Environment, the state's chief minister and forest minister, and other high-ups needed to go personally to the village to force the Maharashtra FD to hand a transit passbook over to the gram sabha, after firmly conveying to the department that the FRA of 2006 over-ruled the Indian Forest Act of 1927. Once authorized, the village earned, in 2011, about Rs 10 million from the harvesting and sale of bamboo, with a large percentage of the money going to villagers as wages. This provides a glimpse of the FRA's potential for enhancing livelihoods and incomes of forest-dwelling communities by restoring their control over forest resources.

Other villages in Gadchiroli district had to contend with illegal conditions being inserted in their CFR titles including that the FD will not be obstructed from implementing its working plan in the CFR areas, which effectively negates the CFR right. Second, most of the other CFR areas are leased to a paper company for bamboo harvesting and the lease had not been revoked despite the recognition of CFR rights and the villagers' demands. However, with persistent struggle and lobbying, more villages have been issued transit permits to transport their bamboo harvests from CFRs to the market with some having undertaken bamboo harvesting. The recently amended FRA rules now empower gram sabhas to issue their own transit permits. During 2013, 18 villages spread across Gadchiroli, Gondia, and Amravati districts have also asserted their right to harvest and sell tendu leaf from their CFRs—an experience which will be monitored by groups from several states to evolve their own strategies for tendu leaf next year.

Inspired by the Mendha experience, on 20 June 2012, Jamgudi village in Kalahandi district in Odisha also harvested bamboo, which had flowered in its CFR area, becoming the first village in the state to do so. With the central minister of Tribal Affairs writing to the state chief minister to instruct concerned authorities to permit the villagers to transport their harvested bamboo outside the village, several other villages in the district which have also received their CFR titles are also planning to harvest their bamboo after the monsoons.

The Soliga tribal community in Karnataka has received community titles over three of the five forest ranges within the core area of the BRT Tiger Reserve, covering a number of forest and cultural rights. The community has also restarted NTFP collection earlier banned by the Karnataka FD (although they still have to deal with the department's resistance). In Uttar Pradesh, land titles have been awarded to the tribal residents of a forest village (which the Uttar Pradesh FD had been trying to evict) in the core area of another tiger reserve and the process of converting it into a revenue village has started.

In response to grassroots mobilization, implementation of the FRA is finally being extended to non-tribal majority districts in Gujarat and Himachal Pradesh only now, five years after the FRA came into force. The pastoral Maldhari community in the Kutch district of Gujarat has challenged the FD's working plan for the Banni grassland misclassified as forest and is in the process of filing a collective community claim for the entire 2,500 sq. kms of their customary pastureland.

Even where the state governments are not being responsive, communities have started asserting their CFR rights. In about 30 villages in the north of West Bengal, the local movement has initiated community institution-building by encouraging the gram sabhas to elect forest governance committees, which are evolving rules for regulating forest use. The gram sabhas have sent notices to the relevant authorities informing them of the constitution of their forest governance committees under the FRA rules for asserting their powers of protection and management. In the Nilgiris in Tamil Nadu, 59 gram sabhas have put up boards outside their demarcated CFRs and banned the Tamil Nadu FD and police from entering them without permission from the gram sabha. The Masinagudi panchayat in the Nilgiris in Tamil Nadu has challenged the legality of the notification of the Mudumalai Critical Tiger Habitat. In most of these areas, in confrontations with departmental staff it is the foresters who have had to back off due to the FRA.

In Jharkhand, grassroots groups have rejected JFM and are demanding restoration of their common property rights under the CNTA and SPTA. Eighteen panchayats mobilized by the Jharkhand Jungle Bachao Andolan in Ranchi district of Jharkhand state filed CFR claims, to reclaim their recorded rights over large forest areas under the CNTA. These forests were vested in the Jharkhand FD by executive fiat after independence without due legal process. The district collector agreed to approve their claims but the Jharkhand FD raised spurious objections. Angered by this, all 18 panchayats have decided to assert their authority over their CFRs on the grounds that their rights stand recognized from the day the FRA came into force.

In Uttar Pradesh, activists and tribal leaders have been made members of the State Level Monitoring Committee and instructions have been issued that panchayat pradhans cannot be the chairpersons of FRCs so as to prevent the elite from sabotaging the claims of the poor. The state government has issued orders for converting several hundred unrecorded *taungya* villages into revenue villages and 4833 individual titles have already been issued in 29 taungya villages spread over five districts of the state. The activists have identified the illegal transfer of over 30,00,000 ha of common land to the FD after independence and are planning to reclaim the same through the FRA.

As awareness about the law and flaws in the initial implementation processes is spreading, demands for re-doing it properly are growing. Many poorly constructed FRCs have been reconstituted and arbitrarily rejected claims are being reviewed. Women's groups are attempting to increase gender-sensitivity of the claim-making process and to ensure that women have a stronger voice in post-recognition management of CFRs.

Although uneven across the country, grassroots assertion of rights under the FRA is beginning to spread. In areas with strong movements, day-to-day rent-seeking and harassment by forest staff has declined. If the FD continues with illegalities, the villagers are now able to implead the nodal tribal departments to take action. Many sub-divisional magistrates and district collectors as chairs of Sub Division Level Committees (SDLCs) and District Level Committees (DLCs) have also started exerting pressure on the FD to follow the law. It remains to be seen how state governments bent upon violating the law's provisions will respond to grassroots assertion. This time, however, the law is on the villagers' side with a perceptible shift in the balance of power between villagers and the forest bureaucracy.

Where to From Here?

Born out of outrage at the sheer brutality of FD evictions of the country's most vulnerable communities from supposedly state forest land, the movement for the FRA has brought the entire edifice of the Indian forestry establishment under scrutiny in a manner never witnessed since the advent of colonial forestry. Although by no means a panacea for the multitude of conflicts and contradictions plaguing India's forested landscapes, especially in the context of the emerging threats posed by globalization, the FRA has sown the seeds for a comprehensive re-examination of the official definition/construction of forests and their boundaries, and the very rationale of current state-controlled, centralized and uni-functional forest management. It has generated the imperative of democratizing forest governance through fundamentally changing the balance of power between local communities and the forest bureaucracy. Through restoring citizenship rights and dignity to

India's indigenous forest-dwelling communities, the FRA can ensure greater livelihood, cultural, and ecological security. By so doing, the new law has the potential of reshaping people–nature relationships that have been badly ruptured since colonial rule, at least in remote and ecologically fragile areas where community cohesion remains strong. This can facilitate restoration of plurality in ecological governance rooted in multi-functional land-uses and values based on indigenous biodiversity knowledge. The fundamental questions of who owns the country's forests and, by whom and for what objectives they should be governed and managed within the country's democratic and constitutional framework, have come centre-stage for the first time.

The rationality of multi-functional community land being kept under centralized state control for uni-functional 'forest' management for the benefit of distant elites is no longer tenable. Prudent extraction for livelihoods by local communities is unlikely to result in the type of systematic forest and biodiversity destruction which has been taking place under the rubric of official 'scientific' forest management.

But major challenges remain. These emanate both from the limitations of the FRA discussed earlier and the external challenges to its implementation. While people are claiming their rights on land already classified as forests, the state forest boundary is continuing to be extended through legally questionable means. The MoEF guidelines now require that non-forest land used for compensatory afforestation must be notified as PF or RF and the records mutated in the name of the FD. The guidelines specifically recommend bringing the few surviving community lands as well as disputed land like Madhya Pradesh's Orange Areas under FD control through notifying them as state forests.

Simultaneously, pressures generated by globalization and the race for a higher rate of economic growth have increased the rate of diversion of forest land for mining, dams, and other development projects. Although it is illegal under the FRA to evict people from forest land till their rights have been examined, and despite MoEF's July 2009 order requiring recognition of rights and community consent before any diversion of forest land, the MoEF has been permitting diversion of such land through the non-transparent and undemocratic process permitted under the FCA. Once gram sabha rights to protect and manage their CFRs are formally recognized, the present process for diverting forest land will no longer be tenable.

Finally, there is the risk posed by the Government of India signing international agreements related to climate change and carbon sequestration, which have major impacts on people's forest rights without any democratic or consultative process. The pressure from industry for the allocation of forest land for dams as well as industrial and mining projects, also remains very high. Although people with recognized rights will be in a stronger position to challenge such impositions, scattered and less organized communities will find them harder to resist. These threats and challenges cannot be addressed till forest management is grounded in holistic natural resource governance by democratic local institutions with secure land and forest tenures.

Notes

1. *T. N. Godavarman Thirumulpad vs. Union of India* WP (C) no. 202 of 1995.
2. Large areas of natural grassland ecosystems have been destroyed by the plantation of exotic tree species in them due to their being classified as forest land and handed over to tree-focused FDs (Sarin 2005a).
3. This still applies to large parts of the Northeastern states.
4. This is a highly misleading and unfortunate term for the uncultivated commons which remains in official use till today. For policy-makers remote from rural realities, the term 'wasteland' triggers the perception of land actually lying waste. Millions of rural livelihoods dependent on such lands continue being destroyed due to their allocation to other uses without taking their existing uses and users into account.
5. The Reserve Forests of north-west Bengal were shifting cultivation land of tribal communities who were initially pushed off their lands and later converted into bonded labour for forestry operations under the *taungya* system (Ghosh 2006).
6. The term nistar means usufruct rights for meeting local households' own needs.
7. Mandated to strengthen the Van Panchayats under the World Bank-funded UP Forestry Project, the state FD revised the Van Panchayat Rules first in 2001 and again in 2005. Despite vehement protests by Van Panchayats, the new rules have reduced their autonomy and made them akin to Joint Forest Management Committees with a forest official now the member secretary-cum-joint-account-holder of every Van Panchayat.
8. South Africa, for example, passed a law after the end of apartheid which enables indigenous African communities to claim restitution of their lands forcibly appropriated during apartheid.

9. Although 4.3 million ha of forest land was diverted to non-forest use between independence and 1980, this was only a fraction of the 26 mha declared to be state forests after independence, often without due legal process.

10. Settlement officers for settling the rights of tribals in over 100,000 ha notified as RF under Section 4 of the IFA decades ago have been appointed in Andhra Pradesh only recently. The Odisha FD is being unable to provide maps of Reserve Forests to gram sabhas during implementation of the FRA as these simply do not exist.

11. See Kumar (2008) for an excellent investigation of the process of disen-franchisement of shifting cultivators in Odisha.

12. See Ghosh (2007) for how the forest rights of tenants in south-west Bengal were illegally extinguished.

13. For examples of similar arbitrary conversion of diverse kinds of land into state forests without due legal process in other states, see Sarin (2003 and 2005a).

14. This case is largely based on Garg (2005).

15. One of MoEF's September 1990 circulars (see section 2.5) required conversion of such leases, pattas, and grants issued by the revenue department under due legal authority into a legal title—but the circular was never imple-mented.

16. Section 3 (i) (g) of the FRA now provides for a right to conversion of such pattas, leases, and grants on forest land into titles but this right has also been ignored during implementation to date.

17. Jharkhand's case is largely based on Upadhyay (2005).

18. However, *bhuinhari* and *mundari khuntkattidari* tenancies were exempt-ed from the ambit of this Act by a 1954 amendment.

19. See http://www.apforests.gov.in/Forest%20Policy-1952.htm.

20. For examples of illegal cancellation of land titles based on the Court's 1996 order in Jharkhand and Maharashtra, see Sarin (2003).

21. Personal interview with the staff of a centrally funded health NGO who were being prevented from delivering basic health services to the villagers living inside the Sunabeda wildlife sanctuary. When they complained to the district collector, she told them that even her staff were being prevented from deliver-ing PDS rations to the villagers. She herself had to complain to her senior to get the PA manager to stop preventing the delivery of essential services to the villagers living inside the PA.

22. Prior to independence, no law could be extended to partially excluded areas (equivalent to present Schedule V areas) without the approval of the Agent of the Crown. The Indian Constitution reversed this provision and, to date, no state governor is known to have exercised her/his power to withhold any law detrimental to adivasis from Schedule V areas. Interestingly, K.C. Deo,

the present Minister for Tribal Affairs, has written to all governors of states with Schedule V areas to exercise their power and prevent the transfer of tribal land to private companies for mining.

23. The apparently constitutionally protected communal tenures under customary laws, even in Schedule VI areas, are easily overruled by statutory laws and Supreme Court orders.

24. According to the 2001 Census, Odisha had 623 settlements on forest land with a total population of 74,047, out of which over 50,000 were STs. These are not 'forest villages' created by the FD but either unsurveyed villages, which have been classified as forests or settlements of the lakhs of those displaced by large dams and other projects without any rehabilitation who have moved to forest land. Residents of such villages are often unable to obtain voter identity cards or domicile certificates to prove their Indian citizenship as these can only be issued by the revenue department, which has no jurisdiction over forest land.

25. Circular no. 13-1/90-FP of Government of India, Ministry of Environment and Forests, Department of Environment, Forests and Wildlife dated 18 September 1990 addressed to the secretaries of FD of all states/UTs. The six circulars under this were:

1) FP (1) Review of encroachments on forest land.
2) FP (2) Review of disputed claims over forest land, arising out of forest settlement.
3) FP (3) Disputes regarding pattas/leases/ grants involving forest land.
4) FP (4) Elimination of intermediaries and payment of fair wages to the labourers on forestry work.
5) FP (5) Conversion of forest villages into revenue villages and settlement of other old habitations.
6) FP (6) Payment of compensation for loss of life and property due to predation/depredation by wild animals.

26. Unofficial estimates suggest their number to be much higher.

27. Only the Maharashtra government issued detailed guidelines and partially implemented these circulars in 2002 after sustained protests against evictions by adivasi movements.

28. In 2009, NAEB revised its operational guidelines for the NAP which generally retain the same structure and have added a stat-level FDA with all division-level FDAs in the state as members. Although the revised guidelines state that different states can use their own JFM guidelines but still provide for the forester/block forest officer to be the member secretary of every JFMC. Totally ignoring the FRA, they aim to promote formation of JFMCs in every forest fringe village in the country (NAEB 2009).

29. As mentioned earlier, about a million families in undivided Madhya Pradesh alone stand to benefit from secure tenures over their cultivated lands under this provision.

30. Given the political economy of law-making and the compromises that had to be made in the enacted version of the FRA, the law has a number of ambiguities and limitations. For critiques of the law, see, for example, Bhullar (2008) and Saravanan (2009).

31. In Odisha, some sympathetic government officials are attempting to regularize pre-1980 forest land occupation by OTFDs displaced by development projects under the MoEF's 1990 circular although such regularization requires approval of the Supreme Court.

32. This was despite the fact that during 2007, the PMO, together with the MoTA and MoPR, had prepared a schedule for the Act's implementation in which four months had been allocated for disseminating information about the FRA among the villagers and training government officials of different departments prior to initiating implementation. (Information obtained from MoTA under RTI.)

References

Agarwal, C. 1996. 'Boundary and Property Rights in Uttarakhand Forests', *Wastelands News*, 11 (3): 4–6.

Anonymous. 2004. 'People Leaving Homes to Fill Empty Stomachs' (original in Oriya). *Dharitiri*, 5 July 2004, translation given at http://vasundhara-odisha.org/DiscussionPaper_eng/Impacts%20on%20Lives%20and%20Livelihood%20of%20People.pdf.

Bera, S., K. Sambhav, A. Pallavi, A. Paliwalm and S. Narayanan. 2011. 'Wealth of Forests Withheld'. *Down to Earth*, 15 September; http://www.downtoearth.org.in/node/33949.

Bhullar, L. 2008. 'The Indian Forest Rights Act 2006: A Critical Appraisal', *Law Environment and Development Journal*, 4 (1): 20.

Campaign for Survival and Dignity (CSD). 2003. 'Endangered Symbiosis: Evictions and India's Forest Communities'. Report of the *Jan Sunwai* (Public Hearing), 19–20 July. New Delhi: Campaign for Survival and Dignity.

Central Empowered Committee (CEC). 2002. 'Recommendations of the Central Empowered Committee in Interlocutory Application No. 703 of 2001'.

Centre for Science and Environment (CSE). 2011. 'Overview (of Forest and Environmental Clearances'; http://www.cseindia.org/userfiles/Overview.pdf.

Council for Social Development. 2010. 'Summary Report on Implementation of the Forest Rights Act'. New Delhi: Council for Social Development.

Diwan, R., M. Sarin, and N. Sundar. 2001. 'Jan Sunwai *(Public Hearing)* on Forest Rights at Village Indpura, Harda District', 26 May 2001, mimeo (later published in *Wastelands News*, May–July 2001, 16 (4) and in *Van Sahyog*, May–July, 3 (1).

Forest Survey of India (FSI). 2005. 'State of Forest Report 2003', Dehradun, India.

Gadgil, M. 2008. *Let Our Rightful Forests Flourish*. Pune: National Centre for Advocacy Studies.

Garg, A. 2005. 'Orange Areas, Examining the Origin and Status, Advocacy Perspective'. Working Paper Series No. 21, Pune: National Centre for Advocacy Studies.

Ghosh, S. 2006. 'Forest/Taungya Villages in Sub-Himalayan West Bengal: A Brief Study in Ecological History, Development Issues and Struggles'. Draft research paper written for the Overseas Development Group, University of East Anglia, Norwich.

———. 2007. 'Commons Lost and "Gained"? Forest Tenures in the Jungle Mahals of South West Bengal,' Overseas Development Group, University of East Anglia, Norwich.

Government of Andhra Pradesh (GoAP), Forest Department. 2002. 'Resettlement Action Plan', AP Community Forest Management Project, Project Monitoring Unit, Hyderabad.

Government of India (GoI). 2007. 'Report of the Steering Committee on the Environment and Forests Sector for the Eleventh Five Year Plan (2007–2012)', New Delhi: Planning Commission.

———. 2002. Circular to all States/Union Territories No. 7-16/2002-FC, 'Eviction of Illegal Encroachment of Forest Lands in various States/UTs—Timebound Action Plan', 3 May 2002, New Delhi: Ministry of Environment and Forests.

Guha, R. 1989. *The Unquiet Woods, Ecological Change and Peasant resistance in the Himalaya*. New Delhi: Oxford University Press.

———. 2001. 'The Prehistory of Community Forestry in India', *Environmental History*, 6 (2): 213–38.

Joint Committee. 2010. 'Manthan: Report of the National Committee on Forest Rights Act'. New Delhi: Ministry of Environment and Forests and Ministry of Tribal Affairs, Government of India. Available at http://fracommittee. icfre.org/FinalReport/FRA%20COMMITTEE%20REPORT_FINAL%20 Dec%202010.pdf.

Khanna, S. 2008. 'Boundaries of Forest Lands: The Godavarman Case and Beyond. Paper presented at the National Workshop 'Beyond JFM: Rethinking the Forest Question in India', organized by the Centre for Interdisciplinary Studies in Environment and Development and Winrock International India in Delhi on 29–30 September 2008.

Kumar, K. 2008. 'Erasing the Swiddens: Constructing Forest Agriculture Dichotomies in Orissa'. Paper presented at National Workshop 'Beyond JFM: Rethinking the Forest Question in India', organized by the Centre for Interdisciplinary Studies in Environment and Development and Winrock International India in Delhi on 29–30 September 2008.

Kumar, K., P.R. Choudhary, S. Sarangi, P. Mishra, and S. Behera. 2005. *A Socio-Economic and Legal Study of Scheduled Tribes' Land in Orissa*. Report of study commissioned by the World Bank. Bhubaneshwar: Vasundhara.

National Afforestation and Eco-Development Board (NAEB). 2002. 'National Afforestation Programme: A participatory approach to sustainable development of forests (centrally sponsored scheme)', Operational Guidelines for the Tenth Five-Year Plan, New Delhi.

———. 2009. 'National Afforestation Programme: Revised Operational Guidelines 2009', New Delhi; http://www.naeb.nic.in/progSchem.html

People's Union for Democratic Rights (PUDR). 2001. 'When People Organise: Forest Struggles & Repression in Dewas'. 'Delhi, June.

Rao, Y.G. 2004. *Livelihood Issues in Satkosiya Gorge Sanctuary, Orissa*. Bhubaneswar: Vasundhara.

Rao, N. 2005. 'Displacement from Land: Case of Santhal Parganas'. *Economic and Political Weekly*, 40 (41): 4439–42.

Samata and CRYNet. 2001. 'Joint Forest Management, A Critique Based on People's Perceptions'. Hyderabad: Samata.

Sarin, M. 1996a. 'From Conflict to Collaboration: Institutional Issues in Community Management', in M. Poffenberger and B. McGean (eds), *Village Voices, Forest Choices: Joint Forest Management in India*, pp. 165–209. New Delhi: Oxford University Press.

———. 1996b. 'Joint Forest Management: The Haryana Experience'. Environment and Development series. Ahmedabad: Centre for Environment Education.

———. 2003. 'Bad in Law. Analysis of Forest Conservation Issues'. *Down to Earth*, July 15: 36–40.

———. 2005a. 'Laws, Lore and Logjams: Critical Issues in Indian Conservation'. *The Gatekeeper Series 116*. London: International Institute of Environment and Development.

———. 2005b. 'The Scheduled Tribes (Recognition of Forest Rights) Bill 2005; Undoing Historical Injustice to Tribals', cover story in *From the Lawyers Collective*, June 2005.

Sarin, M., N.M. Singh, N. Sundar, and R.K. Bhogal. 2003. 'Devolution as a Threat to Democratic Decision-making in Forestry? Findings from Three States in

India' in D. Edmunds and E. Wollenberg (eds) *Local Forest Management: The Impacts of Devolution Policies*, pp. 55–126. London: Earthscan Publications.

Saravanan, V. 2009. 'Political Economy of the Recognition of Forest Rights Act, 2006', *South Asia Research*, 29 (3): 199–221.

Saxena, N.C. 1995. *Forests, People and Profit, New Equations for Sustainability*. Dehradun: Centre for Sustainable Development and Natraj.

———. 1999. *Forest Policy in India*. New Delhi: World Wildlife Fund (WWF)-India and the International Institute for Environment and Development (IIED).

Shah, A.C. 2003. 'Fading Shine of Golden Decade: The Establishment Strikes Back'. Paper presented at the National Seminar on 'New Development Paradigms', 4 to 6 March. Ahmedabad: Gujarat Institute of Development Research.

Srinidhi, A.S. and S. Lele. 2001. 'Forest Tenure Regimes in the Karnataka Western Ghats: A Compendium', Working Paper No. 90. Bangalore: Institute for Social and Economic Change.

Sundar, N. 2001. 'Is Devolution Democratisation?' *World Development*, 29 (12): 2007–24.

Upadhyay, C. 2005. 'Community Rights in Land in Jharkhand'. *Economic and Political Weekly*, 40 (41): 4435–8.

Upadhyay, S. and S. Jain. 2004. *Community Forestry and Policy in North-East India: An Historical Legal Analysis*. California: Community Forestry International.

Vasan, S. 2005. 'In the Name of Law: Legality, Illegality and Practice in Jharkhand. Forests', *Economic and Political Weekly*, 40 (41): 4447–50.

GOVERNING FORESTS
FOR CONSERVATION

NITIN D. RAI[1]

Views from the *Podu*

Approaches for a Democratic Ecology of India's Forests

I can't imagine that a spoken certainty can exist without an unspoken uncertainty behind it...

Josef Skvorecky (1996)

You want to save the elephants in Kenya's parks by having them graze separately from cows? Excellent, but how are you going to get an opinion from the Masai who have been cut off from the cows, and from the cows deprived of elephants who clear the brush for them, and also from the elephants deprived of the Masai and the cows?

Bruno Latour (2004)

On 29 March 2007, the Deputy Conservator of Forests (DCF), Biligiri Rangaswamy Temple Wildlife Sanctuary, Karnataka, arrested 27 Soliga adivasis and kicked and injured a 60-year-old Soliga named Magarikete Gowda for reportedly setting fire to the forest. The injuries sustained by Magarikete Gowda warranted his hospitalization, where he lay when a fact-finding team visited him on the 10 May (Kalpavriksh 2007). The DCF continued to run his 'estate'[2] for more than a year despite protests and appeals by Soligas to state officials, including the Chief Minister of Karnataka, for his removal. While this story has several threads, such as the continued oppression of local communities by the state conservation apparatus, and the denial of justice, I will focus on the production and maintenance of knowledge underlying forest and wildlife management that is so strictly adhered to by its proponents. I suggest that current wildlife management strategies have borrowed heavily from historical forest management practices and that these practices were not necessarily informed by science as is often believed. While there has always been debate both within and outside the administrative apparatus,[3] new knowledge takes enormous time to be incorporated, if at all.[4] The political outcomes of management, rather than the ecological premises of it, seem to justify the actions taken. This explains the one-size-fits-all strategies such as relocation of human settlements, grazing restrictions, ban on forest produce collection, and prevention of fire. If we think of these practices as being justified by their political outcomes rather than the ecological logic, we might begin to understand why such enormous force is used to ensure management practices. The issue is not merely that fires might have adverse effects on forests but that actions such as setting forest fires are seen as protest and are quelled by the use of force.

Thus, in the first half of this chapter, I argue that it is not what is good for the forest or wildlife that is driving current management but state interests in maintenance of control. Once control is consolidated, actual wildlife management gets relegated to such mundane activities as building check dams, roads, and watch towers. Having assumed the symbolic control of forest and the conservation enterprise, these structures on the landscape become the insignia of conservation efforts. Management need not then be based on science or on local knowledge but on pragmatic use of whatever information exists to extend and maintain control.

Indeed, the real power of the forest administration in India is not displayed on the ground where its presence is weak and its ability to be effective compromised by the confused allegiance of its field staff, but in the power to produce knowledge and inform policy to support its continued control over forests (Baginski-Springate and Blaikie 2007). In the second half of this chapter, I argue that much of this knowledge is produced based on received notions of forests as stable and equilibrial structures, derived from colonial and Western notions of pristineness (Lewis 2003). Such assumptions lead to the unquestioned perception that the impact of human use on forests results in 'degradation'. Many studies are beginning to show that the perception of nature as wild, stable, and pristine is a myth (Willis *et al.* 2004). 'Pristine' forests containing high levels of biodiversity have been subsequently shown to have been inhabited, farmed, and burned by humans.[5] Current levels of biodiversity and species composition in most forests are a result of human alteration (see Rangarajan 2003 on the establishment of the Kanha meadows; also Pyne 1990, 1994, and Willis *et al.* 2004). Ecological studies on the impact of local use on forests are often based on the idea that in the absence of human influence, forests will attain ecological climax. This results in undue attention on local users and their impacts on the forest and the convenient neglect of the larger, more damaging impacts of large-scale development projects of the state and private corporations. While it is clear that ecological knowledge on the functioning of systems has changed and that ecologists now acknowledge that systems are dynamic and complex and not necessarily equilibrial, we might ask why management systems are still based on old assumptions. Reasons for the reliance on outdated knowledge could range from the utility such theories hold for continued control of forests, the inertia of state enterprises to adapt to changing knowledge, or the insularity of the knowledge production process.

Conservation Practices in India: A Critique

The Focus on Protected Areas

The notification of forest areas as wildlife sanctuaries and national parks has been the major strategy for conservation of India's wildlife. More recently efforts have been made to identify sets of adjacent protected areas (PAs) to ensure connectivity (Rodgers and Panwar 1988). The focus on PAs, of whatever size, has, however, resulted in the neglect of areas outside PAs that might contain wildlife. Not all areas that have high levels of biodiversity have been protected. While there is increasing evidence that protection does indeed increase diversity levels and decrease rates of deforestation (Nagendra 2008), there is little evidence that non-PAs have consistently low levels of diversity. The presence of a surrounding forest is essential for species within, both as areas into which populations might disperse as abundance increases and as a buffer from more intensive land-use practices that might be biodiversity-unfriendly. The insistence on PAs as the solution is fraught with problems because of the culture of intensification that this encourages. The area under PAs today is simply not enough to conserve India's biodiversity and the intensity of protection within these PAs makes it politically impossible to declare many more. A notion of conservation that looks at 'lesser' areas with as much interest as PAs will ensure that larger areas come under more inclusive conservation regimes. A larger vision of interconnected landscapes that contain human history and experience within the landscape might ensure the maintenance of biodiversity. The insistence, therefore, that the future of India's wildlife is only secure in PAs and conservation outside PA is non-viable comes from a deep allegiance to the idea that humans and wildlife cannot coexist (Karanth 2003).

Wildlife Corridors

The idea of habitat connectivity is key to providing us a way forward in what looks like an intractable and unpassable landscape of jargon, entrenched ideas, and conflicts. Most biologists will agree that habitat connectivity is an extremely important aspect of ecological systems and that it is the fragmentation and truncation of corridors that are a major problem (Borges 2003). Rodgers and Panwar (1988) speak of the need to identify corridors and design PAs based on the idea that connectivity is important. The presence of a matrix of habitats that enable species to move through, even though they might not reside in that habitat, is essential. Thus an agro-forest might be more usable for primates than paddy cultivation. Agriculture, in as much as it may be an unfavourable habitat for most species, also provides forage to many herbivores such as wild boar, *nilgai*, wild ass, and elephants due to its high palatability and nutrition (Sukumar 1994). Chhangani *et al.* (2008) demonstrate that wildlife species that have increased in the Kumbhalgarh Wildlife Sanctuary have been those species that have raided crops and livestock the most while species that have decreased have not been able to avail of the crops on the reserve fringes, suggesting that animals have benefited from agriculture as much as humans have from the forest.

We might then think of landscapes as being a mix of different habitats, all varying in their degree of habitability to different species. This enables us to conceptualize a matrix of biodiversity-friendly agriculture and forest such that the surrounding matrix of agriculture might sustain a greater diversity of species than the current practice of intensification of both conservation and agriculture (Ranganathan *et al.* 2008). We might re-imagine landscapes as being integrated either as they sit beside each other or are nested within each other. Thus cultivated land within PAs could be providing food for raiding animals, providing a matrix for movement to certain species, and even facilitating the survival of several species that are commensal with agricultural systems. Recent work on modelling agricultural systems for biodiversity is showing that intensification of agriculture (the dominant view) might be changed to facilitate biodiversity by making the agriculture matrix more permeable. Conversely, we might even conceive of forest systems as being able to sustain the existence of small patches of cultivation with little impact and even provide forage to enhance animal numbers. Certain kinds of agriculture (shade-grown coffee, diverse agro-forest, mixed crops) could serve as corridors and extension of habitat (Vandermeer and Perfecto 2007).

The Science Behind Local Impacts

The most cited example of the ambiguous role of grazing in PAs is the case of Bharatpur (Lewis 2003; Middleton 2003). The belief that livestock grazing degrades forested ecosystems has a long history and is often cited as one of the major drivers of habitat change. This has led to legislation banning grazing in all PAs in the country. The nation-wide ban was also implemented in Bharatpur with tragic outcomes when nine villagers, protesting the ban, were shot and killed. The ecological effect of the prevention of grazing led to a significant increase in *Paspalum* sp. grass and weeds resulting in choking the waterways, reducing the water surface and nesting area, and even increasing fire due to a biomass build-up (Vijayan 1987). Despite the fact that the scientists of the Bombay Natural History Society suggested that buffaloes be allowed into the park, the forest department (FD) has not relented, claiming that such an act would contravene the Wildlife Protection Act. The management, however, did allow villagers into the park to manually cut grass because they said it would solve the problem of the increase of Paspalum. Lewis (2003), in concluding his account of the Bharatpur case, states that 'the best argument against international ecological advocacy based not on local case studies but on theoretical knowledge, or on the attempt to impose universal conservation models, is that it often has not worked. A people-less, cattle-less park did not help Bharatpur's birds'.

The importance of local solutions to conservation practice is clear, but current laws rarely allow this sort of flexibility. If they did, who better to dictate local efforts than the people living in and around these areas? Universal scientific truths are of little value in managing diverse forests. While Gir and Bharatpur are often cited as examples of this, there are now an increasing number of such cases. The invasion of *Lantana* (*Lantana camara*) in the Biligiri Rangaswamy Temple Wildlife Sanctuary (BRT) and other forests point to the continued lack of on-ground response to such ecological disasters. In the Melkote Wolf Sanctuary, Karnataka, wolf numbers have dropped due to the reduction in livestock as graziers abandoned livestock herding both due to pressure from the FD and the laying of an irrigation canal that changed surrounding land-use (H.N. Kumara, personal communication). A science that seeks to find universal strategies is ridden with problems. The idea that the history of sites vary differently should not be a difficult one to understand. That the use of forests for centuries by cattle and people has determined the current structure of the forest has been demonstrated often. Writing about the Mkomozi Game Reserve in Tanzania, Brockington (2004) states that 'sustained occupation of the reserve for several decades has still left levels of biodiversity exciting to scientists and conservationists six years after the evictions'.

The cases of Gir and Kuno are illustrative of the fact that received wisdom on ecological processes governs management decisions. In the late 1960s, Stephen Berwick and Peter Jordan[6] reported that the diet of lions in the Gir forest comprised predominantly of buffaloes despite there being other herbivores in the forest. They also suggested that removal of Maldhari buffaloes in Gir would not increase wild herbivore numbers as they fed on different plant species, and that the removal of buffaloes might actually affect lions as they would be deprived of buffalo prey. The uproar over this report was predictable and wildlife authorities, with the support of international NGOs, continued with their plans to relocate Maldharis and their cattle outside the reserve. This has led to a serious conflict, with lions leaving the reserve and preying on cattle outside, even killing humans (Saberwal *et al.* 1994). When science provides answers that are inconvenient to the desired political agenda they are ignored (Lewis 2003).

One might argue that the structure of the Gir forest was facilitated by Maldharis and their buffaloes through a process that Jones *et al.* (1994) call ecosystem engineering. Clive G. Jones and his colleagues proposed that many organisms play the role of ecosystem engineers by facilitating the presence of other species. Such 'engineers' could be mega-herbivores, such as elephants or dense assemblages of herbivores such as deer, or even livestock and humans.

Returning to the lion story, we might explore an even more tragic story in the Kuno Wildlife Sanctuary. The Wildlife Institute of India (WII) identified Kuno as a possible second home for lions and suggested that lions be relocated to Kuno (WII 1995). To prepare the ground for the arrival of lions, the report further suggested that people be 'resettled and rehabilitated' to avoid what they saw as problems arising out of conflict among lions, livestock, and humans. That these lions in Gir had lived with livestock and humans was forgotten. When I asked one of the authors of the report why they had so strongly recommended the removal of people, he said that though lions had a history of living with people and cattle, the villagers of Kuno did not have a history of living with lions and it was in their interest that this intervention was proposed. Clearly, it was not science that was behind the decision to move people out of Kuno that led to such dramatic consequences in terms of their livelihood loss (Kabra 2003). Despite the human suffering and close to 20 years after the WII report, lions are yet to be relocated to Kuno. One wonders how different things might have been if the views of others (local dwellers and social scientists) were factored into the initial discussion. Had other views been tabled at the outset, the outcomes of such a large project would have been far less violent.

Tourism and Conservation

The attitudes to tourism differ widely among the state, the scientific fraternity, and the local community in very interesting ways. Tourism is a major activity in PA management. The maintenance of roads, watchtowers, vehicles, elephant rides and wildlife viewing lines are indications of this interest. This is despite revenues from tourism being low in most forests, except in such flagship reserves as Kanha, Corbett, and Periyar. The linkage between reserving areas for biodiversity and enabling tourism becomes most clear to villagers whose lives have been affected by the restrictions on use and residence in PAs. Local people make the connection between their exclusion from the reserve and the increased facilitation of tourists. While there seem to be clear regulations for the use of forests by local people, there are no legal restrictions on the use of PAs by tourists. After decades of use of forests for tourism and the alteration of habitats (creation of road networks, establishment of view lines, waterholes, observation towers, and the movement of hundreds of vehicles) Indian ecologists have still not fully studied the impact of tourism on wildlife and the habitat. Elsewhere, studies have shown that, for instance, tourism affects the behaviour of flamingoes by reducing their time spent on feeding, and subsequently affecting their breeding success (Galicia and Baldassarre 1997).

Tiger Conservation

Project Tiger was established in 1972 to protect and increase tiger numbers in selected forest areas. This flagship project has enjoyed tremendous government backing and has resulted in a network of tiger reserves that receive unprecedented financial support. Greenough (2003) argues that the political leadership during the 1960s and early 1970s was more than willing to listen to international voices that were pushing for wildlife conservation measures. The small and elite group of conservationists (ex-rulers of princely states and urban bourgeoisie) gave their support, which resulted in a resurrection and strengthening of the 'fences and fines' approach that the British had used decades earlier. There was a complete lack of debate in society about any of these measures, and a country that had barely any system of democratic decision-making and debate was taken by surprise as families were evicted by the hundreds in this newfound conservation zeal.

The scientific management of PAs presupposes that benchmarks will be established for subsequent monitoring and evaluation. E.P. Gee's estimate of 40,000 tigers around the turn of the twentieth century was seized upon to demonstrate that tiger numbers had declined rapidly and something had to be done quickly (Greenough 2003). Although there were disagreements with this number, it was widely used by the government to argue for the setting up of Project Tiger. The census of tigers undertaken as part of Project Tiger estimated that there were only 1,827 tigers left in the wild in 1972. Although this census was conducted using techniques that everyone knew were faulty (the use of pugmarks to estimate tiger numbers), these numbers were bandied about with enormous certainty. There was a rush to establish tiger reserves, manage them, and count tigers annually. This flagship project was evaluated by the number of tigers within these reserves and in other PAs. The steady march upwards of this statistic from 1,827 in 1972 to 4,334 in 1989 demonstrated, the authorities argued, the success of the project.

Resentment, however, was beginning to grow among local communities and human rights groups regarding the highhandedness of conservation measures. This led the state to engage in token participatory initiatives such as the World Bank-funded ecodevelopment projects. On the other hand, the scientific establishment was growing increasingly restless about what they felt were lax protection efforts, diluted agendas of the FD (their involvement in 'rural development' activities instead of protection), and the continued marginalization of research in setting management agendas. They berated the faulty pugmark technique and suggested the use of high-tech camera traps and statistically rigorous mark-recapture methods.

All of this fell on deaf ears until the dramatic discovery that there were no longer any tigers left in the Sariska sanctuary. The furore over this was tremendous and galvanized the conservation lobby into pressurizing the government into action. The prime minister set up a task force headed by environmentalist Sunita Narain and had amongst its members a mix of wildlife biologists, ecologists, and former forest administrators. The task force concluded that 'more guns, more guards and... more money... solves nothing' and suggested a series of measures, chief among them being a more inclusive conservation paradigm.

This was at complete odds with the prevailing sentiment of many conservationists. One of the task force members, Valmik Thapar, submitted a dissenting note stating:

> I am constrained to observe that sadly much of the report has become focused on how to improve the life of people inside protected areas rather than protecting tigers inside them. This people focus should have been the job of another task force... this is tragic and if some of the recommendations are endorsed in policy they could have dangerous repercussions for the tiger. (Tiger Task Force 2005: 166)

Thus, the singularity of purpose that marks the conservationist agenda is remarkable and flies in the face of increasing evidence that strict protection and intensification is only possible under fascist and non-democratic regimes. The reliance on external (international) agencies for ratification, funding, and scientific expertise has also resulted in a lack of appreciation for the local: whether knowledge, institutions, culture, or history.

Conservation Attitudes in India

The results of a recent survey by Karanth *et al.* (2008) offers an incisive look at the attitudes of conservationists in India. The authors surveyed 167 people in India who are involved with conservation as practitioners, scientists, or managers. They report that 76 per cent of people surveyed said that force should be used in PAs. Suggestions were also made about the need to have a specialized park protection force and to also use local police, law enforcement agencies, and even guards from the army. Further, only 7 per cent of the respondents said that people could continue to live in PAs; 44 per cent felt that participatory approaches can work; and a mere 2 per cent believed that the Forest Rights Act, 2006 (FRA) should be implemented.[7] These results suggest a hardening of stance amongst the conservation fraternity and that the possibility of building bridges and working together might be getting more difficult. If an overwhelming majority of the respondents believe that people should be moved out and the forests protected by force, the reaction to this from human rights workers could only be more severe.

Interestingly, the authors find that academics were more likely to disagree with the use of force in protecting reserves, and were open to people living inside PAs. They explain these results saying that 'participants engaged in "hands-on" conservation... tended to be more skeptical that parks can achieve their conservation goals with people living in them. Perhaps these differences... reflect a more academic view of human behaviour versus a more practical view informed by first hand experience with people and protected areas'. That the difference in responses might be due to the politics of conservation rather than a 'practical view informed by firsthand experience' seems to be lost on the authors.

The conflict between managers and conservationists is partly explained when we see what happens once control of forests is vested with the managers. Once the PA is declared and made inviolate, the FD largely focuses on activities such as building check dams and roads, clearing vegetation for view lines, and fire retardation. This is clearly a maintenance agenda rather than a management one (Madhusudan and Raman 2003). Ecologists, however, are certain that there is much more to be done and constantly call for greater attention than merely maintenance. The call for better monitoring is resisted by the state as that would mean questioning their established practice and subsequent control. That ecologists and the state share some but not all ideological baggage is clear from what each actor wants: better research based on global science or better control through laws and the selective use of prevailing scientific understanding.

The central issue is: who is framing the problem itself? The notion that grazing and biomass removal is a problem was initially framed by foresters looking for control over forests so as to manage them for timber. For conservationists, it is a problem because it is seen to clash with wildlife management. Are all actors being asked to frame the problem or is the agenda being set by a few? What is the impact of centralized decision-making on local people? How forests are defined might be at the heart of the apparatus of control (Robbins 2003). If local communities are given the opportunity to define forests, the outcomes of conservation efforts might be different from that of state and science-led efforts to define forests (Menon *et al.* 2009). By suggesting that biodiversity is more important than resource-use, we have predetermined the condition of the forest, disregarded the current dependence on the forest by local people, and vested a disproportionate stake for regional groups: whether for cultural, financial, or aesthetic uses.

Displacement for Conservation

Making a case for inviolate areas, Karanth (2003) says 'conservation experiments tend to succeed where participants prefer robust commonsense solutions that keep humans and animals separated at the scale of protected areas, while simultaneously achieving tangible improvements in human welfare at larger landscape scales'. This conservation and development model proposed by one of India's premier large mammal ecologists is the archetypal modernization project of intensification of roles in different landscape types that insists on segregation of outcomes from each landscape type: intensification of agriculture in production landscapes and protection of habitat in conservation areas. This erases the local from all landscapes, whether conservation or agriculture, in that it presupposes scientific management regimes that have little space for local forms of knowledge or for diversity of actor spaces and roles. The presence then of humans and their situated knowledge in any landscape is seen as a nuisance rather than a respectable form of knowledge.

The impact of modernization projects on local communities is staggering, whether in economic (Special Economic Zones, SEZs), agricultural (contract farming and biofuels), or conservation realms (PAs). The number of internally displaced persons in India due to dams is 16.4 million persons, by mines is 2.55 million persons, by industrial development is 1.25 million persons, and by wildlife sanctuaries and national parks is up to 0.6 million persons. (Lama 2000). Such extreme displacement and marginalization does not go unheeded. The state gives out sops to local communities in the form of handouts to placate and assuage. In the case of PAs, this is accomplished by such state led efforts as joint forest management (JFM) and World Bank projects such as Ecodevelopment.

While a lot of attention is paid to the impact of local use by local communities on biodiversity, we seem to ignore large-scale drivers of ecological change: climate change due to energy intensive lifestyles and the modernization agenda of the state that is premised on intensification of land-uses. The reservation of areas for biodiversity might be seen as part of the rationalization project. Conservationists argue that a percentage of land should be made inviolate while the rest might be parcelled off for development and production purposes.[8] The rationale for such a system of intensified land-use flies in the face of everything we know about ecological and social systems. Ecological systems are linked to each other and processes outside PAs will affect ecological processes within. The size of PAs in India is often too small for them to survive as islands of biodiversity in a landscape of intensified production (the average size of 608 NPs and WLS is 258 sq km). Ecosystem matrices outside PAs will be needed to maintain some level of contiguity for animal migration and ecosystem function.

A preservation objective for Indian PAs is not a viable alternative as these forests, grassland, and wetland have been used and transformed for centuries. We, therefore, know little about what these systems look and behave like under a no-use regime. Several systems have changed dramatically under zero management systems: especially due to the spread of invasive species, arguably the biggest threat to forests and wetland today. Our work in BRT suggests that invasive species densities were controlled when Soligas used these forests. When forests were subsequently left untended and fires restricted, invasive species dramatically increased.

The presumed impacts of human use on forests cannot be easily demonstrated due to the range of factors that influence ecological processes. Studies are based on an assumption that there is an inverse relationship between the abundance of wildlife and the presence of people, and this is used to justify the displacement of people from PAs. 'Scientific' managers arrive at explanations that urge a separation of people and forest. A recent sampling of this 'pseudo-scientific' knowledge includes terms such as 'tiger sociology', which is defined as the tiger's need for large *inviolate* space to carry out their 'sociological' activities (cf. Rajesh Gopal in Shankar 2008). Such a premise then leads, not surprisingly, to the conclusion that there is no alternative to inviolate areas to conserve wildlife. The insistence on inviolate areas combined with the recent move by the state to define forests in ways that consolidate its control over 'forest land' is a clear case of the state's self-interest taking precedence over science or social justice.

Writing on the ecological outcomes of the FRA, 2006, a wildlife ecologist states 'research that has carefully examined the effects of chronic resource-use on fragile wildlife has shown that the impacts can be insidious' (Madhusudan 2005). Ecologists are often lulled by the assumption that ecological studies are objective and unbiased. This flies in the face of accumulating evidence that science is not necessarily objective (Latour 1986). It is important to remember that ecological studies in India are conducted by conservationists schooled in the debris of colonial institutions. The author goes on to say that 'we would be ignoring basic ecological facts if we presume that these fragile species, which include tigers, rhinos, elephants and hornbills, can live harmoniously alongside human residence and land-use in their habitats'. The use of language aimed to create images of tigers and elephants withering under 'chronic resource use' has been successful in ensuring support for conservation. The terminology of war and disease is evoked by the use of such words as 'chronic disturbance' (Singh 1999), 'threat' (Barve *et al.* 2005), 'degradation', and 'pressure'. The use of evocative prose has been a tool of the state and its proponents. The conduct of science is in turn influenced by these images and the cycle continues, only to be broken by political efforts such as the FRA or movements that militantly assert their rights over forest resources.[9]

Inserting History into Conservation Practice

Discussions and policy need to historicize conservation in terms of the production of knowledge so as to move away from the artificial divide between scientific conservation and local degradation of the landscape. Several authors such as Gadgil and Guha (1992) and Williams (2003) have provided convincing accounts of the history of Indian forest administration that essentially alienated local knowledge. British forest administrators brought with them ideas of 'wilderness' and based on this their impressions of what local use can do to landscapes. Essentially as highlighted above, they believed that local use was destructive. Shifting cultivation and the extensive use of fire to clear forests was seen to be ravaging lush forests. As a result, draconian forest laws were passed in all the colonies to control forests and to check the itinerant swidden farmers whose cultivation practice was blamed for much of forest degradation. Moreover as Stephen Pyne, writing on the history of fire management in India, argues that the British had to differentiate between primitive practices and their own rational practices, which it did for example by identifying fire, a ubiquitous traditional practice, as 'their chief, almost their only enemy' (Pyne 1994). The immense effort required to prevent fires was daunting, but imperial ambitions could not be seen to be defeated by a primitive practice.

The divide between scientific knowledge and 'local' knowledge that underlies conservation practice today and the presumed superiority of the former is overly simplistic. Colonial foresters, adivasis, and ecologists, for example, all talk about the role of fire in killing the annual shoot growth, resulting in the accumulation of biomass underground year after year that then enables the root mass to put out a robust shoot after some time.[10] Dietrich Brandis noted that while recent transplants died when fires went through a forest, the old ones survived. This is precisely how Achchugegowda, a Soliga elder in BRT, described the effect of fire on plants—the repeated dieback of the shoot and the accumulation in root mass—when he was arguing for a reintroduction of fire into landscapes. Whether Achchugegowda's knowledge comes from pre-colonial Soliga practice or is learned and reinforced by state forest management practices, the point is that this is situated knowledge, however produced. The difficulty in ascribing origins to local and state knowledge makes the separation of knowledge streams a challenging and finally, a polemical exercise. The idea then that situated knowledge is multi-origin and as a result, hybridized, suggests its importance in the management of local ecosystems.

Although knowledge production might have followed multiple pathways, it gets sedentarized in institutions at certain moments in time, based on the political agendas of actors. Let us return to BRT to tell this story. In 1974, the notification of the wildlife sanctuary resulted in the upheaval of Soliga lives and practices. They were displaced from their podus and sedentarized in 'colonies'—a few lost their lives when their podus were razed by FD elephants, then burnt, and their granaries destroyed. These stories have never been recorded until recently. Until they were so violently uprooted, Soligas cultivated the BRT forests for centuries. They used fire extensively to manage their cultivation and the surrounding forest. Fire was set early in the dry season before the build-up of combustible biomass. The practice of early summer fires is called *taragu benki*. Recent efforts at mapping Soliga sacred sites and clan boundaries show that the entire sanctuary was manipulated by Soligas using taragu benki (Mandal *et al.* 2010).[11] This widespread use of fire in past forest management by Soligas has largely gone unmentioned in management plans and ecological studies.

The establishment of the protected area in 1974 resulted in the cessation of taragu benki in BRT. Soligas claim that current PA management is resulting in unprecedented changes to forest vegetation. Soliga elders argue that the control of fire has led to changes in forest structure and wildlife habitat due to the increase in *Lantana*, with serious implications for forest conservation. Mapping of vegetation change in BRT shows that area under *Lantana* has increased dramatically over the last decade (Sundaram and Hiremath 2011).

Soligas offered several reasons for why early-season fires are important: fires control invasive species such as *Lantana*, assist in the regeneration of indigenous species, facilitate the breakdown of seed dormancy, control pests and diseases, increase the availability of fodder for livestock and wildlife, reduce hemi-parasites on amla (*Phyllanthus emblica* and *P. indofischerii*), and increase detection of wild animals. They also claimed that early fires do not kill established seedlings as the rootstock is not affected.

Soliga understanding of the forest contrasts strongly with that of the FD, which continues to be unresponsive to the signals of ecosystem decline that are written large on the landscape due to *Lantana* invasion. In many discussions, Soligas have reiterated their desire to manage a 100-hectare patch of forest, using their practices and by doing so, demonstrate the advantages of fire in rejuvenating forest and increasing wildlife densities. The current atmosphere that privileges a science-based conservation agenda would reject such requests while continuing to search for magic bullets.

Research on NTFP Harvest in PAs

In the BRT Wildlife Sanctuary, ecologists have conducted studies for over 14 years on vegetation dynamics, insect ecology, pollination biology and plant phenology to name just a few, and continue to work there. These studies have resulted in several publications in reputed journals. This is indeed a commendable array and duration of work in one protected area. One major focus of work has been to study the impact of NTFP harvesting so as to evolve sustainable harvest practices. The goal has been to come up with sustainable practices that might be replicated elsewhere. The search for such examples has been necessitated by the constant refrain from conservationists seeking evidence that the 'sustainable use approach successfully conserves mega-fauna such as the tiger and elephant within regimes of extractive use by high-density human populations before this approach is advocated in existing wildlife reserves' (Madhusudan and Raman 2003).

Amla (*Phyllanthus emblica*): Studies on the effect of harvesting amla have examined the size-class distributions to conclude either that resource-use is affecting the (amla) population or that it might in future affect the resource under continued harvest (Murali *et al.* 1996;

Sinha and Bawa 2002; Uma Shaanker *et al.* 2004). The conclusions were based on the premise that in a growing population of a tree species, the number of seedlings and younger individuals should far outnumber adults so that there is adequate replacement of adults when they die. The data showed that populations were being affected variably in different parts of the forest. There were also differences of opinion. Some scholars suggested that the fruit harvest was having an impact (Murali *et al.* 1996) while others suggested that harvesting amla was maintaining populations (Ganesan and Setty 2004). It is also important to remember that a number of other factors other than human harvests of amla were at play. Factors such as rainfall, fruit dispersers, fire, and hemi-parasites were a few other factors that were important. These factors act in tandem to further complicate the picture. Take the case of fire and hemi-parasites. Research showed that fire resulted in the death of small seedlings and hence there were fewer seedlings in dry habitats as opposed to wetter and less burnt-out areas. These findings could have been used by the FD to lend further credence to their anti-fire position and call for greater fire control, but the presence of hemi-parasites queers the pitch. It has been shown that the control of fire increases the density of hemi-parasites on adult amla trees, resulting in low amla tree survival. The more fire is controlled, the more hemi-parasites increase, leading to high tree mortality. It has been widely demonstrated that the largest impact on the demographic growth of plant populations is the removal of adult trees from a population rather than that of seedlings. Clearly, the control of fire has much larger impacts on amla populations than the harvest of fruits (Ticktin *et al.* 2012).

Honey: Researchers have monitored honey bee colony numbers in BRT since 1995. There has been great variation in the number of hives for the whole forest each year (Rai and Setty 2013). Honey is the second most valued NTFP in BRT and has been increasingly harvested by Soligas. The hypothesis that the variation in bee colony numbers is due to the harvest of honey does not stand scrutiny. The ban on NTFP collection imposed around 2003 resulted in dramatically reduced honey collection in the entire study area and, contrary to expectations, a decline in number of bee colonies (Setty *et al.* unpublished manuscript). For some years now, Soligas have maintained that bee numbers in BRT are being affected by pesticide application in floriculture farms outside the sanctuary. Bees are migratory and spend a major portion of the year in the agriculture fields of the plains. Other factors that might impact on the number of bee hives are flowering frequency, average temperature, rainfall that in turn affects flowering, and the availability of nectar.

Both these examples highlight that due to the complexity of ecosystem interactions we are less sure of what might influence biodiversity at local scales. There is also the question of the impact of large-scale drivers such as deforestation and climate change. In the following section I will explore some ideas that are now emerging in ecology with regard to complexity and uncertainty in ecosystems and the problems with managing such complex and uncertain systems.

The New Ecology

Equilibrium, succession, and climax communities have been at the centre of ecological thought over the last hundred years. The notion that ecosystems progress from one stage to the next given prevailing conditions of climate and soils is an old one that is only now being shaken off. Evidence has existed, however, for some time, showing that ecosystems are governed not by equilibrial but by dynamic, complex, and variable processes. This new thinking, however, has not yet pervaded forest management. Most management systems are products of the equilibrium school that assumes that systems progress towards a pristine and climax condition if left untouched. This ignores the fact that ecosystems are constantly affected by a range of disturbances that can never be identified or their frequencies predicted. The inherent disequilibrial dynamics in ecological communities makes the identification of factors responsible for ecological change less certain. Thus the certainty with which most ecological studies hold small-scale anthropogenic use responsible for degradation might be questioned. A recent interdisciplinary study that looked at the hypothesis that Iraqi oil-well fires lit by the retreating Iraqi army during the Gulf war were responsible for the decrease in livestock numbers concluded that a much more complex set of factors including decadal drought, price of barley, and trans-border issues were operating (Gardner 2005).

Assumptions about stability and equilibrium are increasingly being shown to be untenable under new ecological thinking on the function and structure of ecosystems. Recent work has shown that biodiversity at larger scales is maintained by a mosaic of heterogeneous patches within landscapes through spatially and temporally variable disturbance episodes (Pickett *et al.* 1994). This is not to claim that human use does not have ecological effects but that the reliance on ecological science to dictate forest policy is rather like making nuclear scientists the sole decision-makers in detonating nuclear weapons.

The idea that vegetation communities proceeded along successional stages towards a 'climax' formation in the absence of disturbance was mooted by Frederick Clements in the early twentieth century. Evolved from Western notions of stability, this laid the ground for ecological studies in the twentieth century. More recently, ecologists have come to appreciate that disequilibrial and chaotic processes might have a larger role to play in ecosystem structure and that the notion of pristine habitats might not have any real justification. There is now an acknowledgement that there is so much noise in systems that the relationship between local use and degradation might not be as neat as previously assumed.

The study of species diversity has progressed beyond the focus on local processes to determine richness to a more nuanced exploration of historic and regional processes (Ricklefs and Schluter 1993). The work of such authors as Blondel and Vigne (1993) suggests that space and human action has over the last two million years influenced the distribution and occurrence of species. Studying the patterns of bird and mammal distributions in the Mediterranean, they conclude that biodiversity is shaped by disturbances, patchiness, historical events, and the impact of humanity, as much as it is by competition, predation, parasitism, life histories, behaviour, or population structure. Thus the pattern of species that we find at any site or forest is often a result of multiple factors operating variably in time and space. These ecological studies suggest that far more complex processes are structuring our ecosystems than might be explained by current understanding. This belies the certainty with which conservation science lays down management prescriptions that have such far-reaching consequences for both forests and people.

In a discussion on the relevance of non-equilibrium theory to conservation, Begon *et al.* (1990) say that if natural diversity is to be preserved then the non-equilibrium theory tells us not to prevent disturbances that might, in fact, lead to not the destruction but the generation of biodiversity. Although there continues to exist much debate on the role of non-equilibrial processes in structuring communities, ecologists do agree on the role small-scale disturbances have in maintaining diversity. Robbins *et al.* (2006), for instance, suggest that the illegal use of Kumbhalgarh Wildlife Sanctuary by local people might be creating the necessary disturbance required for biodiversity maintenance, especially as these illegal uses are sporadic and spatially variable.

What is meant by degradation is also contested (Lele 2000). There is now a revisionist view that suggests that some levels of landscape change and alteration might facilitate diversity; also that ecological systems are dynamic and thus it is difficult to predict the trajectory of vegetation change (Fairhead and Leach 1996; Robbins *et al.* 2007; Turner 1999). In addition there is also greater appreciation that local knowledge, due to its longer association with the landscape, might have much more to offer forest management efforts than is recognized (Gadgil 2007; Whitehead 2003).

There are forms of local knowledge vis-à-vis ecological functioning and forest management that are now beginning to be asserted (Kothari 2008). Such local forms of knowledge do not necessarily have solutions to issues confronting forest administrations, but management strategies could greatly benefit from both an acknowledgement and an incorporation of it. Communities that have inhabited an area for centuries might know a few things about the area, its forests and biological diversity. Their knowledge might be a little better informed with respect to local context than that of the visiting researcher or the forest administrator whose wisdom is derived from colonial and Western scientific notions of how forests are structured and should be managed—and should be acknowledged.

Current management is based on the understanding that functioning of ecosystems follows equilibrium dynamics that do not take kindly to perturbations; and that if left undisturbed will revert to climax states. Thus any human interventions in forest systems is assumed to be damaging, resulting in renewed state efforts to minimize human use of forest resources. New directions in ecology give us the required framework to move beyond the impasse. This new ecology emphasizes complexity, discontinuities, and uncertainty (Scoones 1999). There is a realization that there are no simple rules that govern ecological systems and that ecological systems do not tend to equilibrium. In many ways, the disequilibrium school mirrors the local knowledge of indigenous people who assess environmental conditions using a complex set of indicators, an understanding of interconnected factors and local context (Sillitoe 1998). The point is illustrated by the contrasting understanding by ecologists and the Soligas of the factors affecting the production of amla fruits in BRT. In their study of the effects of harvesting techniques on fruit production and of the hemi-parasite load on fruit production, Sinha and Bawa (2002) conclude that the presence of hemi-parasites reduces the fruit yield of amla trees and that cutting of branches reduces fruit yield. Soligas, however, have an added explanation for these patterns. Not only do they report that trees with hemi-parasites have high rates of mortality (thus a greater impact than even harvest and branch cutting) but also suggest a link with forest fires. Fires, they observe, kill the hemi-parasites and thus reduce tree mortality, which has led them to define a system of forest management that controls hemi-parasites as well as weeds through periodic burning of the forest.

The phrase 'anthropogenic pressures' appears in the text of many papers published by ecologists working for over a decade on the harvest of NTFPs by Soligas in BRT. These papers have argued that there is indeed a deleterious impact of harvest at every level from the genes to the ecosystem (Uma Shaanker *et al.* 2004). The constant repetition of these conclusions, consolidated by the demonstration of impacts based on the selection of sampling sites that show a gradient of use ('disturbance') has driven home the notion of adverse impacts of local use. When confronted with cases that show no significant impact despite increasing harvests, such as the case of honey collection in BRT (Setty *et al.*, unpublished manuscript), or the case of *Garcinia* fruits in Uttara Kannada (Rai and Uhl 2004), there is the demand for more rigorous methods or long-term data.[12]

Ecological studies that ignore the local, be it local history or local knowledge, often do not engage with the history of forest use and transformation. Studies on the impact of harvest of NTFPs in BRT have based their conclusions on sample sites selected along a gradient of distance from the current settlements, using this as a surrogate for disturbance level (Murali *et al.* 1996; Uma Shaanker *et al.* 2004). It seems to have been ignored that the villages were recently settled as a result of the establishment of the protected area until which time they were moving through the forest either as shifting cultivators or as part of forest department planting operations. In the few cases that studies on the

impact of biomass removal by forest-dwellers do acknowledge historical factors, they assume that such historical processes have played out
evenly across the landscape, trivializing the role of historical landscape
transformations in current forest conditions (Kumar and Shahabuddin
2005). Ecological studies are designed with the assumption that current use regimes, settlement patterns, and management structures are
static. In BRT, for instance, not only have forest management practices
changed in the last few decades but the human settlement pattern has
been highly dynamic due to shifting cultivation. The impacts of these
changes are difficult for existing methods in the ecological sciences to
tease apart. Approaches in ecological sciences do not have relevant
methods that incorporate historical processes. Only contemporary drivers of change are implicated in ecosystem conditions. Inter-disciplinary
approaches, such as those used by Fairhead and Leach (1996), give us
a better understanding than ecological studies that might tell us more
about current ecological processes but little about the history of the
forest and how it came to be in the condition that it is. Conclusions
derived by using ecological methods alone might not provide adequate
management solutions.

Writing for a social science audience, Sivaramakrishnan (1999)
makes a cogent argument for the incorporation of ecological processes
into explanations of landscape change. He urges social scientists and
environmental historians to move away from the essential argument
that power and representation alone might explain forest structure and
practice. He also argues that the dichotomy between local and state
knowledge that comes from power and representation analysis is too
simplistic and that local knowledge informed science and state-making
all along. In other words, he argues that the relationship between state-
making, science, and local knowledge was rather complicated. Add to
this mix the idea that ecosystems are changing and one understands
the complex process of transformation that might not be determined,
predicted, or visualized. This calls for a reframing of the conservation
paradigm based on singular assumptions and static or linear notions of
change. As Visvanathan (2009) observes, 'the complexity of science is
such that old frames of reductionism, productivity, certainty and pre-
dictability no longer work'.

A distinct possibility for the incorporation of local knowledge into
conservation has recently emerged. The FRA, which in addition to

vesting individual rights to land, has provisions for vesting of community rights for conservation and management. The Act empowers rights-holders 'to protect wildlife, forest and biodiversity'. The Act stipulates that conservation measures might be enabled by committees set up by gram sabhas after rights are vested. This applies to all forests, irrespective of their conservation status. The state is mandated to support rights-holders in exercising their rights and duties. There is, however, no explicit mention of the process of how state agencies and communities might harmonize their management plans. An earlier attempt at democratizing sanctuary management was a provision in the Wildlife Protection Act amendment of 2002 which mandated the setting up of sanctuary advisory committees constituted by local and regional representatives. There have, however, been no recorded efforts of any such committees in India. One attempt at establishing a committee in BRT was met with resistance and then wound up after two meetings and no formal ratification of its existence. Often, when confronted with the option of including local people into management committees, it is claimed that they lack capacity. Studies to counter this assumption are numerous and examples exist of local ideas being incorporated into state-driven institutionalized practices (Ribot 2004). Ecologists such as Madhav Gadgil, who have for long argued for an inclusive conservation regime, see the FRA as a 'welcome opportunity to put in place transparent and participatory biodiversity management models' (Gadgil 2007).

Such an effort is underway in the BRT wildlife sanctuary. By the end of 2011, 1516 households have received rights to cultivation on forest land and 25 gram sabhas were awarded community forest rights (CFR). The Soligas won these rights after a four-year struggle and BRT became the first protected area in India in which community rights were awarded to adivasis. Among the several CFRs granted to gram sabhas are rights to harvest NTFP and to conserve, regenerate, and manage the forest according to customary practice. In the same year as the granting of rights under FRA, BRT was also notified as a tiger reserve under the WLPA despite strong protests by the Soligas. This has deepened the conflict between the forest administration and the Soligas. Anticipating possible relocation and hurdles to the exercise of their rights, about 200 Soligas met in July 2011 to produce a community-based conservation plan that they plan to implement in BRT (Kothari *et al.* 2012). The FRA provides them the legal support to do so while the WPA and the forest administration remain silent on such a collaborative approach for the management of BRT. In the two years since the granting of rights, the forest administration continues to snowball attempts by the Soligas to manage the protected area and exercise their community rights. State resistance to local management continues to be based on ecological ideas of stability; centralizing control for usurpation of benefits; and modernization and intensification of conservation through the denial of history, culture, and democratic process.

***

> One reason that science is so often in the firing line in environmental pol-
> itics is that all too often policy decisions are legitimated in purely techni-
> cal terms, leaving opponents with only scientific grounds for contesting
> policies that they oppose for other reasons. (Demeritt 2006)

Following Demeritt, we might argue that there might be social reasons to reject science-driven policy recommendations. For example, even if we assume that the science behind the re-introduction of lions is good, should that be the only information that should be tabled when the proposal to relocate people is made? In other words, when one opposes policy recommendations, does one oppose the science or the assumption that science is central to the exercise? I have argued that the relegation of science to one of several inputs, as opposed to the sole input, for the framing of policy might be a just option.

The claim for a radical reorientation to conservation policymaking cannot be made without taking a good look at the way knowledge is produced by the state and civil society and then exploring why such scientific knowledge might not be appropriate for the task at hand (Chattre and Saberwal 2006). Integrated knowledge for managing forests is required. A singular management regime for all forests is no longer tenable, justifiable, or scientifically valid. The realization that forests have history, are heterogeneous in character, and are immensely complex systems means that we need to use knowledge that has been locally developed. We need what Shiv Visvanathan has called 'cognitive justice' (Visvanathan 2005). Democracy has to be key in the production of knowledge for conservation of PAs. By this I mean a system of knowledge production that is inclusive of local people and decision-making that is flexible and one that periodically incorporates place-based learning.

How does this new ecological thinking affect policy? For one it justifies the push for more decentralized forest management based on local knowledge. It will give local users a bigger say in the way forests might be managed due to their better understanding of local processes. The central point of this new ecological paradigm is that systems are complex and only long-term and multi-perspective approaches will lead to effective management strategies. An appreciation of this approach in policy will enable an inclusive management system that is tolerant of diverse management systems, values heterogeneity in ecosystems, and acknowledges the role of disturbance in maintaining biodiversity. Rather than asking the question 'does the use of forest produce affect wildlife?', one might ask 'how do people co-exist with biodiversity in ecologically friendly ways?' This would result in efforts to manage forest use through a range of knowledge systems whether 'scientific' or 'local'. It is clear that there are no easy answers or certain solutions.

Our desire for rational solutions to perceived problems of biodiversity decline tends to produce simplistic explanations. This is typified by the state's attempts to modernize conservation. In this the state is aided by powerful myths regarding the adverse role of local livelihoods in conservation. Science and its proponents have been complicit in furthering the divide between people and wildlife. Sivaramakrishnan (2003) suggests that the idea of democracy is greater than that of modernity, and that 'Indian democracy necessarily reflects the accommodation of traditional forms of association in the crafting of hybrid political institutions'. Such as effort could bring the sciences together, erode the arrogance of scientific practice, and create space for local voices.

Notes

1. I thank Amita Baviskar and Mahesh Rangarajan for comments on an early draft of this chapter. C. Madegowda and Siddappa Setty have been enthusiastic companions in the evolution and implementation of some of these ideas in Biligiri Rangaswamy Temple Wildlife Sanctuary. I thank Muthatha Ramanathan for many years of critical feedback on my research. Sharad Lele and Ajit Menon have been patient and diligent editors. I alone am responsible for the errors and limitations that persist.

2. In conversation with researchers on 10 March 2006, the DCF referred to Biligiri Rangaswamy Temple Wildlife Sanctuary as his 'estate' and Soliga adivasis as his 'tribals'.

3. Sivaramakrishnan (1999) discusses in detail the debates that ensued in the nineteenth century within the forest administration on the role of fire in eco-sytems. While the upper echelons of the administration argued for fire control, basing it on their assumption that it was bad for forests and associating it with local practice, the field-level officers resisted this as they learnt quickly that fire was essential for forest function. Additionally, field officers were learning from local forest-users about the role of fire and becoming aware of the near impossibility of completely controlling fire.

4. I thank Amita Baviskar for pointing out that rarely is new information incorporated into institutions rapidly. She cites the example of the enormous time that it took the British navy to stock vitamin C that had long been identified as a cure for scurvy that was so prevalent at sea.

5. The literature on this is vast and growing. Willis *et al.* (2004) demonstrate that forests of recent origin are now considered 'pristine'. Pyne (1990, 1994) writes in detail about the history of fire-use in Australia, North America and India and suggests that 'Earth is a uniquely fire planet, and *Homo sapiens* a uniquely fire creature'.

6. Lewis (2003) recounts the events leading from the publication of this report and its being cited in a presentation made by a graduate student, Paul Joslin, at the IUCN meeting in New Delhi in 1969. This invited the attention of a senior Smithsonian official who wrote to Joslin, asking him to reconsider his conclusions. An official from the British Nature Conservancy wrote to Joslin telling him that 'young scientists should devote themselves to their research and not become politicians'.

7. The Scheduled Tribes and Other Traditional Forest Dwellers (Recognition of Forest Rights) Act was enacted to give forest-dwelling tribes and communities rights to their customary agriculture and forest land as well as rights to collect forest produce, conservation, and management.

8. This is best illustrated by the comments of Harish Salve, arguing as *amicus curiae*, in the Supreme Court on 23 November 2007. He stated that forest areas are declared as wildlife sanctuary even if there are four cows and six buffaloes. He suggested that that all such sanctuaries should be denotified, and the land opened up for industrial development, barring four or five sanctuaries in the country, which should be of 'Kenya Class'.

9. Naxalism in Srisailam Nagarjuna Tiger Reserve has ensured that Chenchu adivasis continue to collect non-timber forest produce despite the ban on collection as mandated by the Wildlife Protection Act.

10. Dietrich Brandis (quoted in Sivaramakrishnan 1999), Acchugegowda, a Soliga adivasi, personal communication and more recently, William Bond, an ecologist at the University of Cape Town who argues that major changes in vegetation structure might be linked to high fire activity.

11. The mapping of sacred sites and historic patterns of land-use has evolved as a result of the felt need by Soligas to reassert themselves in the landscape. They believe that these maps could be used to claim rights to the forest under the Recognition of Forest Rights Act, 2006.

12. When the results of a long-term study on honey bee colony numbers that showed no decline in colony numbers due to honey harvest by Soligas were presented, there were vociferous calls from many ecologists who demanded to know if we had looked at this issue thoroughly enough. One ecologist suggested that the 12-year data period was insufficient and that three more years were required before we could draw any conclusions!

References

Baginski-Springate, O. and P. Blaikie. 2007. *Forests, People and Power: The Political Ecology of Reform in South India*. London: Earthscan.

Barve, N., M.C. Kiran, G. Vanaraj, N.A. Aravind, D. Rao, R. Uma Shaanker, K.N. Ganeshaiah, and J.G. Poulsen. 2005. 'Measuring and Mapping Threats to a Wildlife Sanctuary in Southern India', *Conservation Biology*, 19 (1): 122–30.

Begon, M., J.L. Harper, and C.R. Townsend. 1990. *Ecology: Individuals, Populations and Communities*. Boston: Blackwell Scientific Publications.

Blondel, J. and J. Vigne. 1993. 'Space, Time and Man as Determinants of Diversity of Birds and Mammals in the Mediterranean Region' in Ricklefs, R.E. and D. Schluter (eds), *Species Diversity in Ecological Communities: Historical and Geographical Perspectives*, pp. 135–46. Chicago: University of Chicago Press.

Borges, R. 2003. 'The Anatomy of Ignorance or Ecology in a Fragmented Landscape: Do We Know What Really Counts?' in V. Saberwal and M. Rangarajan (eds), *Battles over Nature: Science and the Politics of Nature Conservation*, pp. 56–85. New Delhi: Permanent Black.

Brockington, D. 2004. 'Community Conservation, Inequality, and Injustice: Myths of Power in Protected Area Management', *Conservation and Society*, 2 (2): 411–32.

Chattre, A. and V. Saberwal. 2006. *Democratising Nature*. New Delhi: Permanent Black.

Chhangani, A.K., P. Robbins, and S.M. Mohnot. 2008. 'Crop Raiding and Livestock Predation at Kumbhalgarh Wildlife Sanctuary, Rajasthan India', *Human Dimensions of Wildlife*, 13 (5): 1–12.

Demeritt, D. 2006. 'Science Studies, Climate Change and the Prospects for Constructivist Critique', *Economy and Society*, 35 (3): 453–79.

Fairhead, J. and M. Leach. 1996. *Misreading the African Landscape: Society and Ecology in a Forest-Savanna Mosaic*. Cambridge: Cambridge University Press.

Gadgil, M. 2007. 'Empowering Gramsabhas to Manage Biodiversity: The Science Agenda', *Economic and Political Weekly*, 42 (22, 2 June): 2067–71.

Gadgil, M. and R. Guha. 1992. *This Fissured Land: An Ecological History of India*. New Delhi: Oxford University Press.

Galicia, E. and G.A. Baldassarre. 1997. 'Effects of Motorized Tourboats on the Behavior of Nonbreeding American Flamingos in Yucatan, Mexico', *Conservation Biology*, 11 (5): 1159–65.

Ganesan, R. and R.S. Setty. 2004. 'Regeneration of Amla, an Important Non-timber Forest Product from Southern India', *Conservation and Society*, 2 (2): 365.

Gardner, A. 2005. 'The New Calculus of Bedouin Pastoralism in the Kingdom of Saudi Arabia', in S. Paulson and L.L. Gezon (eds), *Political Ecology across Spaces, Scales, and Social Groups*, pp. 76–93. Piscataway: Rutgers University Press.

Greenough, P. 2003. 'Pathogens, Pugmarks, and Political "Emergency": The 1970s South Asian Debate on Nature', in P. Greenough and A.L. Tsing (eds), *Nature and the Global South: Environmental Projects in South and Southeast Asia*, pp. 201–30. Durham: Duke University Press.

Jones, C.G., J.H. Lawton, and M. Shachak. 1994. 'Organisms as Ecosystem Engineers', *Oikos*, 69 (3): 373–86.

Kabra, A. 2003. 'Displacement and Rehabilitation of an Adivasi Settlement: Case of Kuno Wild-life Sanctuary, Madhya Pradesh', *Economic and Political Weekly*, 38 (29): 3073–87.

Kalpavriksh. 2007. 'Forest Fires and the Ban on NTFP collection in Biligiri Rangaswamy Temple Sanctuary, Karnataka', unpublished report. Pune: Kalpavriksh.

Karanth, K.K., R.A. Kramer, S.S. Qian, and N.L. Christensen, Jr. 2008. 'Examining Conservation Attitudes, Perspectives, and Challenges in India', *Biological Conservation*, 141 (9): 2357–67.

Karanth, K.U. 2003. 'Debating Conservation As If Reality Matters', *Conservation and Society*, 1 (1): 65–8.

Kothari, A. 2008. '*The Cusco Declaration. Infochange*'; http://infochangeindia. org/200807147218/Environment/Politics-of-Biodiversity/Revolutionising-bio-cultural-research.html.

Kothari, A., N. Rai, and C. Madegowda. 2012. 'Green Approach: An Alternative Conservation Model for the BRT Sanctuary is a Step Closer to Becoming a Reality', *Frontline*, 29 (1), January 14–27; http://www.frontline.in/static/html/fl2901/stories/20120127290109900.htm.

Kumar, R. and G. Shahabuddin. 2005. 'Effects of Biomass Extraction on Vegetation, Structure, Diversity and Composition of Forests in Sariska Tiger Reserve, India', *Environmental Conservation*, 32 (3): 1–12.

Lama, M.P. 2000. 'Internal Displacement in India: Causes, Protection and Dilemmas', *Forced Migration Review*, 8: 24–6; http://www.fmreview.org/FMRpdfs/FMR08/fmr8.9.pdf.

Latour, B. 1986. *Laboratory Life: The Construction of Scientific Facts*. Princeton: Princeton University Press.

———. 2004. *Politics of Nature: How to Bring the Sciences in Democracy*. Cambridge: Harvard University Press.

Lele, S. 2000. 'Degradation, Sustainability or Transformation?, *Seminar*, 486: 31–7.

Lewis, M. 2003. 'Cattle and Conservation at Bharatpur: A Case Study in Science and Advocacy', *Conservation and Society*, 1 (1): 1–21.

Madhusudan, M.D. 2005. 'Of Rights and Wrongs: Wildlife Conservation and the Tribal Bill', *Economic and Political Weekly*, 19 November, 40 (47): 4893–5.

Madhusudan, M.D. and T.R.S. Raman. 2003. 'Conservation as if Biological Diversity Matters: Preservation Versus Sustainable Use in India', *Conservation and Society*, 1 (1): 49–59.

Mandal, S., N.D. Rai, and C. Madegowda. 2010. 'Culture, Conservation and Co-management: Mapping Soliga Stake in Biodiversity Conservation in Biligiri Rangaswamy Temple Wildlife Sanctuary, India', in B. Verschuuren, R. Wild, J. McNeely, and G. Oviedo (eds), *Sacred Natural Sites: Conservation of Nature and Culture*, pp. 263–271. London: Earthscan.

Menon, A., C. Hinnewinkel, C. Garcia, S. Guillerme, N. Rai, and S. Krishnan. 2009. 'Competing Visions: Domestic Forests, Politics and Forest Policy in the Central Western Ghats of South India', *Small-scale Forestry*, 8 (4): 515–27.

Middleton, B. 2003. 'Ecology and Objective based Management: Case Study of the Keoladeo National Park, Bharatpur, Rajasthan', in V. Saberwal and M. Rangarajan (eds), *Battles over Nature: Science and the Politics of Nature Conservation*. New Delhi: Permanent Black.

Murali, K.S.U. Shankar, R. Uma Shaanker, K.N. Ganeshaiah, and K.S. Bawa. 1996. 'Extraction of Non-timber Forest Produce in the Forests of Biligiri Rangana Hills, India. 2. Impact of NTFP Extraction on Regeneration, Population Structure, and Species Composition', *Economic Botany*, 50 (3): 252–69.

Nagendra, H. 2008. 'Do Parks Work? Impact of Protected Areas on Land Cover Clearing', *Ambio*, 37 (5): 330–7.

Pickett, S.T.A., L. Kolasa, C.G. Jones. 1994. *Ecological Understanding*. San Diego: Academic Press.

Pyne, S.J. 1990. 'Firestick History', *The Journal of American History*, 76 (4): 1132–41.

———. 1994. 'Nataraja: India's Cycle of Fire', *Environmental History Review*, 18 (3): 1–20.

Rai, N.D. and C.F. Uhl. 2004. 'Forest Product Use, Conservation, and Livelihoods: The Case of Uppage (*Garcinia gummi-gutta*) Fruit Harvest in the Western Ghats, India', *Conservation and Society*, 2 (2): 289–313.

Rai, N.D. and S. Setty. 2013. 'Just Harvest: Ecology and Politics of Forest Canopy Product Use in Protected Areas', in M. Lowman, S. Devy, and T. Ganesh (eds), *Treetops at Risk*, pp. 395–399. New York: Springer.

Ranganathan, J., R.J.R. Daniels, M.D.S. Chandran, P.R. Ehrlich, and G.C. Daily. 2008. 'Sustaining Biodiversity in Ancient Tropical Countryside.' *Proceedings of the National Academy of the Sciences*, 105 (46): 17852–4.

Rangarajan, M. 2003. 'The Politics of Ecology: The Debate on Wildlife and People in India, 1970–95', in V. Saberwal and M. Rangarajan (eds), *Battles Over Nature: Science and the Politics of Conservation*, pp. 189–239. New Delhi: Permanent Black.

Ribot, J. 2004. 'Waiting for Democracy: The Politics of Choice in Natural Resource Decentralization'. Washington DC: World Resources Institute.

Ricklefs, R.E. and D. Schluter. 1993. *Species Diversity in Ecological Communities: Historical and Geographical Perspectives*. Chicago: University of Chicago Press.

Robbins, P. 2003. 'Policing and Erasing the Local/Global Border: Rajasthani Foresters and the Narrative Ecology of Modernization', in Sivaramakrishnan, K. and A. Agrawal (eds), *Regional Modernities: The Cultural Politics of Development in India*, pp. 377–403. New Delhi: Oxford University Press.

Robbins, P.F., A.K. Chhangani, J. Rice, E. Trigosa, and S.M. Mohnot. 2007. 'Enforcement Authority and Vegetation Change at Kumbhalgarh Wildlife Sanctuary, Rajasthan, India', *Environmental Management*, 40 (3): 365–78.

Rodgers, W.A. and H.S. Panwar. 1988. 'Planning a Wildlife Protected Area Network in India'. New Delhi: Ministry of Environment and Forests and Wildlife.

Saberwal, V.K., Gibbs, J.P., R. Chellam, and A.J.T. Johnsingh. 1994. 'Lion–human Conflict in the Gir Forest, Western India', *Conservation Biology*, 8 (2): 501–7.

Scoones, I. 1999. 'New Ecology and the Social Sciences: What Prospects for a Fruitful Engagement?' *Annual Review of Anthropology*, 28 (1): 479–507.

Shankar, M. 2008. 'Relocation of Tigers to Sariska Proceeds, Amidst Caution. Indiatogether, 6 August 2008; http://www.indiatogether.org/2008/aug/env-relocn.htm.

Sillitoe, Paul. 1998. 'The Development of Indigenous Knowledge: A New Applied Anthropology', *Current Anthropology*, 39 (2): 223–52.

Singh, S.P. 1999. 'Chronic Disturbance, a Principal Cause of Environmental Degradation in Developing Countries', *Environmental Conservation*, 25 (1): 1–2.

Sinha, A. and K.S. Bawa 2002. 'Harvesting Techniques, Hemi-parasites and Fruit Production in Two Non-timber Forest Tree Species in South India', *Forest Ecology and Management*, 168 (1–3): 289–300.

Sivaramakrishnan, K. 1999. *Modern Forests: Statemaking and Environmental Change in Colonial Eastern India*. New Delhi: Oxford University Press.

———. 2003. 'Conservation Crossroads: Indian Wildlife at the Intersection of Global Imperatives, Nationalist Anxieties, and Local Assertions', in V. Saberwal and M. Rangarajan (eds), *Battles over Nature: Science and the Politics of Conservation*. New Delhi: Permanent Black.

Skvorecky, J. 1996. *Headed for the Blues: A Memoir. Translated by Kaca Polackova Henley*, p. 16. New York: Ecco Press.

Sukumar, R. 1994. 'Wildlife–human Conflict in India: An Ecological and Social Perspective', in R. Guha (ed.) *Social Ecology*, pp. 303–17. New Delhi: Oxford University Press.

Sundaram, Bharath, and Ankila J. Hiremath. 2011. 'Lantana camara Invasion in a Heterogeneous Landscape: Patterns of Spread and Correlation with Changes in Native Vegetation', *Biological Invasions*, 14 (6): 1127–41.

Ticktin, T., R. Ganesan, M. Paramesha, and S. Setty. 2012. 'Disentangling the effects of multiple anthropogenic drivers on the decline of two tropical dry forest trees', *Journal of Applied Ecology*. DOI:10.1111/j.1365-2664.2012.02156.x.

Turner, M. 1999. 'Spatial and temporal scaling of grazing impact on the species composition and productivity of Sahelian annual grasslands', *Journal of Arid Environments*, 41 (3): 277–97.

Uma Shaanker, R., K.N. Ganeshaiah, M.N. Rao, and N.A. Aravind. 2004. 'Ecological Consequences of Forest Use: From Genes to Ecosystems—A Case Study in the Biligiri Rangaswamy Temple Wildlife Sanctuary, South India', *Conservation & Society*, 2 (2): 347–63.

Tiger Task Force. 2005. 'Joining the Dots: Report of the Tiger Task Force'. New Delhi: Ministry of Environment and Forests, Government of India.

Vandermeer, J. and I. Perfecto. 2007. 'The Agricultural Matrix and Future Paradigm for Conservation', *Conservation Biology*, 21 (1): 274–7.

Vijayan, V.S. 1987. 'Keoladeo National Park Ecology Study'. Bombay: Bombay Natural History Society.

Visvanathan, S. 2005. 'Knowledge, Justice and Democracy', in M. Leach, I. Scoones, and B. Wynne (eds), *Science and Citizens*, pp. 83–94. New Delhi: Orient Longman.

———. 2009. 'Knowledge in Question: The Problem', *Seminar*, 597: 12–15.

Whitehead, P.J., D.M.J.S. Bowman, N. Preece, F. Fraser, and P. Cooke. 2003. 'Customary Use of Fire by Indigenous Peoples in Northern Australia: Its Contemporary Role in Savanna Management', *International Journal of Wildland Fire*, 12 (4): 415–25.

Wildlife Institute of India (WII). 1995. 'Survey of the Potential Sites for Reintroduction of Asiatic Lions'. Dehradun: Wildlife Institute of India.

Williams, M. 2003. *Deforesting the Earth: From Prehistory to Global Crisis*. Chicago: University of Chicago Press.

Willis, K.J., L. Gilson, and T.M. Brncic, 2004. 'How "Virgin" is Virgin Rainforest?' *Science*, 304 (5669): 402–3.

5

NEEMA PATHAK BROOME
SHIBA DESOR
ASHISH KOTHARI
ARSHIYA BOSE

Changing Paradigms in Wildlife Conservation in India

Background

Until very recently, the dominant conservation paradigm in India has been a 'fortress' approach (Brockington 2002) focused on the establishment of a network of wildlife reserves emphasizing law enforcement through 'fences and fines' (Gadgil and Guha 1993). Although the history of competing claims over forest commons may be as old as the history of conservation itself, these contestations were heightened after the creation of state-governed Protected Areas (PAs), a term which gained legal standing and prominence after the promulgation of the Wildlife Protection Act (WLPA) in 1972 (Saberwal *et al.* 2001). This Act (hereafter referred to as the WLPA 1972), and subsequent amendments in 2002 and 2006, allowed for the establishment of PAs of various categories such as National Park, Wildlife Sanctuary, Conservation Reserve, Community Reserve, and Tiger Reserve. Although, control of access and use in these categories varies, with National Parks and Tiger Reserves being the most strictly restricted, the majority of decision-making power across all categories of PAs lies with the state forest department (FD). The under-lying assumption behind strict PAs was that human use is necessarily detrimental to biodiversity/wildlife. However, it can be argued that the rationale to maintain fortress PAs was as political as it was scientific, a form of enclosure and imposed land-use based on a notion of what is desirable by a certain section of society, particularly those who are not directly affected by such enclosures (Saberwal *et al.* 2001).

The social costs of PAs are well documented across the world and in India too, PAs have had severe consequences for communities resi-dent in and dependent on forests and other natural resources (Adams *et al.* 2004; Brockington *et al.* 2006; Ghimire *et al.* 1997; Saberwal *et al.* 2001). Studies suggest that there are three to four million people living inside PAs, and several million more around these, with livelihoods and cultures that are related to the forests and other ecosystems in these (Kothari *et al.* 1995). Many of them have faced physical displacement, or negative social and economic impacts through the loss of access to resources (Lasgorceix and Kothari 2009). The economic, social, and political rights of local communities within PAs have been undermined, usually without consultation, consent, and the provision of adequate alternatives (Wani and Kothari 2007). In a country with widespread hunger and poverty, political marginalization, and overall poor human-development indices, the legitimacy of exclusionary PAs as the primary strategy of wildlife conservation has been strongly questioned by civil society and grassroots social movements (Brechin *et al.* 2002; Wani and Kothari 2007).

The conservation effectiveness of exclusionary PAs and policies is also a highly debated issue. While PAs have been successful to some extent in protecting ecosystems and species, they have also adversely impacted environmental stewardship at a local level as well as the eco-logical security of wildlife. In many PAs in India, there are strong local constituencies against conservation where people have been compelled to engage in activities detrimental to wildlife, either directly through extraction or indirectly through lack of active support for PA manage-ment. The ecological integrity of island PAs and endangered species continues to remain in doubt if conservation efforts do not also address ecosystem conservation at a landscape level, especially the rapid eco-logical degradation outside PAs due to 'development' activities and intense human use. The Global Environment Outlook 5 report (UNEP

2012), has revealed that while globally, PA coverage has gone up in both numbers and spread in the last two decades, bringing under them 13 per cent of the world's land area, global biodiversity has declined at population, species, ecosystem, and possibly genetic levels. The vertebrate populations are reported to have declined by as much as 30 per cent since the 1970s. The report goes on to say that 51 per cent of the sites identified by the Alliance for Zero Extinction as critically important for some endangered species and 49 per cent of Important Bird Areas (IBAs) are still outside PA coverage. The report acknowledges that not all PAs have led to an increase in biodiversity and that not all species may require conventional PAs for protection.

It is increasingly being argued that local stewardship for conservation cannot be built if conservation paradigms do not address the social costs of conservation or take into account traditional indigenous knowledge and common property management practised by local communities for the past millennia. There is now a growing body of knowledge about and political movements in support of what are globally called the territories and areas conserved by indigenous peoples and local communities. These are finding space within the Indian conservation discourse as Community Conserved Areas (CCAs).[1] Numerous examples exist in India where forests, wildlife, and biodiversity are being conserved by people based on their socio-cultural and livelihood relations and dependence on the forests around them (Pathak 2009). In the Ranpur block near Bhubaneshwar, Odisha, 180 villages (many of them adivasi settlements) have conserved forests for several decades, and have come together to form a federation. This is to enable combining their forest conservation initiatives at a landscape level, to minimizing conflicts, and to providing a unified organization. Several hundred Van Panchayats in Uttarakhand have conserved forests for several decades, under state legislation. In Nagaland, the Khonoma Tragopan and Wildlife Sanctuary spread over 2,000 hectares (ha), is an example where through decision-making by communities, hunting and resource extraction is completely prohibited; in another 50 sq. km or so, very minimal resource use for home-use only is allowed. In the nearby Sendenyu village, too, residents have established a Biodiversity Reserve with a complete ban on hunting and destructive resource extraction. Such efforts have historically been ignored in conservation policies and continue to find compromised spaces, if at all, within conservation legislation even today.

This chapter attempts to gain an understanding of the extent to which the idea of democratizing wildlife conservation has actually progressed in India legally and in practice, and the challenges that hinder this progression. We focus on three important new provisions that could potentially lead to democratization of conservation, namely, Critical Tiger Habitats (CTHs) under WLPA 2006; and Critical Wildlife Habitats (CWHs) and Community Forest Rights (CFRs) under the Scheduled Tribes and Other Traditional Forest-Dwellers (Recognition of Forest Rights) Act, 2006 (hereafter the Forest Rights Act or FRA). The chapter attempts to explore answers to the following questions:

1. To what extent do new legislative provisions support possibilities of democratizing and diversifying PA governance in India, in particular, recognizing and vesting access and ownership rights to relevant rights-holders and stakeholders; providing possibilities of inclusive decision-making processes, and creating avenues for the co-existence of humans and wildlife within PAs? How is the actual implementation of these provisions playing out on the ground?
2. What are the wider challenges that need to be tackled while implementing these, which have and would hamper their progression towards realizing true transformation on the ground?

Admittedly, legal provisions are necessary but not sufficient to ensure democratic practices in conservation, as numerous social and political factors also come into play. But we have limited the scope of this chapter to the newly emerging legal spaces to provide a window into the progress towards democratic conservation in India.

Emergence of Democratic Spaces in Laws and Policies in India

In India during the 1980s, questions about exclusionary conservation policies became a more visible and vigorous part of public debate. The mobilization of forest-dependent communities through grassroots social movements, and advocacy by associated social and environmental activists, researchers, intellectuals, and others brought issues about the social impacts of conservation to the forefront of environmental debate. What followed in the next two decades was a highly polarized discussion about the existing fortress approach and whether a paradigm shift towards more inclusionary policies was in order (Kothari *et al.* 1996).

Amongst the first shifts towards such inclusionary policies was the Forest Policy of India in 1988, which prioritized ecological and social functions over commercial ones, and led to schemes such as ecodevelopment (including the World Bank funded India Ecodevelopment Project) on PAs and the Joint Forest Management (JFM) scheme on the rest of the forested landscape. However, because these schemes lacked the necessary legal foothold and democratic vision, and their implementers lacked the intention to relinquish power, they did not fully address many critical issues such as tenure security, access and rights to resources, and community rights to decision-making. On the contrary, the implementation of these schemes has largely meant constitution of local committees to implement activities predetermined by the state, through funds provided by the FD (Das 2007). Till very recently, these were the only legal and policy spaces available to the local communities to voice their concerns in biodiversity management. These schemes have come under criticism on a number of grounds, including that the committees were often undemocratically constituted and suffered from elite capture, and also undermined institutions and initiatives set up by communities themselves (Shahabuddin 2010). Meanwhile the Panchayat (Extension to Scheduled Areas) Act, 1996 was passed. The Central Act provided for the extension of the panchayat local governance system to 'scheduled' areas, with predominantly tribal populations. It required state laws to be made 'in consonance with the customary law, social and religious practices and traditional management practices of community resources'. Gram sabhas were considered competent to protect community resources. They were expected to approve development plans and projects at the village level. This Act was, however, much diluted in state adaptations and limited rights were eventually granted to the communities concerned, thus resulting in much less devolution and fewer benefits to local communities as compared to their expectations (Vagholikar and Bhushan 2000). This once again reflected a lack of willingness of the state and its functionaries to relinquish power.

The decade of 2000 saw further mobilization, including protests and advocacy, in the wake of a directive by the Union Ministry of

Environment and Forests (MoEF) to evict forest-dwelling communities, viewed by the MoEF as 'encroachers'.[2] In the policy arena, a number of interesting changes took place. Recommendations for collaborative management of PAs were contained in the National Wildlife Action Plan (NWAP) 2002, the draft National Biodiversity Strategy and Action Plan (NBSAP) 2004, and the National Environment Policy (NEP). However, the NWAP and the NEP are still at the policy level, with implementation yet to begin, and the submitted draft of NBSAP in 2004 was not accepted by the government, which came up with a much more diluted version of its own (TPCG and Kalpavriksh 2005). The more significant of legislative changes was the promulgation of FRA in 2006. To a lesser extent, but nevertheless important, the WLPA 1972 was amended in 2006 with the inclusion of a section on tiger reserves. Provisions of CFRs and CWHs within the FRA, and CTH within the WLPA 2006 included aspects with potential for greater participation and consultation in part of the formal conservation landscape. Before both of these, the Biological Diversity Act, 2002 offered some possibilities of participation through village-level institutions both inside and outside PAs, though there was little in it to override the alienating provisions of the WLPA 1972 and other forest legislations. It focused more on documenting local traditional knowledge than actually empowering the knowledge-holders and ensuring their continued access to the concerned elements of biodiversity (Pathak Broome *et al.* 2012).

Global Discourse on Democratizing Conservation Laws and Practice

The democratization of conservation laws and practice have been discussed internationally for a few decades, but their acceptance at global conservation forums has been more apparent since 2003. The International Union for Conservation of Nature (IUCN) World Parks Congress (WPC) at Durban, 2003, and the Seventh Conference of Parties of the Convention on Biological Diversity (CBD COP7), at Kuala Lumpur, 2004, have been two major international events to bring these trends into greater focus.

At the WPC, over 3,000 conservation practitioners, policymakers, and others, gathered for what till then was the largest ever gathering of people working on PA issues, and included about 200 representatives of indigenous peoples and other local communities. The presence of the latter was instrumental in the WPC in bringing about the much sought paradigm shift represented by the trends mentioned earlier. This was further pushed by a number of civil society representatives. Elements of new conservation paradigms endorsed by the WPC were included in each of its key outputs: the Durban Accord, the Durban Plan of Action, the Message to the CBD[3] and recommendations on Good Governance of PAs, Diversity of Governance Types of PAs, Indigenous Peoples and PAs, Co-management of PAs, CCAs, Mobile Indigenous Peoples and Conservation, and Poverty and PAs. The Convention on Biological Diversity (CBD) Conference of Parties (CoP) 7 in 2004, heavily influenced by WPC outcomes and civil society organizations (CSOs) and Indigenous Peoples networks mentioned above, adopted a comprehensive Programme of Work on PAs (PoWPA), which included clear goals and actions for moving towards new governance models for PAs, and improving participation, equity and benefit-sharing. A subsequent (2008) review of PoWPA by the CBD Secretariat however showed that progress on these aspects was highly dissatisfactory.

This brings to the fore a global trend, namely the reluctance of state to relinquish their own power and devolve it to other rights-holders and stakeholders. The reasons cited by the signatory states, including India, for lack of implementation included lack of capacity. Consequently, since 2010, a group of agencies including IUCN, Deutsche Gesellschaft fur Internationale Zusammenarbeit (GIZ) GmbH, the ICCA Consortium, and the CBD Secretariat have compiled and published a resource kit to help signatory countries implement governance reforms in PAs more effectively and locate them within the internationally accepted principles of good governance (Borrini-Feyerabend *et al.* 2013).

Primary Values for Democratizing Conservation in the Global Context

The above-mentioned local and global processes have led to the conceptualization of elements that would be crucial for the democratization of conservation globally and within India, including:

1. That territorial and resource rights of indigenous peoples and other local communities that have traditionally lived in or used natural ecosystems, need to be respected in conservation policies and practice, and that the costs and benefits of conservation need to be much more equitably distributed.

2. That governance of PAs needs to be distinguished from management of PAs. The effectiveness of PAs does not merely depend on what decisions are taken but also on *how the decisions are taken, who takes them and what processes and information systems are followed to take these decisions*. PA governance is defined as 'the interactions among structures, processes and traditions that determine how power and responsibilities are exercised, how decisions are taken and how citizens or other stakeholders have their say' (Borrini-Feyerabend *et al.* 2013).

3. That there is not only *one* kind of governance of PAs (by governments), but several kinds; in particular, collaboratively or jointly managed ones, and that local communities and indigenous peoples themselves can and are conserving sites and species across the world. While governance regimes for PAs vary greatly around the world, IUCN and the CBD PoWPA distinguish four broad governance types (Dudley 2008):

   - Governance by government (at various levels and possibly combining various institutions)
   - Governance by various rights-holders and stakeholders together (shared governance)
   - Governance by private individuals and organizations
   - Governance by indigenous peoples and/or local communities[4]

4. It is important to note in this context that there cannot be a standard governance arrangement for all PAs. Governance models are appropriate only when tailored to the specifics of its context and effective in delivering lasting conservation results, livelihood benefits, and the respect of rights. That specific ecological, historical, and political contexts, and the variety of worldviews, values, knowledge (including the local) and outside experts, skills, policies, and practices (including informal and local) that contribute to conservation, should be reflected in the governance regime for each specific PA.

5. That instead of being immutable, the institutions and rules governing PAs must be dynamic and adaptive in response to existing challenges and change. Such adaptive governance should be cautious and well-informed, and nested within a larger vision, developed collectively by all rights-holders and stakeholders.

6. That diversified governance of PAs itself is not enough to achieve democratic and effective PAs. Equally important are the processes by which democratic institutions are set up, those involved in decision-making processes are chosen, the processes by which decisions are made, the processes and knowledge-base which is used to set goals, the fairness with which institutions function, and how effective, transparent, accountable, and well informed the concerned institutions, systems and processes are. The answers to these questions would help determine the *quality of PA governance*. Legal and institutional changes related to governance of PAs alone will not lead to desired results till the actual implementation on the ground is monitored for the *quality of governance* using principles of good governance and parameters of effective management. The governance quality of a PA, or of a PA system, can be evaluated against a number of broad principles of good governance that have been developed by a variety of people, nations, and United Nations agencies, including legitimacy and voice; direction; performance; accountability; fairness and rights (see Box 5.1).

BOX 5.1   Principles of good governance of PAs

Legitimacy and voice

- Legitimacy of a governance arrangement comes from the establishment of institutions with a broad acceptance and appreciation in society; as much as possible attributing management authority and responsibility to the capable institutions closest to natural resources (subsidiarity); ensuring that all mutually agree rules are honoured.
- Voice in a governance arrangement is ensured by making available appropriate and sufficient information to all rights-holders and stakeholders and ensuring that they have a say in advising and/or making decisions; seeking active engagement of all vulnerable groups, such as indigenous

(Cont'd)

peoples, women, youth, and others in decision-making; maintaining an active dialogue and seeking consensus on solutions that meet, at least in part, the concerns and interest of everyone; mutual respect among all rights-holders and stakeholders.

Direction

- Developing and following a consistent strategic vision for the PAs and conservation objectives grounded on values mutually agreed by all rights-holders and stakeholders; ensuring that governance and management practice for PAs are consistent with the agreed values.
- Ensuring governance and management practice for PAs are compatible and well-coordinated with the plans and policies of other levels and sectors in the broader landscape/seascape.
- Ensuring governance and management practice are respectful of national and international obligations (including CBD PoWPA).
- Providing clear policy directions for the main issues of concern for the PA and, in particular, for contentious issues (for example, conservation priorities, relationships with commercial interests, and extractive industries).

Performance

- Achieving conservation and other objectives as planned and monitored, including through ongoing evaluation of management effectiveness.
- Being responsive to the needs of rights-holders and stakeholders by providing timely and effective response to inquiries and reasonable demands for changes in governance and management practice.
- Ensuring that PA staff, rights-holders, and stakeholders, as appropriate, have the capacities necessary to assume their management roles and responsibilities and that those capacities are used effectively.
- Making an efficient use of financial resources and promoting financial sustainability.

Accountability

- Upholding the integrity and commitment of all in charge of specific responsibilities for the PAs.
- Ensuring transparency, with rights-holders and stakeholders having timely access to information about what is at stake in decision-making?
- Ensuring a clear and appropriate sharing of roles for the PAs, as well as lines of responsibility and reporting/answerability.
- Ensuring that the financial and human resources allocated to manage the PAs are properly targeted according to stated objectives and plans.

- Evaluating the performance of the PA, of its decision-makers and of its staff, and linking the quality of results with concrete and appropriate rewards and sanctions.
- Establishing communication avenues (for example, websites) where PA performance records and reports are accessible
- Encourage performance feedback from civil society groups and the media.
- Ensure that one or more independent public institution (for example, ombudsperson, human rights commission, auditing agency) has the authority and capacity to oversee and question the action of the protected areas governing bodies.

Fairness and rights

- Striving towards an equitable sharing of the costs and benefits of establishing and managing PAs and fairness in taking all relevant decisions.
- Making sure that the livelihoods of vulnerable people are not adversely affected by the PAs; that the costs of PAs—especially when borne by vulnerable people—do not go without appropriate compensation.
- Making sure that conservation is undertaken with decency and dignity, without humiliating or harming people.
- Dealing fairly with PA staff and temporary employees.
- Enforcing laws and regulations in impartial ways, consistently through time, without discrimination and with a right to appeal (rule of law).
- Taking concrete steps to respect substantive rights (legal or customary, collective or individual) over land, water, and natural resources related to PAs, and to redress past violations of such rights.
- Taking concrete steps to respect procedural rights on PA issues, including: appropriate information and consultation of rights-holders and stakeholders; fair conflict management practices; and non-discriminatory recourse to justice.
- Respecting human rights, including individual and collective rights, and gender equity
- Ensuring strictly the free, prior, and informed consent of indigenous peoples for any proposed resettlement related to PAs.
- Promoting the active engagement of rights-holders and stakeholders in establishing and governing PAs.

Source: Based on description of the principles by Abrams *et al.* (2003); Borrini-Feyerabend *et al.* (2006); Eagles (2009); Graham *et al.* (2003); Institute on Governance (2002).

Legal Spaces for Democratization of Conservation in India (with a Focus on PAs)

The promulgation of legislations like the FRA 2006, the WLPA 2006, and a number of other legal changes mentioned above, and at the international level, processes within the CBD, have been a conceptual turning-point in the way that forest- and other ecosystem-dependent communities access and interact with traditionally state-governed spaces like PAs. However, the complexity of implementing these legislations on the ground is that the process would involve a paradigm shift, not only in the process of changing words on a piece of paper but also in the historical vision, power dynamics, and mindset of various actors involved. The big question is whether or not the Indian state would be willing to enable the redistribution of power and the building of capacity that is required to implement these legal changes in a meaningful manner.

This section describes the legal provisions within the FRA and WLPA, which have attempted to provide democratic spaces for conservation, their interface, and the manner in which they are being implemented on the ground. What is visible today is a mix of situations. While on the one hand, there is often reluctance in the FD to implement the recent legislation and policy changes, on the other hand, local people, civil society organizations, and conservationists advocating for participatory forest governance do now have some legal provisions in their favour.

The Provisions

The FRA aims to undo historic injustice to tribal and non-tribal forest residents and dependent communities in India by establishing their rights to forest land and resources, including within PAs. In addition to the establishment of rights of ownership and use, FRA also provides for establishment and conservation of CFRs, hence creating a possibility and potential for decentralizing forest governance. As much as possible, it aims to attribute management authority and responsibility to the capable institutions closest to natural resources, that is, to the smallest (recorded or unrecorded) hamlet and settlement or mobile communities. It also vests in the village assembly of such a settlement the right to constitute the governance and management committee, and ensures participation of women and scheduled tribes (STs) in such committees.

FRA also has a provision for creation of CWHs within PAs, where the rights of the local communities can be partially or totally modified, if proven to be irreversibly damaging for wildlife. However, no such modification can be carried out without following clearly laid out steps for doing so in the Act. These include the establishment of rights where they have not been established legally, local consultations with the rights-holders and stakeholders and conducting scientific research to establish impacts of human activities. The WLPA 2006 amendment (coming just two months before the FRA was enacted), introduces the category of CTHs for exclusive protection of tigers in addition to other PA categories. It also supports a participatory process for relocation and modification of rights while creating these CTHs.

Critical Wildlife Habitats and Critical Tiger Habitats

Critical Wildlife Habitat (CWH) and Critical Tiger Habitat (CTH) are, therefore, two similar-sounding concepts introduced by FRA and WLPA respectively,[5] without either of the laws making a reference to the other. Both are special provisions for conservation in PAs, which were introduced into policy discourse in 2005–6, mainly in anticipation of the impacts on conservation of wildlife after recognition of rights under the FRA, which was then being discussed and debated. However, both the laws explicitly support the recognition of the rights process before the creation of these categories and also specify that no relocation can take place without following a process as prescribed in both the laws. These two categories are being used as an example here as they have emerged in a period where recognition of access and rights of local communities have received more priority, and hence carry a greater potential for democratizing PA governance (see Box 5.2).

BOX 5.2   Similarities and differences between CWHs and CTHs

The provisions for CWHs and CTHs are similar in that they are both marked out of PAs; both are defined as areas required to be kept as inviolate on the basis of scientific and objective criteria; both require evidence of irreversible damage being caused, of co-existence not being possible and consent of the village assemblies or gram sabhas before making an area inviolate for wildlife.

There are a few differences though: the purpose of CTHs is tiger conservation whereas the purpose of CWHs is wildlife conservation in general, indicating a difference between a single-species-based and biodiversity-based approach. As a pre-condition for relocation, CWHs mention '*free* informed consent' of the gram sabha obtained in writing, whereas CTHs require 'informed consent' only. For rights modification, a pre-condition for CWH is recognition and vesting of forest rights whereas for CTHs, the pre-condition is recognition, determination, and *acquisition* of land and forest rights. CWHs from which relocation has taken place, cannot be subsequently diverted by the state government, central government, or any other entity for any other uses; this is potentially the most powerful conservation provision in Indian legislation. There is no such restriction on a CTH, which is ironical, given the high degree of attention that tigers have received from formal conservationists compared to other species.

Source: FRA (2006) and WLPA (2006).

In terms of their interface with the local communities, these provisions can be interpreted to allow for the following broad elements:

1. They provide for somewhat broader societal input into the constitution of CWH/CTH areas, as they explicitly require inputs for natural and social scientists.
2. They provide for exploring possibilities of co-existence (which remains legally undefined) between local communities and wildlife, even if, in the case of CTHs, exploration of co-existence is restricted to the buffer zones of CTHs only.
3. They provide for a just process of relocation of communities from the proposed CWHs/CTHs, where their presence is shown to be irreversibly detrimental and they consent to relocate.

Community Forest Resource Rights

As mentioned above, the FRA has certain provisions that entrust the gram sabhas (village assemblies) with the rights and responsibility for sustainable use of their CFR. The CFRs, protection of which is provided as a right (under Section 3[1] i), is traditionally accessed as customary common forest land, and may include such areas within PAs. The gram sabhas are empowered to create mechanisms for the conservation of biodiversity and wildlife, preservation of natural and cultural heritage, for ensuring that internal and external factors do not destroy their community forests, and for maintenance of ecological balance (Section 5). For performing these functions, gram sabhas are to make committees (under Rule 4[1] e). As per the preamble of the Act, these provisions are for strengthening the conservation regime while ensuring livelihood and food security for the concerned community.

Therefore, the CFR provisions could be a powerful basis for initiating processes towards co-existence, co-management, and shared governance resulting in a diversity of PA governance categories and other forest conservation sites, and supporting equitable distribution of benefits thus arising. CFRs, through Section 5 of FRA, also give power to the communities to stop destructive development activities, if they so desire. This has been further strengthened by a circular, issued on 3 August 2009 by the MoEF, stating that all development project proposals requiring diversion of forest land need to enclose evidence that rights of the local people who are likely to be affected have been recognized under FRA, and that consent of the relevant gram sabhas has been obtained, before any clearances are sought under Forest Conservation Act, 1980 (MoEF 2009).

By taking these elements into account, the FRA to a certain extent, establishes the principle of legitimacy, voice, and subsidiarity as mentioned in Box 5.1, as well as certain elements from the principle of fairness and rights. The WLPA is also attempting to move in that direction but falls short by not dealing with a number of contradictions that arise because of the two Acts being silent about each other, and hence not meeting the requirements of *direction*, one of the principles of good governance mentioned in Box 5.1. For example, the WLPA is silent about the settlement of rights process that has been prescribed in WLPA for declaration of PAs, which is in contradiction with the FRA.

This contradiction is further apparent in a lack of clarity on the exact relationship of the two Acts with each other; for example, if local communities claim CFR rights inside a PA, what would be the exact relationship between many such CFRs within a larger PA? On what grounds would the management and governance strategies be decided? Who would decide them and what would be the mechanism to bring various actors together.

In addition, the Acts do not describe the process of identification of CTHs and CWHs other than mentioning that the basis should be 'scientific and objective criteria' although official protocols and guidelines to implement the provisions are present or under preparation. A number of CSOs argue that the seeds of democratic governance of PAs lie in the process by which they are identified and declared. While the importance of wildlife science in this process cannot be overestimated, many CSOs (including those within a national network called Future of Conservation or FoC)[6] submit that the knowledge relevant for such identification also exists amongst amateur wildlife enthusiasts, and even more so with local communities who have valuable traditional knowledge and resource management systems. In particular, given that the CWH/CTH process could involve the modification of people's rights, it is crucial to build a sense of ownership amongst local communities who live in or use sites that are likely to be proposed as CWHs/CTHs, and their say in the decision related to the declaration is a must in creating this sense. This clearly indicates a lack of complete commitment to the principle of voice and legitimacy.

The premise for relocation for creating CWHs and CTHs is to create 'inviolate' zones. However, neither WLPA nor FRA define what 'inviolate' means. In this regard, many CSOs have argued that interpreting a CWH/CTH to be completely human-free (as appears to be in the minds of many conservationists and state forest officials implementing these laws) would lead to only very small areas being notified, whereas interpreting it to mean free of *incompatible* human uses would enable much larger areas to be notified.[7]

All the above-mentioned factors, along with the fact that there are no independent public institutions (with representation of all rights-holders and stakeholders) that have the authority and capacity to oversee the issues and practice related to PA governance, has led to a number of hurdles as described in the process of implementation in the follwing section.

Implementation of the Provisions

This section describes the official guidelines framed for the process of implementing the above provisions, and the status of onground implementation.

Critical Wildlife Habitats

In 2007, a set of guidelines were issued for implementation of CWHs. Some elements of these guidelines had the potential to enhance conservation of biodiversity through more scientific and democratic means. For example, according to this document:

1. The process of identification of CWHs would have required the involvement of experts from both within and outside the government.
2. Section 4 (vii, viii, ix) required that information to be submitted along with the application for CWHs by the state should include a resolution of the gram sabha certifying that recognition and vesting of rights is complete.
3. Section 5 mandated the Expert Committee to engage in an open process of consultations with local communities in areas to be declared CWH and even required a quorum of two-thirds of the adults in the gram sabhas without whose consent a CWH could not have been declared.

These guidelines did have a few limitations (FoC 2007), but in general, these were considered as a good starting-point. However, these guidelines were suddenly withdrawn by the Ministry in 2011 citing 'demand from various quarters'[8] as a reason and a new set of guidelines were issued in February 2011. The new guidelines were criticized for their lack of space for democratic approaches to determining such habitats; and insufficient attention to a proper scientific, knowledge-based approach.

Following public protest, including by FoC members, the February 2011 guidelines were withdrawn and a revised set was made public for comments. These draft guidelines or implementation protocol (as they were called) were a significant improvement over the earlier guidelines. These had a much greater emphasis on public consultations and on acknowledging the possibility of co-existence within CWHs (as mandated by the FRA). With some minor changes, this protocol could be very helpful in furthering the cause of wildlife conservation by respecting people's livelihood rights and hence generating their support and stake for conservation. This protocol, however, has been with the MoEF since March 2011 and a final version had not been issued till the time of writing this chapter.

In the meanwhile the state FDs have approached the identification of CWHs in diverse ways, ranging from total exclusion (contrary to the provisions of FRA themselves) to sticking to the implementation of FRA both in letter and spirit. Assam, on the one hand, and Kerala on the other, represent these two positions with other states located at various points of this continuum. The Assam State FD had planned to declare a total area of 9,67,366.436 ha, consisting of all existing PAs as well as some Reserved Forests (RFs) outside PAs as CWHs. This proposal was not inclusive of the views of the local communities or any scientific report showing whether the impact of local people on these PAs was irreversible. On the other hand, the Kerala FD publicly stated that gram sabhas have a crucial role in the implementation of the FRA.

Proposals for CWHs from different states (Sitamata Wildlife Sanctuary in Rajasthan, Guru Ghasiram National Park, Chhattisgarh, and Gahirmatha Turtle Sanctuary, Chandaka Elephant Sanctuary, and Chilka Bird Sanctuary in Odisha) based on the earlier guidelines of 2007 were reviewed by Kalpavriksh in 2010. Copies of these proposals were obtained through RTI applications. The review revealed the following issues, among others:[9]

Involvement of local communities: Out of all the above mentioned proposals, only one (Sitamata WS) had organized a process that engaged with local communities to some extent. The proposal mentioned that: a) a notice to gram sabhas was sent through the *sarpanchs* of the concerned villages; b) the 2007 CWH guidelines were distributed to the concerned gram panchayats.

Establishing possibilities of co-existence: Little or no evidence is given to establish that co-existence is not possible in proposals which mentioned a need for relocation, such as in the case of Chandaka WS in Odisha. The proposal of the Guru Ghasiram NP mentions that co-existence is not possible as grassland had been converted to agricultural fields and cattle from 78 villages depend on it for grazing, but does not give any scientific proof of the same.

Recognition of rights: In most proposals, there is no mention of recognition of rights under the FRA (except the Sitamata WS where it is mentioned that the verification process under the FRA is under progress). In the Guru Ghasiram NP, the 'recording of rights' had been done by the collector for 29 out of 78 villages, but not under the FRA.

Gram sabha consent for relocation: Consent for relocation by the gram sabha was not attached in any of the proposals. Proposals from the sanctuaries of Gahirmatha and Chilka mentioned that there was no question of rights as there were no human habitation inside, ignoring completely the dependence of a large population on these ecosystems. The proposal of Guru Ghasiram NP mentions eight villages giving their consent, but does not provide copies of gram sabha resolutions.

Some of these issues seem to have been noticed also by the Central Expert Committee constituted in 2007 for evaluating state-level proposals, as indicated in a response by the Minister of Environment and Forests in the Rajya Sabha on 8 May 2012. The response states that the CWH proposals submitted by Odisha to the central-level committee were found to be incomplete and have been sent back for revision and resubmission.[10]

This has not necessarily led to the states following the legally prescribed procedure. The FRA action plan of Tripura presented on 3 December 2012 specifies that a CWH is being established and 2,055 families have been selected for relocation. There is no mention in the plan of the process used for selection of this site or the villages to be relocated. It also does not specify whether or not there has been prior recognition of rights and whether the consent of the gram sabha has been received for relocation.

In Maharashtra, in a response to a circular issued by the department of revenue and forests, the process of identification and declaration of CWHs began in 2012. In some PAs such as the Bhimashankar Wildlife Sanctuary in Maharashtra, the FD held detailed consultations with all the concerned gram sabhas, which rejected the proposal for the creation of CWH, fearing strong restrictions. In others such as the Yawal Wildlife Sanctuary, a process of identification of CWHs was initiated by the FD, but discontinued when in April 2013 the local communities and CSOs raised the issue that implementation cannot be initiated on the basis of draft guidelines, without discussions with the local communities, and without prior recognition of the pending claims under the FRA to the affected villages (Pathak 2013). Thus, CWHs have not been actually implemented anywhere yet, and where proposals are pending, they all fail to follow due process.

Critical Tiger Habitats

Unlike CWHs, which have not been declared anywhere yet, CTHs have been declared all over the country. Available information reveals that the notification of most CTHs in the country has been in violation of the WLPA and FRA. The declaration of CTHs started with rushed notifications of several core/critical tiger habitats in 2007 arguably to create such areas before the implementation of FRA began. On 17 November 2007 the National Tiger Conservation Authority[11] asked all states to set up expert committees to 'finalise and delineate core or critical tiger habitats of tiger reserves… within 10 days of the receipt of this letter'. All relevant states complied by sending in proposals for core or CTHs. As a result, of the total 41 CTHs notified till 2012, 31 were already notified by the end of 2007 with several of them notified on 31 December 2007. It is not a coincidence that this was just one day before the FRA rules were notified, on 1 January 2008. It should be clear from this that no proper scientific or consultative process would have been possible in such a rush.

Indeed, a member of the National Board of Wildlife who was involved with the above notifications of core or CTHs, has recently admitted:

> Declaration of cores was done in a rush in order to insulate our tiger areas against the Forest Rights Act (FRA), which came into being before the end of 2007.… A new core had been created overnight with little basis in science. In Ranthambore, Kailadevi Sanctuary became a core critical habitat encompassing 595 sq km with one tiger, 25,000 people, 40,000 livestock and 44 villages. This makes up 53 per cent of Ranthambore's CTH. (Thapar 2012)

Thus it is apparent that neither the CTHs have been demarcated on the basis of any 'case by case scientific study' as required by Section 38V (4) of the Wild Life (Protection) Act nor have the concerned forest-dwellers been consulted, or their consent taken, in any meaningful manner on this demarcation, nor any attempts made to assess possibilities of co-existence with the local communities. Because of this, many civil society groups and local communities consider these CTHs as illegal.

At the time of writing this chapter, there was no single document consolidating guidelines for notification of tiger reserves and the various issues related to them in a holistic manner. A draft set of guidelines on co-existence and identification of such areas has been submitted by the FoC and could form a useful basis for discussion.[12] It may be useful to add here that in a meeting organized by the MoEF on the revised guidelines for declaration of CWHs held on 4 March 2011,[13] it was decided that the member secretary of the National Tiger Conservation Authority (NTCA) will prepare two protocols related to tiger reserves: one regarding village relocation from tiger reserves and another for 'declaring new tiger reserves after the Forest Rights Act, 2006 has come into effect.' While the protocol on relocation from CTHs has already been prepared, there are still no guidelines on demarcating tiger reserves, securing them for conservation and exploring possibilities of co-existence, particularly in the buffer areas, clearly indicating what the priority for the government is. This issue of a comprehensive set of guidelines has been also raised in the matter of the *Ajay Dubey vs. NTCA and others* case, commonly known as the 'tiger tourism case' (see Box 5.3).

The protocol on relocation from the CTHs has been finalized in 2011 without addressing issues of concern raised by many groups and networks, including FoC.[16] The guidelines do not explain what happens where there is scope for co-existence. Nor does the checklist for relocation require (as should have ideally been the case) a report that would show evidence of irreversible damage and no scope of co-existence. While the protocol mentions that recognition of rights of STs and OTFDs should precede relocation, how such rights and their record in official documents would be useful to the villagers in the place to which they are relocated is left unclear. Even though settlement of rights as a concept and a term is only given in WLPA and not FRA, in many instances 'recognition' and 'settlement' are used in a

Box 5.3   The Ajay Dubey case and conservation through Tiger Reserves

One noticeable development in the tiger conservation scenario has been the case of *Ajay Dubey vs. NTCA and others*. In 2011, a writ petition was filed before the Madhya Pradesh High Court, by Ajay Dubey[14] for stopping 'all kinds of tourism, mining, development or any activity within the core / critical areas of "Tiger Reserves"'. The petitioner also asked for the preparation and implementation of a 'Tiger Conservation Plan' as well as the status of notifications pertaining to core and buffer areas. In response to one of the orders passed on the case, many state governments rushed to notify buffers. The case also led to an NTCA affidavit stating that a 'comprehensive set of guidelines is being framed by the NTCA and the Ministry of Environment and Forests with regard to fixation of core areas, buffer areas, and tourism, including welfare and religious tourism as contemplated, amongst other laws in force, under Section 38-O (c) of the Wildlife Protection Act as well as with regard to the protection of the tigers in forest areas as well as non-forest areas'. It was also submitted that the NTCA 'would consider all aspects while formulating the guidelines after taking the views of Expert Bodies and after letting all stakeholders participate'. However, the guidelines were put in the public domain as a website link for only one week of comments. Moreover, the committee which finalized the guidelines limited its scope to tourism and did not cover issues relating to identification and declaration of core / buffer areas as also issues related to co-existence. Some of its members stated that the committee also had a mandate to formulate guidelines on identification and declaration of core / buffer areas of tiger reserves and that this aspect had not been completed. They believed that this needs to be done, but by following a much wider consultation process. The latter however was neither mentioned in the submission to the Court by the NTCA nor subsequently done.

The NTCA finally did not frame guidelines for the fixation of core and buffer areas. It has also not framed comprehensive guidelines for various aspects of tiger reserves, from identification to demarcation to zonation, management, and governance. What NTCA did submit to the Court (mentioned above) as guidelines are essentially a set of data and principles.[15] These principles have also not gone through the process of consultation with all concerned stakeholders, particularly the local communities. Many CSOs felt that the NTCA had not fulfilled its responsibility and mandate under Section 38Vc of the WLPA and an intervention has been filed by Kalpavriksh raising the above-mentioned issues of concern. There was an interim ban on tourism in the CTHs but the same was lifted with the enforcement of the guidelines. The case is still being heard.

The delineation of buffers without any detailed guidelines for co-existence in buffer zones of tiger reserves, which are inhabited by thousands of tribal and non-tribal forest-dependent communities, makes it appear that reconciling conservation and livelihoods through co-existence is not a priority for the decision-makers.

Source: Authors.

single sentence. For example, 'In case of voluntary relocation also, the rights of people should be recognized and settled before relocation' (page 9). How community rights like intellectual property and traditional knowledge related to biodiversity and cultural diversity, access to sacred sites, to grazing in specific areas, and so on can be 'settled' is unclear. There must be a procedure or clarification on how the more intangible rights, such as use and access rights, are transferred to the relocation site.

Despite the legal provisions and the protocol, a survey carried out in 2011 by Kalpavriksh along with local CSOs in four tiger reserves (the Simlipal Tiger Reserve in Odisha, the Sariska Tiger Reserve in Rajasthan, the Melghat Tiger Reserve in Maharashtra, and the Achanakmar Tiger Reserve in Chhattisgarh) found that legal requirements for creating CTHs or for relocation (even in accordance withthese flawed guidelines) were not carried out.[17] Even the reports of the NTCA monitoring committees[18] point to these violations, as summarized by the authors in the table below.

Even while many CSOs are attempting to draw the attention of the MoEF, Ministry of Tribal Affairs (MoTA), and state FDs about provisions of the law not being fulfilled during relocation from tiger reserves, the NTCA recently has approved a proposal for using Rs 1000 crores per year for the next five years from the funds of the Compensatory Afforestation Fund Management and Planning Authority (CAMPA) for relocation from CTHs and CWHs of PAs (Kalpavriksh *et al.* 2013). It is not surprising that this decision had received the disapproval of many CSOs, who have expressed their concern to the MoEF, urging that relocation should be stopped until there is a detailed investigation on ongoing violations, including prior recognition of rights before relocation, and processes of co-management and participation implemented in PAs.

Table 5.1 Violations and shortcomings in relocation from critical tiger habitats of Tiger Reserves as evident from NTCA relocation committee reports

Legal requirement	Relevant provisions	Relocation violations and shortcomings as evident from NTCA 2010 Relocation Committee report
Prior recognition of rights	38V (5) i of WLPA	No rights' recognition happening. (Achankmar, Panna, Sariska reports mention this explicitly)
Documentary evidence of 'irreversible damage' and that co-existence is not possible	38V (5) ii and iii of the WLPA	No reports of evidence/ basis used for arriving at such a conclusion
Informed consent from gram sabha	Section 38V (5) v of the WLPA	Mostly individual consent, no evidence of 'gram sabha consent' and 'informed' consent (Sariska reports mentions lack of FRA information; Achanakmar report mentions only individual consent)
Compensatory package	38V (5) iv of WLPA, protocol for relocation from CTHs	Land option and compensation for actual assets not being made available (Sariska), Inconsistencies in package (Panna, Ranthambore, and Bandhavgarh), Exclusion from lists (Achanakmar, Panna, and Bandhavgarh)
Stable livelihood and basic facilities in rehabilitated villages	38V (5) iv of WLPA	Insufficient water facilities leading to scarcity for drinking and irrigation (Achanakmar and Panna)

Source: Compiled by authors based on the NTCA 2010 relocation committee reports.

Community Forest Resource Rights

According to official figures (MoTA 2013), 19,680 CFR titles have been recognized over more than 6,50,716.84 acres.[19] Despite the potential of the CFR provisions, it has been noticed, however, that effective implementation is taking place only for a few communities and only in a few states (Desor 2013a). Implementation is especially poor in PAs across the country, where it has actively been discouraged by the FD. Till recently there has also been a lack of clarity among the ground level activists on whether CFR provisions are applicable in PAs.

CFRs, however, have a huge potential in creating a diversity of conservation areas being managed and governed by multiple actors. Such potential is right now visible in a few sites outside PAs and still fewer sites within PAs. Outside PAs, Gadchiroli district of Maharashtra appears to be far ahead of other areas in terms of community claims to forests under the FRA. 3,72,658.17 acres have been claimed and 805 titles handed out as per data presented by tribal commissioner's office during the state-level CFR workshop held in Mumbai on 22 January 2013. Mendha-Lekha, one of the first two villages to receive CFR titles, has subsequently also set an example for a number of processes that can be initiated, using this law towards a more equitable and sustainable local governance and economy. In another village, Murumbodi falling under the Bhikarmaushi gram sabha after recognition of fishing rights over the village lake in 2011, villagers protested against leasing of the lake by the block development officer to the Co-operative Fishing Society of the Dihvar community without any discussions with the villagers. They also demanded benefit-sharing. The Society had to comply to their demand and 50 per cent of the benefits are at present with the villagers (Jathar and Pathak 2012). At the same time, the reasons for such success can be varied, and require an environment of facilitation or a history of community-based efforts. For instance, Mendha-Lekha had moved towards self-rule and forest conservation nearly two decades prior to the Act. Gadchiroli district has also had a history and currently an environment of collective civil society action. A few communities have also mobilized. This, along with a pro-active tribal and revenue department have contributed to effective implementation of the Act (Thatte and Pathak 2013).

In some areas where the process of recognition of rights was essentially government-driven (as in Godda and West Singhbhum districts of Jharkhand) and had taken place without active engagement of gram sabhas as required, not much has changed after recognition of rights (Tenneti 2013).

Slowly, examples are emerging where the CFR provisions are being used in and around PAs to move towards a more democratic resource planning. In BRT Wildlife Sanctuary (also declared a tiger reserve in 2011) where CFRs have been recognized in the CTH area, the Soliga adivasis have developed a three-part plan for collaborative management for conservation, livelihoods, and governance structures, with some landscape-level meetings. However, the villages demand recognition of rights for all the villages of the sanctuary before implementation of the plan (Madegowda *et al.* 2013). In Maharashtra, 45 CFR claims have been filed and titles received in and around the Melghat Tiger Reserve. KHOJ, a local group that has facilitated the process has also moved ahead by drafting and implementing a co-existence plan in some of these villages, which lie in the buffer zone of the reserve. These villages have formed committees for wildlife management under Section 5 of the Act and Rule 4(1)e. In the Yawal wildlife sanctuary in north Maharashtra, the local tribal organization called Lok Sangharsha Morcha (LSM) has used the provisions of the PESA, FRA, and WLPA, to initiate a process of verification of rejected claims under FRA, identification of illegal occupations causing damage to the PA, and micro-planning for social development and conservation in 17 villages inside and around the sanctuary. Although the process has a long way to go, it is a beginning made possible by a strong local resistance movement, using the above-mentioned provisions.

Similarly 33 villages of the Shoolpaneshwar WLS in Gujarat have received CFR titles. Now meetings are being organized for post-recognition conservation and management, with the help of Arch-Vahini, a local CSO. There are also other PAs such as the Tadoba Andheri Tiger Reserve in Maharashtra and the Kumbhalgarh Wildlife Sanctuary in Rajasthan, where although there has been no recognition, people have been filing claims and initiating processes towards such recognition (Desor 2013a).

While CFRs include rights of ownership over NTFP, there have been certain challenges in exercising such rights in PAs. This is evident from the incident in the BRT Tiger Reserve when a range officer confiscated honey collected by the gram sabha of Hosapodu village in Chamarajanagar taluk (Madegowda *et al.* 2013). The village, after receiving CFR rights, had initiated honey-processing and local marketing as an activity independent of the LAMPS cooperative. This happened after discussions with the local NGOs (Zilla Budakattu Girijana Abhivruddhi Sangha [ZBGAS] and Soliga Abhivruddhi Sangha [SAS]) as well as the conservator of BRT. Yet, on 9 May 2013, the range forest officer of the Punanjanur Range seized the honey stored in the village community hall and destroyed the processing equipment, filing a forest offence case under the Karnataka Forest Act, 1963 along with a plea for immediate disposal of the honey. The gram sabha appealed to the Court, claiming ownership of the honey under Section 3(1) (c) of the FRA and requested the Court to stay the disposal of honey and for it to be returned to them. On 23 May 2013, the Yellandur Court ordered that the honey be returned to the interim custody of the gram sabha, stating that the gram sabha is empowered to collect minor forest produce for their livelihoods, leading to subsequent release of the confiscated honey.[20]

CFRs are also being used, especially in the post-recognition scenario, as a tool for demanding more democratic processes of decision-making on forests. An example is the case of local opposition to coupe-felling in forests of Baiga Chak (Dindori district, Madhya Pradesh) using CFRs as a means of assertion. Their argument for these protests is that coupe-felling is leading to degradation of their customary forests and any such felling cannot take place without gram sabha consent. They have been successful in stalling felling operations in some CFR forests (Desor 2013b; Kothari and Desor 2013). The FRA and associated circulars are also being used in some parts of the country by forest-dependent communities to protest against development activities seriously affecting their livelihood base or other interactions with forests, and/or to assert their right for a greater democratic engagement with the process of forest diversion for 'development' projects. The struggle of the Dongria Kondh against Vedanta company's proposal to mine in its habitat (including a sacred mountain) in Odisha is well known and it uses FRA as one of its prime tools for assertion of community forest rights (Patnaik 2013). In Singrauli in Madhya Pradesh, first-stage clearance for coal-mining was given based on what NGOs claim were fake gram sabha resolutions. Simultaneously, the CFR claims process is underway in 62 villages with many villages protesting against neglect of their forest livelihoods due to the diversion (Desor 2013b; Kohli *et al.* 2012). In Thane in Maharashtra, villagers are fighting against the illegal construction of the Kalu dam (being constructed to provide water to Navi Mumbai) with the help of the Shramik Mukti Sanghatna. The dam was being constructed without completing processes under FRA. Many affected villages have filed CFR claims, thus asserting their community rights. Though the project proposal was rejected by the central government, a fresh proposal has been presented by the project proponent to the government in March 2013 and this has been recommended by the FAC (Forest Advisory Committee) on 4 April 2013, despite non-completion of the FRA process. In most instances, forest land continues to be diverted for non-forestry purposes such as mining and power projects without rights recognition and gram sabha consent. Many such instances are also within or around PAs. An example is the forest clearance granted to the windmills project of Enercon-India in 2009. These reserved forests that were cleared were within the boundaries of 14 villages in Pune district and situated within a 10-km radius of the Bhimashankar WLS. Despite this, no consent was taken from the 14 village gram sabhas.[21]

An analysis of official efforts towards implementation of the provisions related to co-existence, and a number of associated events, including judicial processes, indicates an overall lack of seriousness in moving towards co-existence and co-management in PAs and hence a more democratic and diverse PA governance system. This is particularly so in CTHs and CWHs. Even in other PAs or in forests outside PAs where CFR provision has been used by the local communities to assert their rights and responsibility of what they call their community forest resource, there is no linkage between the practices to be followed in these areas and the provisions of the WLPA. Despite the potential of CFR provisions in moving towards shared governance systems within and outside PAs, there is an apparent lack of synergy in implementation of the two laws. Neither of the Acts clarifies the relationship between the people who would govern these areas and the state, which has governed them so far and continues to have certain legal jurisdiction over them. The mindset that conservation cannot take place with people inside PAs continues to be very strong, leading towards prioritizing relocation over exploring co-existence in areas considered important for species and their habitat. Directions in good governance (Box 5.1) include a consistent strategic vision for conservation grounded on values mutually agreed by all rights-holders and stake holders and ensuring compatibility with plans and policies of other sectors in the broader landscape/seascape. Both of these seem to be missing. The following section deals with some of the challenges which have led to the current state of implementation despite the legal provisions being strong, continuous pressure from the ground, regular supporting circulars from the MoTA, among other reasons, which should have led to a diversity of PA and conservation governance models in the country.

Challenges Obstructing Effective Implementation of Such Provisions

Lack of Clear Definitions and Explanation within the Law

The WLPA states that exploring co-existence is a key objective of declaring buffer areas of tiger reserves. However, as mentioned in the sections above, it does not define co-existence. Various CSOs have argued for and the MoEF has agreed to the need for a set of guidelines on co-existence, but this has not been done so far. On the other hand, the protocol on relocation has already been prepared and is in use, clearly reflecting that the priority for implementing agencies is relocation.

Both the FRA and WLPA mention creation of 'inviolate' areas for effective wildlife management. For this purpose, rights of the local communities are to be modified and relocation can be carried out. However, as also mentioned earlier, the term 'inviolate' has not been defined in either of the Acts. There is an urgent need for it to be clarified. Does it mean 'no-use' or 'human-free' areas, or could it also include 'compatible uses' that do not violate conservation objectives. CSOs have given specific suggestions based on wide consultations to the MoEF in which they have suggested broad ecological criteria,[22] which could be used for identifying CWHs. While these criteria provide greater conceptual clarity, many practical complications could arise in implementing these. In a data-deficient scenario, it may be necessary to use thumb rules for decision-making given the urgency of notifying such areas, with the proviso that as data gaps are filled in, adaptive management can review and revise earlier conclusions and conservation recommendations.

There is also a lack of clear direction or guidelines on what should be the progression of the decision-making process while establishing CTHs and CWHs. Should it be first establishment of rights, then exploring co-existence models, and then if proven to be not possible, relocation? If yes, as suggested in the Acts, then the mechanism by which this progression is ensured is not clear. This lack of clarity has led to a situation where it is being assumed that co-existence is simply not possible, and relocation is being insisted upon, at least in the case of CTHs.

Lack of Knowledge, Capacity, and Forums for Monitoring and Evaluation

Lack of adequate ecological and socio-economic knowledge and lack of adequate systems for incorporating traditional and local knowledge is a challenge for identification of areas important for wildlife, and effective conservation management. Government officials, local communities, and CSOs often work in isolation, with little interface to synergize knowledge and experience and develop a long-term vision for governance and management of PAs and other conservation sites. This leads to a situation where the FD identifies an area, decides whether or not co-existence is possible (the knowledge and information basis used for this conclusion is often unclear), and finalizes a management strategy. In the absence of any formal forum for intervention and/or participation, the CSOs and local communities are left to express their views through articles, protest letters, mass movements, rallies, and other such methods.

Officially, none of the CTHs or PAs have carried out an extensive planning process taking into account all the ecological and socio-economic data and all rights-holders and stakeholders. Nor does an inclusive system exist that could constantly monitor the implementation of laws and the progress of jointly established objectives and management goals towards an adaptive management strategy.

Clearly, a one-time planning exercise is not adequate to ensure that implementation of such concepts meets its objectives. Unanticipated problems always crop up. Even the most well-made plans do not necessarily work out perfectly, and local ecological or social situations may change in unexpected ways. It is, therefore, necessary to bring in a continuous monitoring, evaluation, and feedback process, which is fully participatory, and contains independent oversight. Such a process could point to crucial changes in management strategies, governance, boundaries, or other parameters. This also implies that the governance and management institution must be flexible and open to such changes. This is a critical issue and if the process is to move quickly, then adaptive management must be built into the decision-making system.

Ambiguity in Governance Because of Lack of Clarity in the Relationship Between FRA and WLPA

The forest governance regime in PAs is currently ambiguous as the precise relationship of the FRA with the WLPA is unclear, leading to possible confusion on the ground of what action can be taken if a right granted under the Act violates a provision of the WLPA. A conflict could arise in a situation where the management practices/beliefs of the village committee recognized under FRA are in contradiction with the management practices of the FD recognized under WLPA or the other way around. For instance, traditional use of fire, shifting cultivation, and extractive use for commercial purposes are potential points of conflict; these are necessarily detrimental to conservation, as official mindsets would have us believe (nor, of course, can they be unregulated). How this situation will be resolved and what kind of supportive and regulatory mechanism needs to be in place, is not clear from the existing provisions in the two laws and will require further clarification. Additionally, although FRA empowers gram sabhas to ensure conservation and to set up committees for this purpose, it is not clear what happens if, for instance, the rights to harvest of a non-timber forest resource adversely affect its conservation status, in cases where no limits based on ecological criteria are set for resource extraction. Conversely, if the FD imposes conservation or management regimes on communities who demonstrate or feel that such regimes are detrimental to biodiversity (for example, a ban on fire where regulated fire is helpful to local biodiversity), by what mechanism would such feedback and the appropriate action be taken?

There are also some more general post-rights' recognition issues needing resolution, for the forest landscape as a whole. At the policy/ governance level, appropriate institutional arrangements, granting of powers to gram sabhas akin to those of the FD, sharing of such powers between gram sabhas and the FD and the relationship between gram sabha plans and FD's working plans, or other such arrangements, still remain to be worked out (Joint MoEF–MoTA Committee 2010). This is especially relevant in view of the continued operation of FD control and works, even where communities are objecting to these, such as plantations and working plan activities (for example, in Rajasthan and Odisha, the government is collaborating with funders like the Japan International Cooperation Agency (JICA) to implement forestry projects under which plantations are carried out in community land claimed under the FRA).

There is also an absence of planning and institutional structures for conservation and management at a landscape level, that could bring together gram sabhas (or village-level forest management committees), the FD, the tribal department, other relevant departments, and local civil society organizations. Such agencies could monitor and guide forest/ wildlife conservation and enjoyment of CFR rights, facilitate landscape-level planning and implementation, and facilitate convergence of various schemes towards these objectives (Joint MoEF–MoTA Committee 2010). Moreover, there is a lack of convergence between different forest-related laws and policies, partly because the government has not issued any clarification on the relative powers, roles, functions, and responsibilities of the gram sabha and the FD, despite clear recommendations in this regard from a number of sources including the Joint MoEF–MoTA Committee and the National Advisory Committee (NAC).

Conservative Attitudes and Reluctance to Share Power

Anti-democratic attitudes and reluctance to share power amongst government agencies, are one of the key challenges slowing down the decentralization of decision-making powers, or the move towards collaborative or community-based conservation. This is already in evidence, for instance, in the abysmal implementation of the panchayati raj constitutional amendments, especially those pertaining to Scheduled Areas, nearly two decades after they were promulgated. The same is true of the FRA, especially in PAs. State governments and their local departments have simply been reluctant to share administrative and financial powers. There is also a continued belief in conservation by exclusion as indicated by the violation of FRA and in general human rights, in CTHs mentioned above. There have been no fundamental changes in the Indian Forest Services curriculum to cover legislative developments like the FRA, which could support a more democratic model of conservation (Kothari 2013).

This difficulty in implementing the FRA can be understood by recognizing that forests and wildlife habitats in India are spaces that have been largely controlled by the Indian state, first colonial then independent. It could be argued that the purpose of implementing the FRA within this state-centred context is a struggle over governmentality, within which certain types of exclusionary conservation models have been historically deployed to further the government's larger aim of managing the lives of its constituents (Foucault *et al.* 1991). Such centralized control gets internalized both within the constituents of the state, and also amongst many people outside the state, creating enormous resistance to paradigm shifts requiring decentralized redistribution of power (Agrawal 2005; Bryant 1998; Foucault 1991).

However, despite the larger challenges of moving the state, there have been a number of officials within the environment bureaucracy who have shown different ways of doing things on the ground (for example, in promoting tribal livelihoods linked to the Periyar Tiger Reserve in Kerala, or providing employment options to pastoralist communities in conservation areas of Sikkim). This has resulted in policy level changes with more conviction. Additionally, other wings of government, such as the MoTA in the case of the FRA, in the period from 2011 onwards have also taken more proactive role in influencing conservation policy.

Outside of the state, there is an almost equally powerful force of resistance: a strong section of formal sector conservationists continuing to believe in or espouse exclusionary, top-down conservation. The debate on tiger protection is dominated by such people, as they dogmatically hold to the assertion that only 'inviolate' (read: human-free, except tourists) areas will do if the tiger has to be saved. This automatically leads to one of the problems mentioned above, that even where the law mandates an exploration of co-existence, it is relocation that gets all the attention, budgets, and political will. Even increasing scientific evidence of the possibilities of co-existence within specified limits of human activity, and the ethical imperative of democratic decision-making (which they may even assert in other contexts), has not substantially shifted this mindset amongst a small but powerful section of the conservation community.

Development Context

It is important to look at the larger landscape (and seascape) within which each PA or CFR is located. Such areas will not survive long as islands within a landscape of unsustainable development, for sooner or later the forces of demand and conversion, and long-range phenomena like pollution and climate change, will also enter these. Possibly the greatest threat to both wildlife and forest dependent livelihoods in India today is India's economic growth model. In the pursuit of a double-digit growth rate, environmental priorities have been brushed aside in the last few decades. Since 1991 in particular, with economic liberalization and globalization, this process has greatly intensified (Shrivastava and Kothari 2012). There is pressure for dilution of norms of gram sabha consent of affected villages in the process of forest clearance to speed up clearances and remove 'bottlenecks' to the inflow of huge investments. Such a pressure has led to the formation of a Cabinet Committee of Investments and recent exemption (through an MoEF circular on 5 July 2013) from the need of gram sabha consent for linear projects such as roads, highways, and canals (Kothari 2012).

Such policies, lacking a focus on areas outside of the PAs, create an ever-increasing competition between the local people, urban needs, industrial needs, and needs of wildlife, particularly the less charismatic ones, for the fast-reducing ecosystems and resources therein. As fuel and fodder resources outside a PA diminish, people inevitably try to enter into the PA; as mining and hydro-electricity sites reduce outside, commercial agencies inevitably ask for the opening up of PAs. In a situation where no local community stake has been created in the continuation of the PA, government agencies and conservation NGOs will find it impossible to contain these forces. Already several PAs are threatened with such forces (for example, several dozen have mining going on or proposed within or just adjacent to their boundaries). Many PAs, from where traditional communities are being moved out, are being opened up for large-scale commercial tourism in the name of ecotourism. Such pressures will only increase as globalization makes further inroads into India unless seriously challenged by both environmental and human rights groups working with local communities. Also, focusing on merely the PA network for conservation, as seems to be the case in official implementation, could leave out many crucial wildlife habitats that are currently outside the PA network, such as most of our marine ecosystems. It will also impact migratory species and species with large home ranges. Therefore, what is lacking and is urgently required, is planning at landscape (and seascape) level where natural resources and biodiversity both within and outside of PAs are conserved and managed (TPCG and Kalpavriksh 2005). Yet, the larger developmental process has no interest in or time for conservation, and even less for participatory conservation.

***

The FRA 2006 and WLPA 2006 (through its newly inserted provisions), provide significant legal opportunities for Indian conservation to take steps towards more democratic and effective conservation. This includes participatory and knowledge-based ways of identifying and managing special areas for wildlife, strategies for a wide range of governance arrangements depending on the local context, a focus on inclusion and co-existence, strong legal and social protection of critical conservation areas against destructive development processes, and enlarging the scope of conservation to include entire landscapes and seascapes rather than only islands of protection. These could also help lead to greater livelihood security, and more democratic and robust governance, across such scapes. But to enable this potential to be met, civil society organizations and those within the government who are sensitive to such issues, will need to join hands, and also win the trust of and help empower local communities who have been thus far at the receiving end of both the 'development' and the 'conservation' sticks. Further legislative changes are needed, especially to strengthen (with reference to Box 5.1) the direction, accountability, and performance. There will need to be continued shifts in mindset and attitudes, through dialogue, assertion of community controls, changes in the training of the forest bureaucracy, and greater public participation. Perhaps most important, the conservation 'community' will need to seriously challenge the currently dominant economic development model in the search for sustainable, equitable alternative models.

This is a very difficult task, but not impossible. In many ways India's conservation story has been one of swimming against the tide with some remarkable stories of reversing processes of extinction. We need a similar resolve, this time with a much stronger knowledge and democratic base, to achieve lasting conservation and livelihoods security.

Notes

1. Globally these are now termed Indigenous Peoples' and Local Community Conserved Territories and Areas (ICCAs), mirroring (and having emerged from) the concept and practice of community conserved areas in India; for conceptual treatment, case studies, and analytical reports on ICCAs, see www.iccaforum.org. See www.iccaconsortium.org for more information on ICCAs and Pathak Broome and Dash (2012) and Pathak (2009) for Indian examples.

2. Circular no. 13-1\90-FP of Government of India, Ministry of Environment and Forests, 18 September 90. Addressed to the forest secretaries of all states and union territories.

3. See http://www.iucn.org/about/work/programmes/gpap_home/gpap_capacity2/gpap_parks2/?2137/2003-Durban-World-Parks-Congress, accessed January 2014.

4. The WLPA amendment introducing the category of CTHs was dated 4 September 2006 and the FRA introducing the category of CWHs was passed on 29 December 2006.

5. The Future of Conservation in India is 'a network of ecological and social organizations and individuals committed to effective and equitable conservation of biodiversity. FoC is not an organization, but a forum where organizations and individuals can meet, dialogue, and take joint actions.' (See http://kalpavriksh.org/index.php/conservation-livelihoods1/networks/future-of-conservation.html.)

6. See 'Proposed Guidelines on Identification of Critical Tiger Habitats, Co-existence and Relocation related to Tiger Reserves (in pursuance of the WLPA as amended in 2006)', submitted in September 2007, by Ashoka Trust for Research in Ecology and the Environment (Bangalore), Council for Social Development (Delhi), Himal Prakriti (Munsiari), Kalpavriksh (Delhi/Pune), Samrakshan (Delhi), SHODH (Nagpur), Vasundhara (Bhubaneshwar), Wildlife Conservation Trust (Rajkot), WWF-India (Delhi).

7. MoEF circular F. No. 1-39/2007 WL1 (pt) dated 7 February 2011 by Deputy Inspector General Prakriti Srivastava to all chief wildlife wardens.

8. Information based on a summary of RTI data from 2010 received by Sreetama Guptabhaya, member of Kalpavriksh, Delhi/Pune.

9. Response by MoEF minister, Rajya Sabha on 8 May 2012 to question no. 3,439 asked by Ramachandra Khuntia on the Notification of Critical Wildlife Habitats in Odisha.

10. Vide its circular No. 1501/11/2007-PT (Part) to all relevant states.

11. See FoC, 2007, 'Proposed Guidelines for Identification of Critical Wildlife Habitats in National Parks and Wildlife Sanctuaries Under Scheduled Tribes and Other Traditional Forest Dwellers (Recognition Of Forest Rights) Act, 2006, submitted on December 2007 to NTCA.

12. See MoEF, 2011, 'Summary records of the meeting held on 4 March 2011 to discuss the issue of revised guidelines for declaration of Critical Wildlife Habitats', 23 March.

13. SLP (C) no. 21339/2011 (*Ajay Dubey vs. Union of India & Ors*).

14. In August 2012, Kalpavriksh filed an intervention in the *Ajay Dubey* case regarding violations of the FRA and WLPA being caused by the rushed process of notifications of buffer areas of tiger reserves. It intends to extend the intervention to cover overall issues of violations in the process of notification of tiger reserves.

15. See http://kalpavriksh.org/images/CLN/FOC/Relocation%20protocol_Comments.pdf for FoC's comments on the draft relocation protocol. The protocol was finalized without incorporating most of these points.

16. See Kalpavriksh, 2011, 'Recognition of Rights and Relocation in Relation to Critical Tiger Habitats'.

17. Unpublished reports provided by NTCA, Government of India, in 2011.

18. However, it should be noted that there are many gaps and ambiguities in relation to these official figures (Desor 2013a).

19. Information on the Hosapodu incident and the court order was provided by Archana Sivaramakrishnan (Keystone Foundation) and Mahadesha on behalf of the Hosapodu gram sabha.

20. From the letter submitted to Minister of Tribal Affairs on 19 September 2011 by members of Kalpavriksh (Neema Pathak, Saili Palande, and Pradeep Chavan) on the subject 'Permission granted to Andhra Wind Power Project Enercon-India, Maharashtra, based on misrepresented facts and in violation of provisions of the Forest Rights Act, 2006'.

21. See report of the *National Workshop on Critical Tiger Habitats and Critical Wildlife Habitats*, May 2008, http://kalpavriksh.org/index.php/conservation-livelihoods1/networks/future-of-conservation.html.

References

Adams, W.M., R. Aveling, D. Brockington, B. Dickson, J. Elliott, J. Hutton, D. Roe, B. Vira, and W. Wolmer. 2004. 'Biodiversity Conservation and the Eradication of Poverty', *Science*, 306 (5699): 1146–9.

Agrawal, A. 2005. 'Environmentality: Community, Intimate Government, and the Making of Environmental Subjects in Kumaon, India', *Current Anthropology*, 46 (2): 161–90.

Borrini-Feyerabend, G., N. Dudley, T. Jaeger, B. Lassen, N. Pathak, A. Phillips, and T. Sandwith. 2013. *Governance of Protected Areas: From Understanding to Action*. Best Practice Protected Area Guidelines Series No. 20, Gland, Switzerland: International Union for Conservation of Nature.

Brechin, S.R., P.R. Wilshusen, C.L. Fortwangler, and P.C. West. 2002. 'Beyond the Square Wheel: Toward a More Comprehensive Understanding of Biodiversity Conservation as Social and Political Process', *Society & Natural Resources*, 15 (1): 41–64.

Brockington, D. 2002. *Fortress Conservation: The Preservation of the Mkomazi Game Reserve, Tanzania*. Bloomington: Indiana University Press.

Brockington, D., J. Igoe, and K.A.I. Schmidt-Soltau. 2006. 'Conservation, Human Rights, and Poverty Reduction'. *Conservation Biology*, 20 (1): 250–2.

Bryant, Raymond L. 1998. 'Power, Knowledge and Political Ecology in the Third World: A Review'. *Progress in Physical Geography*, 22 (1): 79–94.

Das, P. 2007. 'The Politics of Participatory Conservation—The Case of Kailadevi Wildlife Sanctuary, Rajasthan', in G. Shahabuddin and M. Rangarajan (eds), *Making Conservation Work*. New Delhi: Permanent Black.

Desor, S. (ed.). 2013a. *Citizens' Report 2013 on Community Forest Rights under Forest Rights Act*. Pune: Kalpavriksh and Bhubaneshwar: Vasundhara, with New Delhi: Oxfam India, on behalf of Community Forest Rights Learning and Advocacy Process; http://kalpavriksh.org/images/LawsNPolicies/Community%20Forest%20Rights%20under%20FRA%20Citizens%20Report%202013.pdf.

———. 2013b. 'Baiga Chak,' in S. Desor (ed.), *Citizens' Report 2013 on Community Forest Rights under Forest Rights Act*. Pune: Kalpavriksh and Bhubaneshwar: Vasundhara, with New Delhi: Oxfam India, on behalf of Community Forest Rights Learning and Advocacy Process: 41–9.

Dudley, N. (ed.). 2008. 'Guidelines for Applying Protected Area Management Categories'. Gland, Switzerland: International Union for Conservation of Nature (IUCN); http://data.iucn.org/dbtw-wpd/edocs/PAPS-016.pdf.

Future of Conservation (FoC). 2007. Comments on 'Guidelines to Notify Critical Wildlife Habitats including Constitution and Functions of the Expert Committee, Scientific Information required and Resettlement

Matters Incidental thereto'. New Delhi: MoEF; http://kalpavriksh.org/images/CLN/FOC/foc20075.pdf.

Foucault, M., G. Burchell, C. Gordon, and P. Miller (eds). 1991. *The Foucault Effect: Studies in Governmentality*. Chicago: University of Chicago Press.

Gadgil, M. and R. Guha. 1993. *This Fissured Land: An Ecological History of India*. Berkeley: University of California Press.

Ghimire, K.B., K.B. Ghimire, and M.P. Pimbert (eds). 1997. *Social Change and Conservation: Environmental Politics and Impacts of National Parks and Protected Areas*. London: Earthscan.

Graham, J., B. Amos, and T. Plumptree. 2003. 'Governance Principles for Protected Areas in the 21st Century', Discussion Paper, Phase 2, Institute on Governance in collaboration with Parks Canada and Canadian International Development Agency (CIDA).

Grazia, B.F., N. Dudley, T. Jaeger, B. Lassen, N. Pathak Broome, A. Phillips, and T. Sandwith. 2013. *Governance of Protected Areas: From Understanding to Action*. Gland, Switzerland: International Union for Conservation of Nature [IUCN]).

Institute on Governance. 2002. 'Governance Principles for Protected Areas in the 21st Century'. Discussion paper for Parks Canada, Ottawa, Canada.

Jathar, R. and N. Pathak. 2012. 'Maharashtra', in S. Desor (ed.) 2012. *A National Report on Community Forest Rights under Forest Rights Act: Status and Issues* for Community Forest Rights Learning and Advocacy Process. Pune: Kalpavriksh; Bhubaneshwar: Vasundhara and New Delhi: Oxfam India, pp. 41–9.

Joint MoEF–MoTA Committee. 2010. 'Future Structure of Forest Governance', in *Manthan: Report of National Committee on Forest Rights Act*. New Delhi: Ministry of Environment and Forests and Ministry of Tribal Affairs.

Kalpavriksh. 2011. 'Recognition of Rights and Relocation in Relation to Critical Tiger Habitats. Pune: Kalpavriksh; http://www.kalpavriksh.org/images/Documentation/Advocacy/Recognition%20of%20Rights%20and%20Relocation%20in%20relation%20to%20CTHs.pdf.

Kalpavriksh *et al.* 2013. Open letter to MoEF, 'No more funds for relocation without establishing a democratic process!' (cited1 July2013); http://kalpavriksh.org/images/LawsNPolicies/letter%20of%20protest%20against%20more%20funds%20for%20relocation%20%20july%202013.pdf.

Kohli, K., A. Kothari, and P. Pillai. 2012. *Countering Coal? Community Forest Rights and Coal Mining Regions of India*. Delhi/Pune: Kalpavriksh and Bengaluru: Greenpeace India.

Kothari, A. 2012. 'National Investment Board: One More from Pandora's Box'. *Economic and Political Weekly*, 47 (45): 10–13.

———. 2013. 'Forestry Education and Training: Time for Major Reforms. *Indian Forester*, 139 (4): 324–30.

Kothari, A., N. Singh, S. Suri. 1996. *People and Protected Areas: Towards Participatory Conservation in India*. New Delhi: Sage Publications.

Kothari, A. and S. Desor. 2013. 'Baiga's battle', *Frontline*, 1 May: 93–5

Lasgorceix, A. and A. Kothari. 2009. 'Displacement and Relocation of Local Communities from Protected Areas in India: A Synthesis and Analysis of Case Studies', *Economic and Political Weekly*, 44 (49), 5 December: 37–47.

Madegowda, C., N. Rai, S. Desor. 2013. 'BRT Wildlife Sanctuary, Karnataka', in S. Desor (ed.), *Citizens' Report 2013 on Community Forest Rights under Forest Rights Act*. Pune: Kalpavriksh; Bhubaneshwar: Vasundhara and New Delhi: Oxfam India on behalf of Community Forest Rights Learning and Advocacy Process.

Ministry of Environment and Forests (MoEF). 2009. 'Diversion of Forest Land for Non Forest Purposes under the Forest (Conservation) Act 1980—Ensuring Compliance of the Scheduled Tribes and Other Traditional Dwellers—Recognition of Forest Rights Act, 3 August, F.No. 11-9/1998 FC- PT.

Ministry of Tribal Affairs Government of India (MoTA). 2013. 'Status Report on Implementation of the Scheduled Tribes and Other Traditional Forest Dwellers (Recognition of Forest Rights) Act, 2006 [for the period ending 30 June 2013]'. New Delhi: Ministry of Tribal Affairs Government of India.

Pathak, N. (ed.). 2009. *Community Conserved Areas in India—A Directory*. Pune/Delhi: Kalpavriksh.

———. 2013. 'No Basis Yet for Declaration of Critical Wildlife Habitat: A Report from Yawal WLS', *Protected Area Update*. 19 (3): 19.

Pathak Broome, N. and T. Dash 2012. 'Recognition and Support of ICCAs in India' in H. Jonas, A. Kothari, and H. Shrumm (eds), *Recognising and Supporting Conservation by Indigenous Peoples and Local Communities: An Analysis of International Law, National Legislation, Judgements, and Institutions as They Interrelate with Territories and Areas Conserved by Indigenous Peoples and Local Communities*. Bengaluru: Natural Justice and Pune/Delhi: Kalpavriksh.

Pathak Broome, N., S. Bhutani, R. Rajagopalan, S. Desor, and M. Vijairaghavan. 2012. 'An Analysis of International Law, National Legislation, Judgements, and Institutions as They Interrelate with Territories and Areas Conserved by Indigenous Peoples and Local Communities', Report No. 13. Bangalore: Natural Justice and Pune/Delhi: Kalpavriksh.

Patnaik, S. 2013. 'Dongria Kondhs Have Shown the Way'. *The Hindu*, 21 August; http://www.thehindu.com/news/national/other-states/dongria-kondhs-have-shown-the-way/article5044669.ece.

Saberwal, V.K., M. Rangarajan, and A. Kothari. 2001. *People, Parks and Wildlife: Towards Co-existence*. New Delhi: Orient Longman.

Shahabuddin, G. 2010. *Conservation at the Crossroads: Science, Society and the Future of India's Wildlife*. New Delhi: Permanent Black.

Shrivastava, A. and A. Kothari. 2012. *Churning the Earth: The Making of Global India*. UK: Penguin.

Tenneti, A. 2013. 'Jharkhand', in S. Desor (ed.), *A National Report on Community Forest Rights under Forest Rights Act: Status and Issues for Community Forest Rights Learning and Advocacy Process*. Pune: Kalpavriksh; Bhubaneshwar: Vasundhara and New Delhi: Oxfam India.

Thapar, V. 2012. 'Tourism Did Not Kill the Tiger', *The Indian Express*, 29 August.

Thatte, M. and N. Pathak. 2013. 'Maharashtra' in S. Desor (ed.), *Citizens' Report 2013 on Community Forest Rights under Forest Rights Act*. Pune: Kalpavriksh, and Bhubaneshwar: Vasundhara, with New Delhi: Oxfam India on behalf of Community Forest Rights Learning and Advocacy Process.

Technical and Policy Core Group (TPCG) and Kalpavriksh. 2005. 'Securing India's Future: Final Technical Report of the National Biodiversity Strategy and Action Plan'. Prepared by the NBSAP Technical and Policy Core Group. Delhi/Pune: Kalpavriksh.

United Nations Environment Programme (UNEP). 2012. *GEO5: Global Environment Outlook Environment for the Future We Want*. Nairobi: United Nations Environment Programme; http://www.unep.org/geo/geo5.asp.

Vagholikar, N. and S. Bhushan. 2000. Panchayat (Extension to Scheduled Areas) Act, (1996). A study of complementarities and contradictions with laws and policies relating to natural resource conservation. Kalpavriksh, unpublished.

Wani M. and A. Kothari. 2007. 'Protected Areas and Human Rights in India—The Impact of the Official Conservation Model on Local Communities', *Policy Matters*, July, 15: 100–14.

GOVERNING THE FOREST–NON-FOREST BOUNDARY

SHOMONA KHANNA

Boundaries of Forest Land

The Godavarman Case and Beyond[1]

This chapter sets out to examine the manner in which the Supreme Court of India, through the Godavarman case[2] and the World Wildlife Fund (WWF) case,[3] has impacted the discourse as well as the praxis regarding the boundary between forest areas and non-forest areas. Both these litigations, filed almost contemporaneously in 1995, have been treated by the court as 'continuing mandamus' and come up for monitoring and hearing at regular intervals. The Godavarman case, in particular, has resulted in numerous innovations being made by the Court itself, both in terms of setting up implementation structures, as well as in the area of legal interpretation. The sitting of the Special Forest Bench at regular intervals for the last 19 years in the Godavarman case has changed the dynamics of the operation of forest law, and consequently of forest governance, in this country. A detailed critique of such innovations, and the obvious incursion of the judiciary into the functioning of the executive and the legislature, is beyond the scope of the chapter. I focus here on the question of how forest boundaries are defined and understood as a result of the Godavarman case, with some reliance also being placed on the WWF case, together referred to as the forest case(s).

One point for analysis is how the judgment dated 12 December 1996[4] and the numerous interim orders passed subsequently in the forest cases have resulted in the physical expansion of the forest boundary and, therefore, of the control of the state Forest Departments (FDs) and of the Supreme Court over larger areas of land. Two other processes are equally important: the process of forest 'conversion' under the supervision of Courts and the expansion of the hegemony exercised by the Supreme Court and the state executive, through the Union Ministry of Environment and Forests (MoEF) and the state FDs, within the existing and expanded forest boundaries. The chapter will attempt to examine these three processes. It relies upon many years of observation and information gathering. Unfortunately, access to court records that would provide a wealth of information is next to impossible as such records are not public documents.

At the outset, it must be mentioned that the forest law regime in India, as articulated in the Indian Forest Act, 1927, and the numerous state-level legislations flowing out of it traces its roots to a colonial government which was focused on maximizing Britain's industrial needs. It is, therefore, characterized by exclusion, extraction, and centralization of power in the state FDs, which themselves carry a colonial lineage. The Forest Conservation Act, 1980, (henceforth FCA) purportedly enacted to control the massive depletion of forests taking place in the country due to unchecked commercial activities in forests, carried this command and control approach to forests forward and intensified centralized decision-making. The centrality of the issue of exclusion and inclusion, resonating almost immediately in illegality and legality, to the subject of forest law and governance cannot be overemphasized. This is why the creation of boundaries in the Godavarman case is so important. It is this underlying context within which the impact of the Godavarman case on forest boundaries is examined here.

The December 1996 Judgment

The FCA made it mandatory for every proposed conversion of forest land for non-forest purposes to obtain a clearance from the central government-appointed committee specially set up for the purpose. This was no doubt a laborious and time-consuming procedure. It was, therefore, natural that many state governments attempted to evade the operation of the FCA by arguing that the land being diverted was not covered by the statute for one reason or another.[5] Rampant felling of trees was also taking place in forests that did not, at that time, fall within the definition of forest land, either because of incomplete forest settlement or because in areas such as the Northeast, the forests were largely community-owned.

Stepping into this perceived gap, on 12 December 1996, the Supreme Court passed a landmark judgment[6] in the Godavarman case. The focus of the Court was on what are the 'forests' and 'forest land' covered by Section 2 of the FCA, the provision which stipulates that any conversion of forest or forest land to a 'non-forest purpose' must first be cleared by the central government in what is commonly known as 'forest clearance'. The judgment laid down the following:

> the term 'forest land' occurring in Section 2 will not only include 'forest' as understood in the dictionary sense, but also any area recorded as forest in the Government record irrespective of the ownership... The provision enacted in the Forest Conservation Act, 1980 for the conservation of forests and matters connected therewith must apply clearly to all forests so understood irrespective of ownership or classification thereof.[7]

Therefore, the Court held that conversion of any forest land as defined above, would necessarily invite the provisions of the FCA. Directions were also issued to all the state governments to set up committees to identify all forest land according to the aforesaid definition.

In addition, the Court issued certain state-specific directions[8] relating to existing forest lands, avowedly with the purpose of protecting them. These are summarized as follows:

1. *Jammu and Kashmir:* complete ban on felling of trees in forests; state government may remove fallen trees, or fell and remove diseased or dry standing timber from forests other than Protected Areas (PAs);
2. *Tamil Nadu:* complete ban of all felling of trees in forest areas with the following exemptions:
   - Felling in lands given to tribals under the tree *patta* scheme;
   - Plantations were granted certain exemptions;
   - In Janmam land, the ban on felling will operate subject to the final orders passed by the Supreme Court in appeals challenging the constitutional validity of the Janmam Lands Abolition Act.[9]
3. *Himachal Pradesh, West Bengal, and the hill areas of Uttar Pradesh (now Uttarakhand):* ban on felling of trees in forests whether public or

private, with an exemption being made for timber from plantations for the personal use of local populations.

The Court also specifically observed that these directions will operate and will be implemented notwithstanding any order of government, authority, or tribunal or any court including the High Courts. Significantly, while passing this judgment, the Supreme Court had made no effort to examine the complex rubric of state laws, government orders, and court precedents, apart from customary practices, that form the substance of forest law and its operation at the ground level. At the same time, the inclusion of a *non obstante* clause in a judgment of the Supreme Court seems a bit superfluous, since it is recognized as precedent overriding all existing law by the Constitution itself. This observation seems to indicate that the court was well aware that it was stepping into an arena that was complex and far from homogenous, and that for its directions to operate in fact, no room for 'interpretation' must be allowed to remain.

As a result, the December 1996 judgment has continued to operate in spite of glaring contradictions with innumerable state-level laws,[10] government orders, court decisions and practices. Any variation which may be required has to be sought from the Supreme Court itself by way of filing intervention applications (IAs) seeking a modification. In the last 19 years, the number of such applications filed has crossed 3,700, and at least for the last eight years the Special Forest Bench hearing the matter sits for two hours once every working week. Finding that it was inundated with litigation it was ill-equipped to handle, the Court began to take the assistance of specially constituted committees in specific cases.[11] The Central Empowered Committee was set up initially for a period of five years by an order of the Supreme Court on 9 May 2002,[12] where it was described as 'a Committee for all the forest areas in the whole of India with power to give directions, hear objections and take decisions so that there is no need to approach this Court from time to time.' Since then it has seen several extensions. While one *amicus curiae* was originally appointed in 1995, at present there are four designated senior advocates and four advocates on record. The time and resources dedicated to the forest cases by the Supreme Court Registry are nothing short of astounding.

The issue of 'forest protection' has, in other words, been seen by the highest Court through the narrow lens of whether 'forest land' falls within the purview of the FCA or not. Flowing from this, the Court has examined whether such land attracts the numerous consequences of such classification, such as payment of compensatory afforestation and net present value, the necessity for forest clearances, and so on. In the process, the discourse within the courtroom has lost sight of the complex web of laws, rights, land-use, and relationships within which forests are in fact interwoven with people's lives, in particular, the lives of forest-dwellers.

As we will see later in this chapter, the reality of what forests are and mean outside of the narrow lens of what the Supreme Court had defined for itself in the December 1996 judgment, became blurred, and that lens has continued to remain the same even while the issues and debates that have come before the Court have continued to expand. It is the fundamental starting-point from which all applications are articulated and solutions addressed, the microscopic truth which permeates all argumentation, as well as the premise on which the directions of the Court are founded. This entry point into the issue has, in fact, become a boundary within which the Court itself has chosen to operate, compelling those that approach it to infuse their argumentation with other 'truths' that supplement or complement this boundary, rather than challenge it at their own peril.

Extending Existing Forest Boundaries

The first and most obvious impact of the December 1996 judgment was the outward expansion of the forest boundary, the area over which the FD and the MoEF claim their jurisdiction. In practice, the application of the December 1996 judgment happened in several different ways, and led to all kinds of confusion and contradictions, which are explored in this section.

Forest-like Areas

The December 1996 judgment resulted in committees established in several states to identify 'forest-like areas', that is, areas which may not be classified as forests in the land records, but are forests according to the 'dictionary meaning' of the word. One of the early challenges to this expanded definition of forest land came from certain mines in

Lalitpur, Uttar Pradesh, where it was argued by the miners that the state government had arbitrarily included the land on which they held valid mining licences as 'forest-like areas' and thereby directed them to stop mining operations. The matter was referred by the Supreme Court to the Central Empowered Committee, which submitted a detailed report on 10 September 2003,[13] stating that there appears to be some merit in the argument made by the mining companies. It observed:

> In view of the above, the CEC is of the view that the identification of 'forest like area' has not been done in a proper, unbiased and systematic manner in the Lalitpur district... No guidelines have been prepared on the basis of which proper and objective field verification could have been done... Some of the mines have been closed down arbitrarily whereas others have been allowed to continue. The guideline regarding no mining up to 100 m from the forest area has not been applied uniformly.[14]

In the normal course of a litigation, such a finding by a court appointed committee would result either in the petitions of the miners being allowed so that their licences are restored, or the matter being referred back to the concerned authority for reconsideration. The Supreme Court did neither, but rather, based on the recommendation made by the CEC, directed the state government to come up with a set of guidelines on how 'forest-like areas' are to be identified. Once this was done and approved by the Court, they should be implemented.[15]

The matter is still on board, and the state government has, under the sustained monitoring of the Court, devised a set of guidelines and undertaken a time-bound exercise of identifying the 'forest-like areas' according to the said guidelines. This process will be reported back to the Court. The issue relating to the legality of the particular mine, and whether it can continue its mining activities, remains unsettled, much to the chagrin of the applicants![16]

Another example of how the December 1996 judgment impacted the lives and livelihoods of ordinary citizens comes from Punjab. The Punjab government in its initial enthusiasm, while filing an affidavit in response to the December 1996 judgment, included in 'forest-like areas' all lands notified under Sections 4 and 5 of the Punjab Land Preservation Act, 1900. This is a statute designed to give certain powers to the local administration over lands which are subject to soil erosion, regardless of whether they are private property or state-owned, so that the government can take steps towards soil conservation. Having included such land in the list of 'forest-like areas', the state government found itself compelled to enforce the provisions of the FCA, and restrict non-forest activity on such land. Large numbers of farmers across the state found themselves on the wrong side of the law for cultivating their private land, which they had been doing for more than 50 years. When they approached the Supreme Court, a report from the CEC was called for. In its report[17] the CEC took a view that the farmers had been cultivating the said lands for 50 or more years, and the state government as well as the central government had given their no-objection to the exclusion of lands notified prior to 12 December 1996 under the Punjab Land Preservation Act, 1900 from the purview of the FCA. However, instead of leaving the matter at that, the report observed that 'the CEC is of the view that deletion of areas, which were under cultivation/habitation prior to 25.10.1980, i.e, enactment of the FC Act, would not be against the spirit of the FC Act and this Hon'ble Court's order date 12.12.1996, if such areas were included in the "list of forest areas" on technical reasons alone.' It, therefore, recommended that a clarification be issued by the Court on these lines.

Accepting the report of the CEC, when the matter came up for hearing on 9 September 2005, the Forest Bench directed the MoEF to consider the proposal of the state government for deletion of these lands from the purview of the FCA, with a rider that '(t)he lands to be considered would be such lands which are under bona fide agricultural use.'[18]

Interestingly, applications filed before a different bench by mining lease-holders from Haryana, argued that their lands would not fall under the purview of the FCA since these had been notified under Sections 4 and/or 5 of the Punjab Land Preservation Act, 1900, and that mining ought to be permitted to continue without approval of the central government.

The Supreme Court however, rejected this argument[19] on the ground that the state government has been showing the area as 'forest' in its own affidavits before the Court in the Godavarman case, and cannot now be permitted to make a complete somersault in these proceedings, contending that its own earlier stand that the area was 'forest' was somehow erroneous. Therefore, the area will be treated as forest and the FCA shall apply, which means that for any non-forestry use procedure under the said Act must be adhered to. It was held, 'On the facts of the case, the mining activity on areas covered under section 4 and/

or 5 of the Punjab Land Preservation Act, 1900, cannot be undertaken without approval under the FCA.'[20]

These are two instances which were agitated before the Court in a sustained way and where CEC reports were submitted and released as public documents. However, the bulk of the documentation being filed by various state governments regarding the identification of 'forest-like areas' and the expansion of the control of the state FDs over such lands remains inaccessible as part of the court record. It, therefore, is difficult to speculate on whether arguments have been placed before the Court, drawing its attention to the inherent absurdity of applying the same approach to land mistakenly classified as 'forests' by an over-enthusiastic state government with respect to agricultural farmers and commercial activities such as mining.

Orange Areas in Madhya Pradesh and Chhattisgarh

The status of Orange Areas[21] in Madhya Pradesh and Chhattisgarh, has always been tenuous in law, being classified as 'undemarcated protected forests' in forest records, and also as revenue land in revenue records. For years these areas have been the subject matter of dispute between the FD and the revenue department (RD), each claiming control over them. The disputes have been compounded in several areas where one or the other department has transferred the land to third parties, including forest-dwelling communities, through the issue of pattas or grant of leases. It is not surprising, therefore, that the issue has found mention in several pending matters in the Godavarman case consequent upon the December 1996 judgment.

Some time in 2001, an application was filed by the state of Madhya Pradesh challenging the decision of the state-empowered committee that *chote jhad ka jungle* and *bade jhad ka jungle* constitute 'forest' and, therefore, any diversion of such land for non-forest use must be approved under the FCA. The application came up before the Supreme Court on 1 August 2003 and was dismissed, with a further observation that 'it is open to the State of Madhya Pradesh to approach the Central Government for their exclusion from the purview of the definition of 'forest' under the provisions of the Act.'[22]

In another independent proceeding, the Court was approached by a private applicant challenging the allotment of forest land in Korba to a private company, Messrs Maruti, to establish a coal-washery plant, without seeking approval under the FCA.[23] The state of Chhattisgarh opposed this plea on the ground that the allotted land is not recorded as Reserved Forest (RF) or Protected Forest (PF), and that there is no actual notification to the effect that the land is an Orange Area. The CEC examined the matter and submitted a report that notwithstanding the non-availability of any notification showing that the area was a notified RF or PF, this area was continuously being considered by the Chhattisgarh FD as a 'forest'. The report cites a letter of the district forest officer, Katghora, which mentions that the land comes under the category of *bade jhar/chote jhar ka jungle* (that is, Orange Area/ undemarcated PF) and contains about 200 trees per ha. Further, the report cites the findings of the Orange Area Survey and Demarcation Unit, Bilaspur that the allotted area is suitable for notification as demarcated PF. Therefore, the land should be treated as 'forest' for the purpose of the FCA.

Meanwhile a consolidated application related to the legal status of 'Orange Areas' in Madhya Pradesh and Chhattisgarh was filed by Ekta Parishad before the CEC itself.[24] After detailing the uncertainty around the legal status of these lands, and the direct and adverse impact this has on the fundamental right to life of the people who are dependent on such land, the application prayed for:

- appropriate directions to settle the exact legal status of Orange Areas;
- constitution of a high-level committee, comprising representatives of the central government, state governments of Madhya Pradesh and Chhattisgarh, including revenue and forest officials, and representatives of the affected people, to draw up a time-bound action plan;
- cessation of all evictions of valid patta-holders and occupants of land in Orange Areas, as well as those areas where the process of settlement of demarcated forests is not complete; and
- issuing of appropriate directions to the concerned state governments, to ensure bona fide requirements of people, including their *nistar* rights, are met.

Although notice was issued to the respondents, the application remained pending for several years before the CEC without any substantive hearing. Finally, the applicants approached the Forest Bench on the ground that no progress is being made in the matter by the CEC, which needs a quietus as soon as possible.[25] The Court initially sought a response from the CEC, and upon its submission through the amicus that a joint demarcation exercise be conducted by the CEC, the RD and the FDs, the Court directed the state government to file a response.

At the time of writing, the Orange Areas matter remains undecided. In fact, the case has not been argued at length even once even though over a decade has passed since it was first brought to the attention of the Forest Bench. Perhaps the challenge lies in the fact that the unidimensional approach to forest boundaries articulated in the December 1996 judgment and the subsequent orders are unable to comprehend the complex lived reality of those who people these boundaries.

Buffer Zones and Eco-sensitive Areas

There has also been a systematic effort to extend the control of the Court over areas surrounding forest areas, ostensibly to curb uncontrolled commercial activity such as mining. This includes the creation of buffer zones around PAs and the declaration of 'eco-sensitive zones'. While these classifications are made under the Environment Protection Act, 1986 and are not technically within the regime of forest law, the trend is clearly to extend the control of the Court as well as the state FDs over these areas. While some part of this process has taken place within the Godavarman case, the main developments took place in a case entitled *Goa Foundation vs. Union of India*,[26] which has been pending since 2004, and has only recently been shifted to the same Forest Bench which hears the Godavarman case.

Filed in 2004, this writ petition points out that there are a large number of mining and industrial units operating in the vicinity of National Parks and Sanctuaries without the requisite clearances under the Environment Protection Act, 1986, and particularly without an Environment Impact Assessment (EIA).[27] The petitioner sought urgent directions for closure of all units operating without the requisite EIA clearance in order to prevent destruction of wildlife and environment.

While the Court initially passed a blanket order that all defaulting units operating in violation of environmental laws be closed,[28] it subsequently passed an order directing the closure of some units and the continuation of others based on a status report filed by the MoEF in

May 2005.[29] The vicinity of the units to PAs was a major consideration in this classification. The overwhelming majority of the units that were close to PAs were mining operations. The Court took note of a decision taken on 21 January 2002 by the Indian Board for Wildlife on 21 January, 2002 to notify the areas within 10 kilometres of the boundaries of national parks and sanctuaries and the wildlife corridors as 'eco-sensitive areas' under the Environment Protection Act, 1986, and directed as follows:

> The MoEF is directed to file an affidavit, within three weeks, placing on record their stand in respect of the decision taken in the meeting of 21st January, 2002 and on the issue of grant of clearances for mining in areas in close proximity of the sanctuaries and the policy, if any, as to the distance of the area from the boundaries of the sanctuaries for the purpose of considering the application for grant of mining lease.[30]

Since the issue of declaration of eco-sensitive areas or buffer zones around national parks and sanctuaries has also been pending before the Forest Bench in the Godavarman case, in early 2008 the Court transferred the *Goa Foundation* case to the Forest Bench. On several occasions when the applications came up for hearing, the Court was pressed by the amicus curiae in the Godavarman case to pass directions concerning the rampant mining taking place in the country in the close vicinity of PAs. While the declaration of buffer zones around PAs, ostensibly to prevent mining and other commercial activity, remains undecided till date, the Court has chosen to pass directions on a case-by-case basis.[31] Here again, it has been observed that issues regarding the complex user and ownership rights in the lands surrounding PAs are not addressed, even as vociferous arguments are repeatedly addressed before the Court by affected mining lease-holders.[32]

While examining each of the above processes towards expansion of forest boundaries subsequent to the December 1996 judgment, it becomes apparent that the Supreme Court has made huge inroads into the legislative domain by re-writing laws enacted by Parliament, and also into the executive domain by centralizing to itself functions which are the responsibility of the MoEF and the state FDs, among others. This has been done at enormous cost, both in terms of the resources the Court has requisitioned for this purpose and in terms of the opportunity cost of dedicating a three-judge bench of the Supreme Court, numerous court staff, and a wide array of lawyers, to these proceedings on a weekly basis. What is also evident is that the voices of forest-dwelling communities who depend upon the forests for their livelihood and survival have either been absent, or where present, have been unable to articulate the substantive issues emerging and the consequences flowing from the Godavarman case on their lives. Commercial interests such as mining lease-holders, on the other hand, have taken every opportunity to create a formidable resistance to the avowed objectives of these orders, that is, the protection of forest wealth from untrammelled exploitation.

It is, therefore, important to ask whether a simplistic binary classification of forest–non-forest can capture the inherent complexity of forested ecosystems and their use in countries like India, where the rights and uses of forest land vary in type and intensity as do the ecological conditions. The next section attempts to examine whether the forest regime put in place by the Supreme Court in the Godavarman case at such enormous cost, has been able to achieve its purported objective.

Inside Existing Forest Boundaries

Even as the processes for expanding the forest land over which the Court, and through it the MoEF and state FDs, would have control expands, parallel processes have been underway to increase the quantum of control of the Court over such forest land. While this process has been far slower and hard to track, it has nonetheless been inexorable.

Rigidity Regarding Alteration of Boundaries

As mentioned earlier, while the proceedings in the Godavarman case continued before the Forest Bench of the Supreme Court, another public interest litigation (PIL) relating to protection and conservation of wildlife in the PAs of the country was initiated in 1995. The WWF case[33] has resulted in certain pivotal interim orders that have impacted the operation of forest law. A cryptic one-line order was passed by the Supreme Court on 13 November 2000 in the WWF case as follows: 'Pending further orders, no dereservation of forests/sanctuaries/national parks shall be effected.'[34]

Since there is no judgment accompanying this order, it is impossible to determine the context in which it was passed. One may hazard a guess that the Court was responding to attempts by user agencies and state governments to get past the requirement of a forest approval under the FCA by declassifying the concerned forest land altogether.

Similarly, another key order was passed in the WWF case on 9 May 2002, again without a judgment explaining the reasons and the legal foundations on which it is based. This order said the following: 'In the meantime, no permission under Section 29[35] of the Wild Life Act should be granted without getting the approval of the Standing Committee.'[36]

These orders have dramatically altered the statutory framework of law relating to forests and PAs as follows:

1. State governments were wrested of their statutory power[37] to alter the land classification of forests (through dereservation). This power was taken over by the Supreme Court.
2. State chief wildlife wardens were wrested of their power to grant permission for any non-forest activity inside a national park or a wildlife sanctuary (WLS) that disturbs wildlife, its habitat, forest produce, or water courses. These powers were given to the Standing Committee of the National Board for Wildlife.[38]

It is important to point out that there is a distinction between dere-servation of forest land (which involves a declassification of the land as forest), and the grant of approval under Section 2 of the FCA for non-forest use of forest land (which need not involve a change in the legal status and classification of the land). While both require a 'forest approval' under the FCA from the Forest Advisory Committee, it is the former which now also requires specific permission of the Supreme Court, and therefore, of the Central Empowered Committee.

The orders of 13 November 2000 and 9 May 2002 were passed by the Supreme Court in the WWF case without any accompanying judgment explicating the reasons why it felt compelled to issue directions contrary to statutory law enacted by Parliament. It is curious that although numerous applications are filed seeking modification of these orders in the form of case-specific exemptions for developmental projects, the orders remain intact. Moreover, no party has come forward raising substantive legal and constitutional arguments regarding their validity and the clear incursion they make on the domain of the executive and the legislature (Khanna 2008).

Diversion and Net Present Value

There already existed a statutory requirement in the Forest Conservation Act, 1980 and its rules that diversion of forest land for non-forest purposes must be accompanied by compensatory afforestation, usually in an area up to double of the diverted land. During the course of hearings in the Godavarman case,[39] it was pointed out to the Court that the utilization of funds collected as compensatory afforestation was dismal, with 50 per cent or less of the amount received actually being spent by state governments. Taking *suo moto* note of this fact, the Supreme Court treated it as an independent interlocutory application (IA),[40] issued notice to the defaulting States, and called for a CEC report (Ranthambore Foundation 2005a).

The CEC recommended, among other things, that while granting approval under the FCA, in addition to the funds realized for compensatory afforestation, the net present value (NPV) of the forest land diverted for non-forestry purposes shall also be recovered from the user agencies. Although the term NPV has not been specifically defined, in a subsequent order the Court did attempt to articulate its meaning as 'the amounts... required to be used for achieving ecological plans, and for the regeneration of forest and maintenance of ecological balance and eco-system. The payment of NPV is for protection of the environment and not in relation to any proprietary rights'.[41]

The CEC further recommended the creation of a 'Compensatory Afforestation Fund' in which all funds received from the user agencies towards compensatory afforestation, additional compensatory afforestation, penal compensatory afforestation, net present value of forest land, catchment area treatment plan funds, and so on, shall be deposited.

The Forest Bench, in October 2002,[42] accepted this report in its entirety, and directed the central government to frame rules for creation of a body to manage the compensatory afforestation fund. Regarding the concept of NPV, the Court directed:

> The net present value is to be recovered at the rate of Rs. 5.80 lakhs per hectare to Rs. 9.20 lakhs per hectare of forest land depending upon the quantity and density of the land in question converted for non-forest use. This will be subject to upward revision by the Ministry of Environment and Forests in consultation with Central Empowered Committee as and when necessary.[43]

Following upon these orders, a complex set of directions and orders issued by the MoEF have been put in place, including a direction constituting the Compensatory Afforestation Fund Management and Planning Authority (CAMPA).[44] These directions have been challenged by user agencies as well as state governments, claiming that in certain situations, payment of NPV for diversion of forest land is an onerous burden. It was argued that the diversion of forest land is a necessity in certain cases to enable the state to fulfill its welfare functions and for economic development. In such cases, NPV should not be levied. The MoEF as well as some of the state governments have gone on record to state that no NPV should be payable while regularizing land rights of tribals and forest-dwellers, as well as for rehabilitation of persons displaced from developmental projects in other areas.[45] Some state governments also took issue with the centralization of control of the funds collected, arguing that when the funds are collected with respect to diversion of forest land at the state-level, there is no reason for it to be centrally controlled by CAMPA.

In an attempt to settle the issue, the Court in September 2005 consti-tuted what has come to be known as the Kanchan Chopra Committee to look into issues relating to NPV, which submitted its report in April 2006 (Chopra *et al.* 2005).[46] The Committee in a formidable report drafted by economists from the Institute of Economic Growth after numerous public hearings and consultations with a variety of user-groups, proposed a detailed economic methodology and formulae for calculation of the value of a forest. The value of a forest should include:

- the value of the goods and services which flow from a forest area, such as timber, carbon storage value, fuelwood and fodder;
- non-timber forest produce (including grass), ecotourism, and water-shed services.

The Committee proposed that this method of calculation be applied on a case-by-case basis, and accompanied by public hearings with local forest-users to apprise them of the intention of diverting forest land. The rights, privileges, and concessions existing in the forest area also should be settled by the district collector prior to diversion, again on a case-by-case basis. The Committee went on to make detailed recommendations regarding activities and projects that should be exempted from payment of NPV and to what extent. In addition, it recommended that diversion of national parks and sanctuaries must not take place at all.

One recommendation of the Committee that struck at the core of the Godavarman case relates to the constitution of CAMPA. The Committee recommended that since the goods and services which flow out of forest areas are of a varied nature, and the users of these goods and services likely to be adversely affected by its diversion for non-forest purpose are also diverse, there is no justification for consolidation of the compensation received for such diversion (the NPV) in a centralized fund such as CAMPA. Instead, the stakeholders who stand to lose the most, that is the local communities, must receive the larger share of the compensation, through the panchayat, with the state government receiving a second share, and the central government a much smaller share. It, therefore, proposed the distribution of the NPV collected at the panchayat, state, and central levels in separate funds.

It was little surprise that this report did not meet with the approval of the Court or the amicus curiae. A detailed note was filed by the CEC on 17 October 2006[47] where it disagreed with practically every recommendation, especially that NPV should be calculated on a case-by-case basis. Instead, the CEC proposed an alternative methodology for calculation of NPV for each state based on the classification of forest (such as evergreen forest, semi-evergreen, moist deciduous, dry deciduous, Himalayan pine, Himalayan broad leaf, mangrove, thorn scrub, shoal grassland) and density of tree cover.

The CEC also disagreed with the recommendation of the Kanchan Chopra Committee that there should be no diversion of PAs, and stated that in exceptional circumstances, diversion of national parks, WLSs, biosphere reserves, and other eco-sensitive areas is being permitted by the National Board for Wildlife (NBWL) and the Supreme Court through a well-defined procedure based on the principle of economic deterrence. Therefore, in national parks and WLSs, NPV should be calculated at 10 times and five times respectively of base value of NPV subject to a minimum of Re 1 crore. This will act as a monetary disincentive, while at the same time not make essential diversions impossible.

Finally, the recommendation regarding distribution of NPV to local, state and central level funds was also found unacceptable by the CEC, for the reason that this issue has already been settled by a judgment of the Supreme Court,[48] in which constitutional validity of the CAMPA rules had been upheld. In a detailed judgment dated 28 March 2008, the Supreme Court laid the entire debate to rest by accepting the CEC's recommendations, with some modifications based on the arguments addressed by the MoEF and counsels for various project proponents.[49]

The question remains whether economic deterrence in the form of NPV has been a successful tool in preventing the diversion of forest land for commercial purposes, large or small. Although no consolidated data exists on how much forest land has been approved for diversion, available data suggests that the diversion of forest land has, in fact, increased rapidly. According to one estimate the quantum of forest land diverted under the FCA (approved and in-principle) went from 3,499.22 ha in 1982 to 31,836.88 ha in 1996 (when the 1996 judgment was passed), and a shocking 108,680.20 ha in 2006.[50] This data suggests that the enactment of the FCA has, far from resulting in a decline, coincided with a ten-fold increase in diversion till 1996. And subsequent to the December 1996 judgment of the Supreme Court, forest diversion has increased at the confounding rate of 340 per cent.

A corollary of this astonishing fact is that the CAMPA fund, in which the money collected as NPV and compensatory afforestation are located, has crossed Rs 30,000 crores. The fund is administered by an ad-hoc committee, which was not empowered to utilize it due to pending applications by various state governments insisting that the funds collected from diversion of their forest land must remain within their control rather than be centralized in CAMPA. Even as a Bill was pending before Parliament to address this intractable issue, in a surprise move the amicus curiae announced that the matter had been settled amicably between the CEC and the forest minister, and placed a proposal before the Court which was approved the same day.[51] The Supreme Court directed the following:

- setting up of state-level CAMPAs;
- release of Rs 1,000 crores from the CAMPA fund to the state CAMPAs in a proportionate manner, after approval of the state-level plan for utilization of this money;
- five per cent of the total amount released will be available with the national CAMPA Advisory Council for monitoring, evaluation, and implementation.

Once this order was passed, it was with a sense of immense self-congratulation that the amicus curiae pointed out to the Court that the amount released from the fund exceeded the total annual budget for 2009 of all the state FDs put together. When success is measured in terms of the quantum of money collected through CAMPA because of large projects such as mining that divert large areas of forests, we can safely conclude that the original goal of economic deterrence of such diversion has long been abandoned.[52]

Encroachments[53]

The repeated use of the term 'encroacher' to club all forms of extra-legal possession and use of forest land, including that of forest-dwelling communities, which have a prior legal right over the land, has served to create a bias in the mind of the Court. Such a bias first came to the fore in July 1998 in an order passed by the Supreme Court regarding encroachments in the Athol Reserve Forest, Chikmangalur, Karnataka. The order restrained all encroachers from changing the nature of the encroachments during the pendency of these proceedings.[54] A court commissioner was appointed to conduct a site visit, and subsequently, reports were also submitted by the Survey of India and the CEC. No attempt was made to differentiate between encroachments by large plantation owners, and those by marginal and small farmers or even the landless.

The judgment of the Court on the subject of encroachments in Thatkola Reserve Forest[55] also deserves scrutiny. After setting out the entire history of the litigation, the Court noted that 611.23 acres of forest land had been encroached inside the RF, of which 556.04 acres had been encroached for coffee cultivation and 55.19 for 'other purposes'. The arguments of some of the larger plantations were heard, but were rejected. The Court held:

> There can be no manner of doubt that any land which forms part of the Tatkola Reserved Forest could only belong to the Government. Once the forest was established in the year 1936, all other rights therein came to an end.... From the aforesaid, it is quite clear that all encroachers in to the Tatkola forest have to be evicted. It is no doubt true that according to Section 64A[56] show cause has to be issued. But that can only be with a view to enable the person to whom notice is issued to show that his

land does not fall within the boundaries of the forest as drawn up by the Survey of India. If the land is identified as falling within the Survey of India boundary then there could be no other defence open to the person concerned and the State would be under an obligation and duty to evict the encroacher, by force if necessary.[57]

The Court directed immediate removal of these encroachments, if necessary with the use of force. In the event the encroachers did not vacate the land within the given time-frame, they would be liable to pay damages.[58] An application seeking a review of this judgment was dismissed.[59]

In another proceeding[60] emerging from applications filed by three environmental organizations regarding indiscriminate deforestation in the Andaman and Nicobar Islands, and its impact on the primitive tribes, the Court again appointed a court commissioner,[61] who submitted a detailed report in January 2002. On the basis of these recommendations, the Court passed the following order: '… Regularisation of encroachments on forest land in any form, including allotment/ use of forest land for agricultural or horticultural purpose, shall be strictly prohibited…. All post 1978 encroachments shall be completely removed within three months.'[62]

The CEC also submitted a report recommending that the 'removal of encroachers should be implemented forthwith'.[63] Unlike in the Tatkhola case, the directions of the Court and the CEC recommendations have led to a complicated array of litigations by a range of interest groups, asserting their right to be treated differently from others. These are summarized below.

- Environmental groups[64] have challenged the failure of the Andaman and Nicobar Islands administration to implement these orders, espousing the cause of the primitive tribes.
- The Andaman and Nicobar Islands administration has sought dilution of these directions, pleading impossibility in implementation.[65]
- Immigrants who are tribals from mainland India, relocated and settled by the Indian government, are seeking recognition of their rights to the lands they occupied.
- Descendents of prisoners sentenced to imprisonment in the Islands during colonial rule, describing themselves as 'Local Borns', are also seeking recognition of their land rights.[66]

- Owners of sawmills, timber traders, and other groups who have benefited from the commercial exploitation of the forests, are seeking their exclusion from the definition of 'encroachers'.

It is clear from the experience of the litigation around encroachment of forest land in Tatkhola and the Andaman and Nicobar Islands that the issues emerging are complex and intractable. Yet, when the amicus curiae filed an application on 23 November 2001 before the Forest Bench seeking similar directions regarding encroachments on forest land across the country, the Court immediately took the opportunity and passed the following interim order:

> An application has been filed by the ld. Amicus Curiae in Court against the illegal encroachment of forest land in various States and Union Territories is taken on board. Let the same be registered and numbered. Issue notice to the respondents returnable after six weeks. There will be an interim order in terms of prayer.[67]

Since the above order is incomprehensible by itself, the prayer (a) made by the amicus in IA 703 is reproduced: '(a) Restrain the Union of India from permitting regularisation of any encroachments whatsoever without the leave of this Hon'ble Court'.

This historic order has become part of the legend of the tribal rights movement in India, and its repercussions resounded in places perhaps the Court had never anticipated.[68] The history of opposition to the order need not be gone into here. Suffice it to say that the FDs of numerous states have used the order to launch massive eviction drives of tribal and forest-dwelling populations residing inside forest land across the country on the ground that the Supreme Court has passed orders for eviction of encroachers.

A report submitted by the CEC shortly after its constitution in August 2002 also supported the eviction of 'encroachments'.[69] This led to a spate of interventions being filed in the Supreme Court by representative organizations of forest-dwelling communities challenging the approach taken by the Court, the CEC report, and also terming the eviction drives as illegal. State governments also filed affidavits in response, and some even filed similar applications seeking rejection of the CEC report. A strong affidavit in opposition was also filed by the Union government (Khanna and Naveen 2005).

An examination of these applications and affidavits indicates that an effort was made to articulate the perspective of tribal and forest-dwelling communities. The applications highlighted the 'symbiotic relationship' between forest-dwelling communities and the forests and the 'historical injustice' of the non-settlement of rights initially by the British colonists and later by the Indian government. It was also argued that the forest-dwelling communities actually protect the forest and, therefore, should be allowed to continue to do so. Finally, it was pleaded that the report of the CEC recommending eviction be rejected by the Supreme Court.

However, it is clear that even the most bravely articulated arguments have been constrained within the boundaries of forest law as redefined by the Supreme Court in light of the FCA. The applicants took great pains to assert that people were not 'encroachers' at all, relying heavily upon the 1990 guidelines,[70] which themselves were an exercise in drawing boundaries between 'eligible' and 'ineligible' encroachers. Furthermore, they relied upon the 1988 Forest Policy to substantiate the claim that forest-dwelling communities are not detrimental to the forest ecology, and in fact, protect the forests. Some applicants bent backwards to make their case by saying forest-dwelling communities could be used to supplement the severe manpower shortage being faced by State FDs through rapid expansion of the JFM programme.[71] Notably absent was a substantive articulation of rights based on the constitutional framework, including the special provisions for scheduled tribes (STs) and scheduled areas, the panchayats, and separation of powers between the judiciary, , and legislature.

Since February 2004, these applications have not been listed for hearing. There have, however, been other applications filed by various state governments seeking permission to 'regularize encroachments', which have been found to be 'eligible', and these have received uneven treatment both from the CEC and the Court. While in some cases, the Court has permitted the regularization,[72] in other cases the matter has remained unresolved for the reason that the state governments have not filed consolidated proposals for eviction of all ineligible encroachments along with regularization of all eligible encroachments.[73]

With the enactment of the Scheduled Tribes and Other Traditional Forest Dwellers (Recognition of Forest Rights) Act, 2006 (hereafter FRA) much of these debates have become academic, and the enforceability of many of the Court's orders will have to be thoroughly examined in light of the dramatic change brought about by this law in the power relationship between the state machinery and forest-dwellers.

Nevertheless, the manner in which the Godavarman case has dealt with the issue of encroachments remains an important area of study and concern.

Exercise of Usufruct Rights

Without a robust challenge to the fundamental premises on which the Supreme Court's intervention in forest law and governance was based, the Godavarman case rapidly made incursions into areas of exercise of power inside already declared forest areas. This resulted in serious curbs being placed, in particular on the exercise of usufruct rights of forest-dwelling communities in PAs (Khanna 2008).

In early 2000, the amicus curiae filed an application[74] against commercial exploitation of national parks and sanctuaries. Instead of transferring the application to the Bench hearing the WWF case, since it related to a subject matter squarely falling within the domain of that case, the Forest Bench chose to retain the matter with itself.[75] No explanation for this procedural innovation was given. While issuing notice to the respondents (which included only the state and union territory governments) on 14 February 2000, the Court passed the following interim order:

> In the meantime, we restrain respondents No. 2 to 32 from ordering the removal of dead, diseased, dying or wind-fallen trees, drift wood and grasses, etc., from any National Park, Game Sanctuary or forest. If any order to this effect has already been passed by any of the respondent-States, the operation of the same shall stand immediately stayed.[76]

Soon after, a clarificatory order was passed deleting the word 'forests' from this order and thus restricting its operation to PAs.[77] Not surprisingly, the 14 February 2000 order has been the subject matter of a number of challenges by a variety of interest groups which have met with uneven and inconsistent results. These include:

- Upon a representation by the state of Rajasthan, the Court held that: 'it is clarified that the said interim order will have no application in so far as plucking & collection of tendu leaves is concerned.'[78]
- By an order dated 11 May 2000, the Court granted exemption to 'cut, transport and sell' eucalyptus, subabool and casurina by farmers/tribals under the Farm Forestry Programme to Orient Paper Mills, Amlai, Shadol (Madhya Pradesh).[79]

- In a subsequent clarification[80], the Court stated that:

  It is clarified that the order of this Court prohibiting cutting of trees does not apply to bamboos including cane, which really belongs to the grass family, other than those in the national parks and sanctuaries. In other words, no bamboos including cane in national parks and sanctuaries can be cut but the same may be cut elsewhere.

- Where the rights of pastoralists and graziers is concerned, the prohibition of grazing rights in Ranthambore National Park has on several occasions led to forcible entry into the park by armed graziers along with livestock, and assaults on the FD staff.[81] Rather than lead to any re-examination of the Court's orders, the CEC has issued directions seeking ehnanced security and action against the graziers who violated the Court's orders.

- A joint application was filed by Vasundhara and Kalpavariksh,[82] placing evidence before the Court pertaining to the havoc created by this order, compounded by subsequent 'clarificatory' orders issued by the CEC. However, when this application was listed before the Forest Bench, instead of referring the matter to an independent committee for examination, or even applying its own mind to it, the Court directed the CEC to submit its response. Till date arguments on this application have not been heard, although it has been listed on occasion.

- The Odisha government sought permission for collection of kendu leaves from three sanctuaries in the state: Sunabeda, Badrama, and Satkosia.[83] The CEC recommended[84] that kendu leaf collection be permitted in Satkosia WLS and that for the other two sanctuaries permission be sought from the Standing Committee of the National Board for Wildlife. Although the Court has not approved these recommendations, the CEC has gone ahead and forwarded them to various functionaries of the state FD, resulting in considerable chaos.

Although the order dated 14 February 2000 was meant to be of an interim nature, it has become entrenched over the years, thus turning established law on its head. Even subsequently, a notice was not issued to any of the other Union ministries concerned, nor to any representatives of tribal or forest-dwelling communities or to any persons whose rights to forest produce in these areas had been summarily stopped. What is most startling is that at no point of time has the

Court considered that the order of 14 February 2000 is contrary to the provisions of the Wildlife Protection Act, 1972 itself. The 1972 Act provides for extraction of minor forest produce from existing PAs for bona fide livelihood needs, and also lays down a detailed procedure for the settlement of rights of persons living in and around PAs *prior to* final notification and declaration of the boundary of the protected area. Despite orders in the WWF case as far back as August 1997,[85] directing time-bound settlement of rights, and numerous subsequent orders, the process of settlement is incomplete even today, with state governments expressing their helplessness to settle rights and relocate populations on such a massive scale.

This order, being an unreported order, did not gain any public attention until state FDs began to curb the legal rights of local communities to extract forest produce from national parks and sanctuaries, whether under customary law or under pre-existing settlements and awards under the Wildlife Protection Act, 1972. In the Nilgiris Biosphere Reserve, made up of a number of national parks and WLSs, a large number of tribes originally resided. Many of them were already facing displacement from their traditional habitats as a result of the declaration of these PAs under the 1972 Act. The Supreme Court's order resulted in further harassment in terms of accessing minor forest produce. The result has been hunger deaths and malnutrition.[86] In several states, the extraction and collection of *tendu patta*, a critical source of cash income for forest-dwelling communities, has come to a complete standstill, devastating the lives of populations already battling poverty.

Rather than speedily apply its mind to these critical issues, the Supreme Court kept the matter in abeyance for several years, even while the CEC itself issued its own directives to the State FDs in purported clarification of the Court's orders. In November 2004 the CEC submitted a report[87] recommending that clarifications be made by the Court exempting certain conservation and wildlife protection-related activities for better management of the PAs, as well as clarifying which activities are prohibited. It further recommended that certain small public utility projects of a non-commercial nature, such as the laying of drinking water pipelines, electricity distribution lines for rural areas, telephone lines and optic fibre cable be exempted from the cumbersome approval procedure currently in place.

The Forest Bench chose to hear and dispose of the issue on 14 September 2007 with reference only to the report of the CEC, without any of the other pending applications relating to the same subject matter being listed on that date. Hearing only the submissions of the amicus curiae, the Court accepted the recommendations of the CEC. The only concession to the existence of forest-dwelling communities was an exception: 'wells, hand pumps, small water tanks, etc., for providing drinking water facilities to villagers, who are yet to be relocated from the protected area'.

Meanwhile, the original objective of IA 548 (to prevent commercial exploitation of PAs) remains unrealized, with applications being filed in a mechanical fashion by governments and project proponents, who readily agree to the 'conditional approval' recommended by the CEC, including payment of NPV. Recently, debates raged within the courtroom over several hearings on a national highway being constructed through the Rajaji National Park which crossed the traditional route of the elephant population.[88] Arguments flew thick and fast on whether an overpass should be built for the elephants to walk over, or whether a flyover for the vehicles would be more conducive to the elephants' habits. The matter was deemed important enough for a personal site visit by the amicus himself.

Of Boundaries Forgotten

In the above analysis, we have seen some of the consequences of the unidimensional approach of the Forest Bench in the Godavarman case commencing with the December 1996 judgment. The discourse on boundaries, which began with the December 1996 judgment, appears to have become established in the repeated use of negatively loaded terminology such as 'encroacher' by the Court, the CEC and the Amicus Curiae in their treatment of forest-dwellers' habitation, cultivation, and exercise of usufruct rights. The repeated use of the term 'encroacher' has resulted in diminished space for articulation of alternative perspectives within the courtroom, such as the symbiotic relationship between tribals and the forest, the interdependence of forest ecology and the village economy, and the inter-relationships between community governance structures and the state.

This bias was apparent in an early case filed by the state of Madhya Pradesh seeking permission to regularize encroachments by forest-dwellers for the period 1 January 1977 to 25 October 1980, where a telling order was passed by the Supreme Court:

> Experience has shown that whenever regularisation takes place subject to imposition of conditions such as compensatory afforestation, the regularisation becomes effective without the conditions ever been fulfilled.
>
> In our opinion, it will be more appropriate that the conditions imposed in relation to regularisations are required to be fulfilled first before any regularisation is granted. The result of this would be that the regularisation would be deferred but the fulfillment of the conditions ensuring inter alia compensatory afforestation would be ensured.[89]

Paradoxically, the Court is careful not to use similar negative terminology when referring to commercial activities inside forests. Thus, mines with expired leases are granted 'temporary working permits', saw-mill owners operating illegally in relation to forest land are granted 'rehabilitation packages', and many illegalities committed by large industry in use of forest land long before any statutory approvals are obtained, are covered by the fig-leaf of NPV and ex post facto forest approvals.

This chapter has looked at the life-threatening challenges of expanding forest boundaries on the rights of agriculturists, tribals, and forest-dwelling communities. Reports have been received from states such as Madhya Pradesh and Odisha about the mysterious and inexorable shifting of 'pillars', which demarcate the boundaries of forest land, expanding the forest land under the control of the state FDs. Whole villages suddenly find themselves inside forest boundaries and being declared 'illegal occupations', and farmers wake up one morning finding a brand new 'pillar' in the middle of a field with a standing crop; no explanations, no hearings, no process, no appeal.[90] Can such inexplicable shifting of boundaries on the ground be a reflection of shifts that have already taken place inside the closely guarded paper trail within the Supreme Court registry in the Godavarman case? On the other hand, commercial interests have been able to launch a formidable opposition within the courtroom to any such inroads into their domain, and where direct opposition has failed, they have found ways to 'work the system'.

Along with the voices of the dispossessed, the constitutional framework itself, which establishes a representative democracy committed to certain fundamental rights and directive principles of state policy based on universal ideals of liberty, equality, and distributive justice has also become invisible. It bears reminding that the Constitution of

India affirms the role of the panchayats as instruments of local self-governance,[91] and further creates a special dispensation with regard to STs, scheduled areas, and tribal areas,[92] creating spaces for participation at the grassroots level for decision-making vis-à-vis developmental processes. This constitutional design has been further affirmed in numerous legislations which flow from it, such as the Panchayats (Extension to Scheduled Areas) Act, 1996, as well as in judgments of the Supreme Court itself, the most well-known among them being *Samatha vs. State of Andhra Pradesh & Ors.*[93] Yet, in a Court where even attempts to point out conflict with statutory law are met with indifference, attempts to raise issues of violation of constitutional principles are rare indeed.

It is unfortunate that even in the few instances where those whom these constitutional provisions seek to benefit approached the Forest Bench, the environment inside the courtroom was not perceived as conducive to the articulation of substantive arguments based on constitutional provisions and principles. Arguments have not been raised against the blatant abuse of the power of eminent domain by the state within forest land, which has belonged to tribal and forest-dwelling communities for centuries. Nor do we find any articulation of the state as encroacher through force, fraud, and the law to take over the ancestral domains of the indigenous people of this country. Clearly, neither the time nor the place was right.

We further observe a process of centralization of forest control through the use of an expanded definition of forests, increased monitoring over what forests are used for, and disempowerment of the administrative and statutory implementation machinery under the existing statutory forest laws. The Forest Bench relies heavily upon the CEC and the office of the amicus curiae for its recommendations, not having the expertise nor the time to examine the minutiae of the thousands of applications that come before it. This has meant a concentration of power in the hands of a committee that is not bound by any statute, except the notification setting it up, and in a small body of lawyers who are unaccountable to anyone except the Court.

The constitutional principle of separation of powers between the judiciary, the executive, and the legislative wings of the state, upheld by the Supreme Court itself in numerous decisions, has completely been disregarded. The intermittent conflicts between the judiciary and the executive (represented in Court through the Union MoEF), are in fact, nothing more than mere skirmishes. Let it not be forgotten that when the Forest Bench, and with it the CEC, took over the statutory functions of the Forest Advisory Committee under the FCA for a period close to two years, the protest by the MoEF was quickly withdrawn through the discharge of its advocate on record and the replacement of its arguing counsel.[94]

Clearly, the Godavarman case, for all its avowed commitment to the protection of forests and environment, in fact, represents a space where the judiciary and the executive have joined hands with vested interests represented by large capital. The modus operandi has been an age-old legal tool of setting false boundaries to the discourse within the domain of the courtroom. This litigation provides a convenient space to re-write laws through interim court orders without having to go through the democratic rigour of the legislative process, while at the same time undermining the role of the legislature in the long term through the combined force of the political executive and judiciary. This process has been commented upon at length in the context of the dismantling of the protective labour law regime and the welfare state, in order to create an environment conducive to resource extraction in post-liberalization India. This could never have been achieved through Parliament, subject as it still is to checks and balances inherent to the democratic process. It was the Supreme Court which stepped into the gap and over the last 10 years has changed the face of labour law in the country (Breman *et al.* 2009).

The political economy of the Godavarman case in particular, and the forest case(s) in general, has been obscured by the discourse and posturing of environmental protection. While there has been a nagging discomfort among many who have been observing these proceedings closely about the amount of time spent by the Court in debating and deciding issues relating to the 'rights' of mining companies, saw-mills, and timber traders, apart from private contractors constructing dams and so on, little time is afforded to the marginalized, whose petitions are, in any case, few and far between.

The prism of political economy, like in the case of labour laws, unravels the manner in which the demolition of existing forest law regimes, biased as they already were against the poor and marginalized, has served the purpose of aggressive capitalist advancement in the country. However, unlike in the arena of labour law, the legislature has been compelled to enact a historic statute in the FRA. A close analysis of this law reveals that while 'recognizing and vesting' forest rights in forest-dwelling communities, which may or may not have been recognized in earlier statutory law, it strikes a body blow to the command and control approach which suffuses the regime established by the Godavarman case. For the first time, independent India has given statutory recognition to the notion of the ancestral domain of forest-dwelling peoples over forests. How far the textual promises that this law makes will actually be realized remains to be seen. A lot depends on how (and where) a resolution will be found between the narrow worldview adopted inside the rarefied atmosphere of the courtroom in the Godavarman case and the historically appropriate, constitutionally sound, and internally robust approach adopted by the FRA.[95]

Notes

1. The author acknowledges the support and direction provided to her by the Society for Tribal Rural and Urban Initiative (SRUTI) in the monitoring of the forest case(s) over the last five years, and the detailed comments on an earlier draft of this chapter by Sharad Lele, C.R. Bijoy, and Shankar Gopalakrishan. After this chapter was finalized, the Forest Bench delivered judgements in a few important cases, including the Vedanta bauxite mining project in Niyamgiri, mining in Bellary, and the Lafarge case. The chapter does not reflect the implications of these cases.

2. *T.N. Godavarman Thirumalpad vs. Union of India & Ors*, Writ Petition (C) 202 of 1995, Supreme Court of India, pending.

3. *Centre for Environmental Law, WWF-India vs. Union of India & Ors*, Writ Petition (C) 337 of 1995, Supreme Court of India, pending.

4. (1997) 2 SCC 267.

5. See, for instance, a spate of litigations that followed the enactment of the statute, including *State of Bihar vs. Banshi Ram Modi* (1985) 3 SCC 643, *Ambika Quarry Works vs. State of Gujarat* (1987) 1 SCC 213, *Rural Litigation and Entitlement Kendra vs. State of Uttar Pradesh* 1989 Supp (1) SCC 504.

6. Henceforth referred to as the 'December 2006 judgment'.

7. (1997) 2 SCC 267, para 4.

8. Interestingly, the Northeast region, which constituted a major area of concern in the early years of the litigation, did not find mention in this particular judgment.

9. The correct citation of the relevant statute is the Gudalur Janmam Estates (Abolition and Conversion into Ryotwari) Act, 1969. Although this Act was

included in the Ninth Schedule of the Constitution, it has been mired in litigation for several years. As a result, of the 32,000 ha of land covered by the statute, only about 11,000 ha could be properly demarcated and settled in different categories, including forests. The uncertainty in the classification of the Janmam lands into different categories, including forest land, has led to considerable insecurity of tenure for forest-dwellers and tribals in the area for decades. The issue remains to this date unsettled, as the constitutional validity of the Act of 1969 is pending examination by the Supreme Court in *IR Coelho vs. State of Tamil Nadu* (C.A. 1344-5 of 1976). See also (1999) 7 SCC 580.

10. According to a study conducted by this author, Odisha alone has at least 18 state-level statutes that govern various aspects of forest rights and forest management. See Khanna (2005).

11. For instance, a high-powered committee was set up in 2000 to oversee the implementation of orders relating to operation and licensing of sawmills, plywood and veneer factories, and transportation of timber, assisted by a special investigation team for investigation into violations. In 2001, State Empowered Committees were set up for the states of Madhya Pradesh and Chhattisgarh.

12. 2002 (5) SCALE 6.

13. Recommendation of the Central Empowered Committee in IA nos 442–6 in WP (C) 202 of 1995 (10 September 2003); Ranthambore Foundation (2005b), p. 141.

14. Ibid., p. 144.

15. Order dated 11 January 2008 in I.A. no. 979 in I.A. nos 443, etc., in WP (C) no. 202/1995. Interestingly, the Court escalated the pressure on the state government prior to this hearing by listing the matter at regular intervals, and then directing the principal secretary (forests), Government of UP to be personally present. It was only after this that the state government finally notified the guidelines, and commenced the process of identification of 'forest-like areas'.

16. As recently as 10 January 2010, when this matter was argued at length and no solution could be arrived at, the counsel for the applicants articulated his dismay at the unending process by asking for the disbanding of the CEC itself. The Court adjourned the matter. See order dated 10 January 2010 in WP (C) 202 of 1995, unreported.

17. Report dated 16 September 2003 in IA no. 727 in Ranthambore Foundation (2005b), p. 153.

18. Order dated 9 September 2005 in IA no. 976 in IA no. 727 in WP (C) 202 of 1995, unreported.

19. *M.C. Mehta vs. Union of India* (2004) 12 SCC 118.

20. Ibid., para 95 (3).

21. The reason these lands are called 'Orange Areas' is because they are coloured orange in the official state government maps, marking their disputed nature.

22. Order dated 1 August 2003 in IA nos 791–2, *T.N. Godavarman vs. Union of India*, unreported.

23. IA nos 857–8 filed by Deepak Agarwal in Writ Petition 202 of 1995.

24. IA no. 198 before CEC.

25. IA nos 2000–2000A in WP (C) 202 of 1995, pending. This application last came up for hearing on 20 February 2009.

26. Writ Petition (C) no. 460 of 2004, Supreme Court of India.

27. Ministry of Environment and Forests, Environment Impact Assessment Notification, SO 60 (E) dated 27 January 1994.

28. Order dated 21 February 2005 in *Goa Foundation vs. Union of India*; 2005 (5) SCALE 276.

29. *Goa Foundation vs. Union of India* 2005 (5) SCALE 287; The Court examined the status report filed by the MoEF relating to a total of 292 units. Of these, 121 were industrial units, and were allowed to continue functioning. Of the remaining 171 mining units, two were denied clearance, 18 units were granted clearance, 35 coal mines were allowed to continue functioning, and 64 units were found to be operating prima facie without environmental clearance.

30. Order date 30 January 2006 in *Goa Foundation vs. Union of India*, Writ Petition (C) 460/2004.

31. An example is the recent approval granted to the Panna diamond mines where mining operations had been stopped for several years on the ground that it lay in the buffer zone of the Panna National Park. The Forest Bench has permitted the mines to start operating, under the supervision of a committee headed by Belinda Wright.

32. This petition was disposed of by the Supreme Court in April 2014 with a direction to MoEF to notify eco-sensitive zones around PAs in Goa within six months. *Goa Foundation vs. UoI*, 2014 (5) SCALE, 364.

33. *Centre for Environmental Law, WWF-India*, Writ Petition (C) 337 of 1995, Supreme Court of India.

34. Order dated 13 November 2000 in I.A. no. 2 in WP (Civil) no. 337 of 1995; 2000 SCALE (PIL) 325.

35. Section 29 of the Wildlife Protection Act, 1972 prohibits the destruction of wildlife, habitat, forest produce, or diversion of water courses inside a sanctuary without a permit granted by the chief wildlife warden. Section 35 (6) makes a similar provision for national parks.

Interestingly, with regard to removal of forest produce, both these provisions contain a proviso that 'where the forest produce is removed from a sanctuary the same may be used for meeting the personal bona fide needs of the people living in and around the sanctuary and shall not be used for any commercial purpose'. Section 35 (6) makes a similar provision for national parks.

36. Order dated 9 May 2002 in I.A. no. 18 in WP (Civil) no. 337 of 1995; 2002 SCALE (PIL) 174. The Standing Committee referred to here is the Standing Committee of the National Board for Wildlife.

37. See for instance, Section 27 of the Indian Forest Act, 1927.

38. The National Board for Wildlife is a high-powered body constituted under the Wildlife Protection Act, 1972 and chaired by the prime minister. It has constituted from within itself a Standing Committee which meets at regular intervals and deals with such permissions.

39. Order dated 17 April 2000 in IA nos 419 and 420 in WP (C) 202 of 1995; 2003 SCALE (PIL) 104.

40. IA no. 566 in WP (C) 202 of 1995.

41. Judgment dated 26 September 2005 in *T.N. Godavarman Thirumalpad vs. Union of India & Ors;* (2006) 1 SCC 1.

42. Vide order dated 30 October 2002 passed in I.A. 566 in WP (C) 202 of 1995; 2002 (9) SCALE 81.

43. 2002 (9) SCALE 81, p. 88.

44. Some of these directions are contained in letters dated 17/18 September 2003 and 19/22 September 2003, F.No.5-1/98-FC (Pt II), among others, and include:

- NPV is payable in all cases that have been granted in-principle approval after 30 October 2002;
- NPV must be realized before Stage II (final) approval is granted;
- NPV must be recovered in all cases where Stage I approval has been granted after 30 October 2002, even if Stage II approval has also been granted;
- State governments should charge NPV within the range of Rs 5.80 lakhs to Rs 9.20 lakhs per hectare depending upon the quality of forest, density, and the type of species in the area.
- The funds thus recovered should be transferred to CAMPA.

45. For instance, in the affidavits filed by the Union of India and the state of Tripura in IA no. 703 in WP (C) 202 of 1995.

46. Of course, with the enactment of the FRA, payment of NPV for lands where forest rights are vested has been obviated. See Section 4 (7) of the Act.

47. *Observation of CEC on Report of Expert Committee on NPV (NPV Committee)*, filed on 17 October 2006 by M.K. Jiwarkja, member secretary, Central Empowered Committee.

48. Judgment dated 26 September 2006 in IA no. 826 in IA no. 566 in *T.N. Godavarman vs. Union of India;* (2006) 1 SCC 1.

49. *T.N. Godavarman vs. Union of India;* (2008) 7 SCC 126. This judgment was modified subsequently by an unreported order dated 9 May 2008 in the same case, to the extent that some of the exemptions to NPV were clarified.

50. All figures from MoEF, as analysed by the Campaign for Survival and Dignity (2003).

51. *T.N. Godavarman vs. Union of India;* order dated 10 July 2009 in IA no. 2143 in WP (C) 202 of 1995, unreported.

52. In March 2014, the Court permitted the ad-hoc CAMPA to annually release upto 10 per cent of the principal amount lying to the credit of each State/UT (2014 [3] SCALE 566). The judgment makes no mention of the report of the Comptroller and Auditor General drawing attention to the gross mismanagement of this fund on numerous fronts (Report no. 21 of 2013).

53. For a detailed examination of the implications of the Godavarman case on rights of forest-dwelling communities, see Khanna and Naveen, (2005).

54. *T.N. Godavarman vs. Union of India* (2000) 10 SCC 494, order dated 29 July 1998 in IA no. 276 in WP (C) 202 of 1995.

55. *T.N. Godavarman vs. Union of India* 2002 (9) SCALE 81, order dated 29–30 October 2002.

56. Of the Karnataka Forest Act, 1963.

57. 2002 (9) SCALE 81, paras 17 and 18.

58. Ibid.—at para 21. In a follow up of the judgment, the CEC presented a report dated 6 October 2002 where it pointed out the blatant non-compliance of these orders of the Court, in that no penalties had been collected by the state government from anyone and the structures and standing trees in the coffee plantations were left untouched. It completely ignored the state government's submission that the encroachments still remaining to be removed were by poor and landless forest-dwellers. Instead, the CEC pointed out that on its site visit it had found large-scale encroachments in adjoining Reserve Forests of Sargod and Maskali which needed the Court's intervention. An order was passed by the Court on 25 August 2005 (in IA no. 990 in 860 and 818) taking note of the affidavit filed by the state government that removal of encroachments in Sargod and Maskali RF was underway, and directing it to file complete details after removal of all encroachments and taking physical possession of the land. It is not known what has happened subsequently.

59. Order dated 14 July 2003 in IA 860 in WP (C) 202 of 1995, unreported.

60. IA no. 502 in WP (C) 202 of 1995. These organizations had initially approached the Port Blair bench of the Calcutta High Court bench and on its advice approached the Supreme Court.

61. Shekhar Singh of the Indian Institute of Public Administration.

62. Order dated 7 May 2002 in IA no. 502 in WP (C) 202 of 1995, unreported.

63. Recommendation of the Central Empowered Committee in Special Leave to Appeal (Civil) no. 18030 of 2003 (dated 18 March 2004).

64. Such as IA no. 1024 in IA no. 502 in WP (C) 202 of 1995 filed by SANE, BNHS, and Kalpavariksh.

65. Such as IA no. 918 in IA no. 502 in WP (C) 202 of 1995 filed by the Andaman &Nicobar Islands Administration.

66. *Local Borns' Association and ors. vs. The Chief Secretary, Andaman & Nicobar Islands and ors;* SLP (C) no. 18030 of 2003, pending.

67. Order dated 23 November 2001 in IA no. 703 in WP (C) 202 of 1995, unreported.

68. See Chapter 3 in this volume. See also Campaign for Survival and Dignity (2003).

69. Recommendations of the Central Empowered Committee in IA no. 703 of 2001 (dated 5 August 2002), ibid.

70. Campaign for Survival and Dignity (2003) cites the '1990 Guidelines' in full. These are a set of six circulars issued by the MoEF on 18 September 1990 regarding regularization of encroachments on forest land, settlement of disputed claims, forest villages, and so on. The circulars are numbered 13-1/90/FP (1) to (6).

71. IA no. 830-831 in IA 703 in WP (C) 202 of 1995 filed by Paromita Goswami of Elgar Pratishthan, Maharashtra.

72. IA no. 1252 in WP (C) 202 of 1995, being an application filed by the state of Odisha for regularization of encroachments on 3328.4151 ha of forest land in favour of 3754 families under FP (1) of 18 September 1990, on which a report was submitted by the CEC (numbered as IA 1345) recommending that the regularization be allowed on the condition that ineligible encroachers will be evicted. An order in these terms was passed by the Supreme Court on 13 April 2006.

73. IA no. 54 in WP (C) 337 of 1995 (pending before the Forest Bench) filed by the state of Maharashtra for permission to grant *ek saali pattas* to eligible encroachers in Reserve Forests in the state.

74. IA no. 548 in Writ Petition no. 202 of 1995.

75. Since 2011 the WWF case is also being heard by the Forest Bench, although this was not the case at the time the order dated 14 February 2000 was passed.

76. Vide order dated 14 February 2000 in IA no. 548 in WP (Civil) no. 202 of 1995; unreported.

77. Vide order dated 28 February 2000 in IA no. 548 in WP (Civil) no. 202 of 1995, unreported.

78. Vide order dated 3 April 2000 in IA no. 548 in WP (Civil) no. 202 of 1995, 2000 (4) SCALE 168.

79. Vide order dated 11 May 2000 in IA no. 539 in WP (C) no. 202 of 1995. This application was not specifically related to IA 548.

80. Vide order dated 18 February 2002 in IA 707 in order dated 11 May 2000 passed in IA no. 539, as quoted in Ranthambore Foundation (2005b).

81. See for instance IA no. 418 before CEC.

82. IA no. 1355 of 2005 in IA no. 548 in WP (C) no. 202 of 1995. When the matter first came up on 16 September 2005, a report from the MoEF was called, and the application was tagged with IA 703 relating to encroachments.

83. IA no. 2069 of 2006 in IA no. 548 in WP(C) no. 202 of 1995, filed by the state of Odisha.

84. Recommendation of the CEC in IA nos 2069 of 2006 filed by the state of Odisha seeking permission for the collection of kendu leaf from three WLSs in the state (dated 14 October 2009).

85. *Centre for Environmental Law, WWF-India vs. Union of India*, WP (C) no. 337 of 1995, order dated 22 August 1997 as follows:

Even though notification in respect of sanctuaries/national parks have been issued under section 18/35 in all States/ Union Territories, further proceedings as
required under the Act i.e. issue of proclamation under section 21 and other steps
as contemplated by the Act have not been taken. The concerned State Govern-
ments/ Union Territories are directed to issue the proclamation under section 21
in respect of the sanctuaries/national parks within two months and complete the
process of determination of rights and acquisition of lands or rights as contemplated by the Act within a period of one year.

86. *Nilgiris Biosphere Reserve: Fading Glory*, Equations, 2004.

87. The interim report in IA no. 548 for clarification/modification of the Court's order dated 14 February 2000, being IA no. 1220.

88. IA nos 2147–8 in WP (C) no. 202 of 1995, pending.

89. Vide order dated 22 September 2000 in IA 424 in WP (C) no. 202 of 1995, as reported in Ranthambore Foundation (2005a).

90. Personal observation of the author during field visits. See also Campaign for Survival and Dignity (2003).

91. Part IX read with the Eleventh Schedule of the Constitution of India.

92. See in particular Article 244 read with the Fifth Schedule and the Sixth Schedule of the Constitution of India.

93. (1997) 8 SCC 191.

94. Order dated 4 April 2008 in WP (C) no. 202 of 1995, unreported.

95. A significant beginning has been made in the Niyamgiri judgment (*Orissa Mining Corporation vs. MoEF* [2013] 6 SCC 476). Passed by the same Forest Bench, it has proceeded, correctly so, without reference to the restricted world view adopted in the forest case(s) in the past.

References

Breman, J., I. Guerin, and A. Prakash. 2009. *India's Unfree Workforce: Of Bondage Old and New*. New Delhi: Oxford University Press.

Campaign for Survival and Dignity. 2003. *Endangered Symbiosis: Evictions and India's Forest Communities*. Report of the Jan Sunwai (Public Hearing), July 19–20. Delhi: The Delhi Forum and the Other Media.

Chopra, K., G.K. Kadekodi, and V.B. Eswaran. 2005. *Report of the Expert Committee on Net Present Value*, submitted to the Supreme Court of India; http://www.fedmin.com/html/npvk.pdf; accessed 11 February 2014.

Khanna, S. 2005. *Proposal for Reform of Laws Relating to Rights of Tribals and Forest Dwellers in the State of Orissa*. New Delhi: Society for Rural, Urban and Tribal Initiative (SRUTI).

———. 2008. *Exclude and Protect: A Report on the WWF Case on Wildlife Conservation in the Supreme Court of India*. New Delhi: Society for Rural, Urban and Tribal Initiative (SRUTI).

Khanna, S. and T.K. Naveen. 2005. *Contested Terrain: Forest Cases in the Supreme Court of India*. New Delhi: Society for Rural, Urban and Tribal Initiative (SRUTI).

Ranthambore Foundation. 2005a. *Saving India's Forests and Wildlife: The Pioneering Role of the Supreme Court of India, Vol. I, Selected Orders/Judgments*. Mumbai: *Sanctuary* Magazine.

———. 2005b. *Saving India's Forests and Wildlife The Pioneering Role of the Supreme Court of India, Vol. II, Reports of the Central Empowered Committee*. Mumbai: *Sanctuary* Magazine.

7

KANCHI KOHLI
MANJU MENON

The Making of Forest (Re)Publics

Popular Engagement with Official Decision-making on Forest Conversions

Understanding Forest Contexts

Forests have been at the frontier of progress since the beginning of human history. Progress has often meant that forests had to give way, whether it was for the transformation of human communities from hunter-gatherers to settled agriculturists or from that to modern industrial societies. Forests have been pushed further back, circumscribed to particular areas or cut down and made to disappear altogether. Forests have represented wild nature that stands in opposition to civilization, culture, and human progress (Merchant 1980). But although the predominant gaze of human society has increasingly become utilitarian the world over, there is a qualitative difference in the 'objective' (for the larger good of the nation) decision-making process of the state with regard to forests and the experiential and localized forms of conserve and use regimes followed by forest-dependent communities.

In India, the 'management approach' to forests dates back to the colonial period (Rajan 2006; Rangarajan 1996). The possibility of turning tropical forests into an economic resource led the colonial government to systematize record-keeping on forests, protect forests from unregulated use, harvest scientifically, and regenerate methodically. During the late nineteenth century, many forests of central and eastern India were zoned off through 'protection' clauses of colonial legislations. Several of these forests were turned from 'native' multiple-use forests into economically viable, single-use timber plantations. In regions where villagers resisted such moves because they affected local-use regimes, access or control of forests was handed over or shared between the local administration and village communities through complex legal arrangements. At the same time, the colonial government also encouraged the expansion of settled agriculture, as this enhanced land revenue. Towards this end, it liberally handed out forest land in many parts of the country (Das 2009).

Following independence, in the interest of establishing both significant growth and food security, forest conversions were undertaken primarily for the expansion of agriculture on the one hand and the creation of infrastructure like roads and transmission lines and the opening up and building of new mines and hydroelectric projects on the other. In the three decades from the 1950s onwards, newly formed state governments collectively engaged in large-scale conversions of forests for these uses, something which had started in the colonial period itself. The National Commission on Agriculture reported that from 1950 to 1976, approximately 4.3 million hectares (ha) of forest land was diverted for non-forestry use, much of it (2.6 million ha) as part of the government's 'grow more food' policy (Das 2009).

Following this rapid deforestation phase, a major piece of national legislation, namely, the Forest Conservation Act (FCA) was passed in 1980 to regulate forest diversion. How effective has the FCA been in actually conserving forests and checking conflicts around their conversion to industrial uses? In this chapter, we examine the processes of the interpretation and implementation of the FCA. We find that the FCA created new power tussles of a federal, judicial, and scientific nature. Through the analysis of some real-life cases that spilled out of the otherwise strictly departmental process of 'forest clearance' due to challenges by forest-dependent communities and/or advocates of wildlife conservation, we argue that despite this Act, the objective of economic growth takes precedence over local socio-ecological uses of forests.

In this chapter, we first look at how forest land can legally be converted to non-forest use. We then attempt to understand the scale of forest conversion, keeping in mind the socio-ecological context in which conversion takes place. The specific stories and histories of impacts of 'forest clearances' are looked into to understand the design of the FCA. In a substantial portion of the chapter, we examine the systems of compensation and valuation, which have been arrived at as a matter of practice in the process of permitting forest land diversions under the FCA. What were the tussles between the Supreme Court of India and the Ministry of Environment and Forests (MoEF) that gave final shape to these methods and whether they were at any point of time seen as deterrents towards forest conversions for industrial purposes? In an important section further on in the chapter, we look at the FCA in the light of a new legislation, the Scheduled Tribes and Other Traditional Forest Dwellers (Recognition of Forest Rights) Act, 2006 and how this has influenced decision-making around the conversion of forests to non-forest land uses. We finally outline some of the systemic changes that would be needed to make forest regulation work within an adversarial political and economic context. This final section is more policy reflective than policy prescriptive.

Studying the process of forest conversion allows us to explain the increasing challenges that policy faces. This space that is routinely dominated by experts, official knowledge, and decontextualized decision-making may not be able to address localized concerns of balancing ecological concerns and livelihood needs. One aim of the chapter is to explore the nature of knowledge-making on forests in the legal and bureaucratic realms that feed into decision-making on clearances. We also point out that rather than sporadic efforts at opening up these processes to community voices, these decision-making processes will benefit immensely from participatory approaches that are truly open to other priorities of forest-use/conservation. In our analysis, we use methods of social theory to understand the relationship between human society and nature and to explain the problem of forest clearances through the axis of fact versus value, a distinction critical to the study of society and the environment (Barry 2007).

Making of the Forest Regulation

In 1977, following an amendment to the Indian Constitution, forests were made part of the Concurrent List, indicating that its administration and management would now be the joint responsibility of both the central and state governments (http://www.constitution.org/cons/india/shed07.htm, accessed 4 January 2014). It was at this juncture that the Union government began playing a more significant and decisive role in matters relating to forests, especially when it came to determining non-forest use.

Since 1980, the process of forest conversion for development has been popularly referred to as 'diversion of forest land for non-forest use'[1] or 'forest clearance' as put forth under the FCA. Decisions pertaining to freeing up land for non-forest use, de-reserving forests under the Indian Forest Act (IFA), 1927 and felling of trees, all began to be governed through decisions made by the Ministry of Environment and Forests (MoEF).[2]

Although the title of the law and its preambular text indicated that it was a law meant for conserving forests, the procedures it laid out were more directed towards putting into place a legal system whereby forests can be converted or diverted to non-forest uses such as industry, plantations, and infrastructural development. The logic driving this was that regulated use of forests would necessarily lead to protection. In the sections further on in this chapter, we attempt to present evidence and deliberate upon the fact that although regulating the use of forest land is critical to the larger schema for conservation, it may have counter-effects in light of the contradictory priorities of national and state governments. It might also privilege systems that are more expert-driven and less publicly accountable.

Each time forest land is diverted, it must be seen in the wider context of economic growth and development. It is also not devoid of the impacts that decisions around development would have both on forest-dependent communities and the ecological system. In India, where forests are integrated with human use and dependence, any decision about changing forest land use or ownership has a direct bearing on the people living in and around the area. It also has strong ecological consequences as many forest areas act as migratory routes and corridors for wildlife species. Decisions to divert land result in both fragmentation of forests as well as a severing of its relationships with dependents. While there may be differing views on whether forest diversion should take place or not, our main concern here is see how decision-making around conversions take place and what factors ultimately determine whether conversion happens or not.

It is also important to reiterate that before the enactment of the FCA, the powers to reserve or de-reserve any land as forest or allow for the felling of trees were vested in the state governments who carried out these activities in consonance with the respective state forest acts. While the IFA, 1927 framed the broad legal contours of forest management, many state governments also established their own state legislation that at times elaborated upon the IFA (for example, the Karnataka Forest Act, 1963). Therefore many problems of decision-making are linked to both the purposes with which the FCA was created and what impact it had on state–centre relations.

Supreme Court's Intervention in Forest Decision-making

Since 1995, the Supreme Court of India has played a central role in matters of forest policy and governance as a result of the ongoing matter being heard in the *T.N. Godavarman Thirumulpad vs. Union of India* [W.P. (Civil) no. 202 of 1995] case, popularly known as the *Forest* case or the Godavarman case. One of the initial orders in this case substantially changed the manner in which forests were viewed and decision-making around it determined. The order of 12 December 1996 expanded the meaning of the word 'forests' to its dictionary meaning. What this meant was that any area which fit the dictionary definition of a forest could be converted to non-forest use or felled only after approval under the FCA. Forests, in other words, went beyond land recorded as forests in government databases.

Over the years, interventions in the Godavarman case have had a significant bearing on decisions around forest land (see Chapter 6 in this book). This has been both with reference to forest land requirements for industrial and/or infrastructure projects as well as overall policies on compensatory afforestation (CA), forest valuation, and institutional frameworks for the implementation of these policies. The case also resulted in the monitoring and enforcing of procedural requirements of the FCA being influenced by deliberations and decisions in the Godavarman case.

This post-1996 phase of forest diversion added to the complexity of the ongoing tussle between central and state governments and resulted in tussles between the bureaucracy and judiciary in FCA implementation.

The Practice of Forest Diversions: A Private Affair[3]

The Forest Conservation Rules, 2003 (earlier 1981), specify the process by which forest diversions, dereservation, or felling of trees can take place. The procedure does not involve a direct application to the MoEF by the user agency. Whenever a project proponent, either government or non-government, requires the use of forest land for the construction of an industrial or infrastructure facility, the person or agency needs to approach the state government through the relevant Forest Department (FD). There are separate forms for first-time approval and renewal of forest diversions. This form is to be submitted to the concerned nodal officer authorized by the state government (which in most cases is the divisional forest officer (DFO) of the forest division from where the forest diversion is being sought). Amongst other things, this form has to include a detailed cost-benefit analsysis for the use of the forest land. It is only once the state government is satisfied through site visits and the 'application of mind' on the documents submitted that the proposals which require prior approval of the central government, based on Section 2 of the Act, are sent to the MoEF.

In order to carry out the task of forest clearances, the FCA and its rules envisage the role of a Forest Advisory Committee (FAC). The FAC which comprises four non-official expert members, other than the MoEF representatives, has the responsibility to screen these applications, seek additional information, and subsequently recommend the grant or rejection of clearance. The MoEF refers every proposal with complete documents and site inspection report (wherever required) to the FAC. The committee takes a view based on parameters that include the project's location around a protected area (PA), whether the use of the forest land is for agricultural purposes linked to rehabilitation or whether alternatives to the use of this land have been considered or not. The FAC is also to ascertain whether the state government has considered feasible alternatives and the forest land proposed for diversion has been minimized. Finally, it ascertains if the area for CA (see next section for more details) has been acquired. The FAC has the responsibility to screen these applications, seek additional information and subsequently recommend or reject the grant of clearance. Once the FAC has given its opinion, the MoEF decides whether the final forest clearance can be granted or not.[4]

The forest clearance process as envisaged under the FCA and implemented through various guidelines and procedures (MoEF 2004) relies on the interaction between project authorities, state FDs, and relevant offices of the MoEF. At no given point of time does the procedural design of the FCA necessitate the participation of and official dialogue with either forest-dwelling communities, environmentalists, or larger civil society groups seeking opinions on whether the decision to divert forest land would impact forest-dependent communities and conservation initiatives. If at all such an interface takes place, it is at the discretion of the FAC members or the minister. The FCA does not require it to be done. Therefore, while groups and individuals have continued to engage with members of the FAC and send in written observations, they remain external to the official process of decision-making. The insitutionalization of the public interface prior to the grant of forest clearances has remained unaddressed.

This purely internal procedural format of decision-making has led to several disingenuous and furtive practices. An example of such disingenuous and furtive practices was witnessed when forest clearance was given for the USD 12 billion steel plant and captive port of M/s POSCO India Limited. The clearance, and clarifications sought on phone, were preceded by an aerial survey carried out by the DFO instead of a ground verification. The reason cited for this was that since there was local resistance and the villagers had barricaded the area, forest officials decided not to go to the site for ground verification. The DFO, nonetheless, recommended the diversion of the forest land (Mining Zone Peoples' Solidarity Group 2010). This was in violation of the procedures laid out under the FCA wherein the site inspection report of the DFO needs to be appended to the application for forest land diversion. The MoEF subsequently approved the diversion of 1,253.225 ha of forest land for non-forest use in December 2009, specifically for the steel plant. The lack of a full local assessment of both ecological and social implications as is required to be done along with a cost-benefit analysis did not adversely impact the approval given. This suggests two things. First, that the legal requirement of a site inspection report was never completed; and second, that the baseline which was to inform both the state government and MoEF following the ground verification was never undertaken. The approval was granted based on secondary information put forward by the FD. Since these documents were not required to be made public prior to a decision on forest diversion, the contents of the site inspection and subsequent recommendation of the state FD were never known to the affected people.

Section 4.4 of FC Guidelines: All Command and No Control

There are several instances where industrial and infrastructure projects require the use of both forests as well as forest land. Normally, the project authorities initiate construction activity on non-forest land while they await forest clearance to be issued under the FCA. Their justification is that work is being carried on while the approval is pending. The assumption clearly is that the approval will be granted or that the project could go on irrespective of the decision.

It is these specific scenarios that Section 4.4 of the Guidelines for the FCA, 1980, seeks to address. The section is self-explanatory and unambiguous:

> Some projects involve use of forest land as well as non-forest lands. State governments/project authorities sometimes start work on non-forest land in anticipation of the approval of the central government for the release of forest land required for the project. Though the provisions of the Act may not be technically violated by starting work on non-forest land, expenditure incurred on works in the non-forest land may prove to be infructuous if diversion of forest land is not approved. It has therefore been decided that if a project involves forest as well as non-forest land, work should not be started on non-forest land till the approval of the central government for the release of forest land under the Act has been given. (MoEF 2004)

Yet, the implementation of Section 4.4 guidelines has been marred with controversy and conflict of interpretation for a long time now. Several project authorities have chosen to disregard its existence and others consider it a voluntary requirement as it is part of a set of guidelines and not a legal framework. The MoEF on 21 March 2011, keeping many such instances in mind, wrote to all chief secretaries and administrators and reiterated the importance of Section 4.4. It also clarified any possible confusion in interpretation.

Despite clarifications from the MoEF, Section 4.4 has often been loosely applied. One clear example of this was the case of bauxite mining in the Niyamgiri Hills stretching across Kalahandi and Rayagada districts of the eastern Indian state of Odisha. The mining proposed by the Odisha Mining Corporation (OMC) was for the purpose of supplying Vedanta Inc.'s requirements for its alumina refinery located in Lanjigarh. In late 2009, the MoEF issued a press statement during the winter session of Parliament stating that construction activity at the proposed site was carried out on non-forest land without the final approval under the FCA. This was because at that stage the company had received an 'in-principle' (Stage I) approval, not the final approval (Stage II) to initiate mining activity. The MoEF's statement was based on a site inspection report by its regional office. It contended that the construction activity was in contravention of Section 4.4 guidelines (MoEF 2009; Kohli 2010)

This admission of the MoEF in 2009 also needs to be viewed in light of the history of litigation against mining in Niyamgiri Hills. Section 4.4 had previously been invoked in three petitions filed before the Central Empowered Committee (CEC) of the Supreme Court in 2004. These petitions highlighted the illegality of initiating construction activity on the Lanjigarh refinery, primarily on non-forest land, by Sterlite Industries (a subsidiary of Vedanta) without having the forest clearance for mining operations. At that point of time neither the refinery nor the mining project had received forest clearance under the FCA, which made the construction of the refinery even in non-forest land illegal. However, during the course of the hearing the refinery received its forest clearance.

As matters progressed in the Supreme Court, both the project authority and the government of Odisha conveniently delinked the mine and the refinery projects and, therefore, the applicability of Section 4.4 to this case by saying that the construction of the refinery was not dependent on the mining. If this had not been done, the construction of the refinery, too, would have been rendered illegal because as a separate project it would have needed its own environment and forest clearance. What was pending was the bauxite mining in Niyamgiri Hills, which by virtue of this delink, was treated as a separate project with separate clearances. It was also done with the presumption by the project authorities that approval for mining bauxite in Niyamgiri hills could be procured over the next few years (Kohli 2006). The refinery was allowed to be constructed and operate with a separate forest clearance in anticipation of the approval for mining in Niyamgiri Hills.

In October 2012, Vedanta shut down its refinery at Lanjigarh citing the scarcity of bauxite. The mining project has not been granted forest clearance so far (Satapathy 2012). The company was able to get around Section 4.4 to get clearance for the refinery and the MoEF made it possible for them to do this through the artificial delinking of this project from the mine. However, the MoEF itself used Section 4.4, along with environment and social impacts, as reasons for why mining operations should not be allowed in Niyamgiri Hills. Political intervention, local struggle, and international lobbying had a significant role to play in the ministry's decision (Sahu 2010). What this example illustrates is the selective manner in which the MoEF implemented its own guidelines. Such failure to regulate creates difficulties for project proponents whose investments are put at risk by their own actions and also compromises the environmental and social relevance of the decision itself.

Another example of Section 4.4 being circumvented was with respect to the OPG Power Gujarat Pvt Ltd in Bhadreshwar, Mundra taluka, an ecologically fragile area in Kutch district of Gujarat. The company initiated construction activity of an intake/outflow pipeline after they received environmental clearance for the plant and approval under the Coastal Regulation Zone (CRZ) notification, 1991. However, they applied for the diversion of 3.6768 ha of forest land on 17 October 2011 only after the environment and CRZ clearances were in place. Neither of these clearances had disclosed the presence of forest land in the project area. When OPG initiated construction activity, local villagers brought the fact that forest land was involved to the attention of the MoEF. They also raised the issue with India's newly set up National Green Tribunal (NGT). When the NGT heard the case, the legal arguments raised by the company questioned the mandatory nature of the guideline and succeeded in convincing the NGT that they be allowed to carry out construction activity at their own risk (Interim Judgement of the NGT dated 10 January 2012 in M.A. no. 32 of 2011 arising out of Application no. 32 of 2011). This was problematic and against the spirit of Section 4.4 guidelines that warn against carrying out construction activity. This was significant because once the construction was over, the land-use even on the non-forest land underwent transformation and the social impact, if any, was irreversible. Moreover, the construction of part of the project became an argument to prove a fait accompli and to claim that forest land diversions were essential for sustaining project activity.

This guideline has not been upheld by the courts or the MoEF either. Project proponents ignore Section 4.4, continue with business as usual, and cry foul each time action is sought to be taken against them for having started construction without receiving forest clearance. There are only two ways to resolve such situations. One is for the MoEF to stick to the guideline as if it is indeed the law and implement it in all cases irrespective of whether the investor is willing to risk the consequences of ignoring the guidelines. The other is to forsake Section 4.4 altogether and take decisions only on the merit of the case for forest diversion. Between the two, the first makes more sense as a precautionary approach. It helps to maintain a clear regulatory condition, offers the possibility of better and holistic decision-making vis-à-vis the project as well as protect investment, which is in any case the reason why the guideline was introduced. The latter option makes little economic sense. Furthermore, denial of permission to a project already underway has little purchase from politicians, the judiciary or investors. A regulation that has so many people working against it stands a small chance of success.

State of Our Forests: Lost in Numbers

In recent policy debates on forest governance, the actual numbers of forest diversion cases and the physical area they cover have assumed great significance (CSE 2011; Rajshekhar 2012). Yet they tell only a small part of the story of forest diversions. When the Right to Information (RTI) Act, 2005 was first enacted, it became a critical tool through which environmental and human rights groups got access to official data on rates of diversion of forest land from governments. The new opportunities, made possible by the RTI Act, to access this data, were a novelty and a big change from earlier times when all records on decisions related to forest land circulated only between forest offices. But numbers cannot be an end in themselves, especially when the MoEF itself is not shying away from them and making it public through websites and press statements. Today, when the MoEF puts out figures about forest land diversions, it is often to prove that the forest clearance mechanism is functioning and not coming in the way of economic growth as is often alleged by project authorities (Rajshekhar 2012).

When we sought data for this chapter (through the RTI) regarding rates of diversion in 2007 (Kohli and Menon 2009), the figures obtained indicated that since 1980, the MoEF had allowed for the diversion of 11,40,176.86 ha of forest land for non-forest use. These figures also indicated that due to the up-scaling of forest clearances granted to dams, mines, industries, or infrastructure projects like roads, a quarter of these diversions were between 2003 and 2007. The numbers showed that in this period, forest diversion rates matched the rates prior to the promulgation of the FCA. During this period, India diverted an average of almost 1 lakh ha annually (MoEF 2004). Going by the numbers alone, we were as badly off as before the legislation.

In a recent media release of the MoEF in October 2012, former Minister of Environment, Jayanthi Natarajan disclosed that as of 2012, a total 11,44,861 ha of forest land have been diverted in the last 32 years since the FCA came into force.[5] If the figures we obtained through RTI are accurate, it would imply that between 2007 and 2012 only 4,684 ha of forest land have been officially diverted under the FCA. This figure does not seem accurate at all when read with a government press release that states that 15,639 ha of forest land had been diverted between 13 July 2011 and 12 July 2012 (term of Jayanthi Natarajan) The release further adds that the area of forest land diverted during this one year period is 44 per cent of the average annual rate of diversion (35,775 ha per annum) during the 32-year existence of the FCA (MoEF 2012).

Another set of forest diversion figures were presented by the Centre for Science and Environment (CSE) in August 2011. According to their figures, the MoEF granted approval for the diversion of 11,98,676 ha of forest land between 1981 and 2011 (CSE 2011). It is not clear if the figures for forest diversion given by the MoEF under RTI are different each time or if aggregate figures of forest diversion for various kinds of projects calculated by different researchers varies. In a situation such as the above, it becomes clear that the annual or periodical diversion figures are not useful to assess the actual effect of the FCA. The centralization of permissions may have also pushed under the table the rate and extent of illegal use of forest land (Kohli and Menon 2009). And for this, there are no statistics available.

The above inconsistencies raise a number of problems related to the extent of forest diversion. First, the MoEF's own records of forest diversion rates are inconsistent and do not render confidence that there is actually one credible figure that can be relied upon. Second, these numbers do not give a sense of the amount of forest land that might have been converted to non-forest use without permission from the state governments and the MoEF. At the least there should be a centralized record of forest clearance-related violations and the extent of illegal conversion. This is particularly important as there are instances where encroachment on forest land by project authorities has been condoned and land handed over to them post facto (Anonymous 2004; Menon and Vagholikar 2004).

There are also larger questions to reckon with in this puzzle of forest data. There is an overemphasis on numerical data to establish the efficiency or the lack of it of the FCA. This trend is not just in accordance with how the MoEF functions, but also how other research groups and advocacy organizations operate. While it is important to keep an eye on numbers, there are relatively fewer research efforts to critique the functioning of the FCA on substantive procedures and ultimate outcomes. These figures do not reveal the kind and extent of impact such land-use change would have had or what the qualitative nature of the forest that has been lost is.

The Decisive Powers to Divert Forests: Who has the Last Word?

An area of consistent negotiation at the political, judicial, and administrative levels has been with regard to who has the power to decide about forest diversions. Different actors situated within these domains have mobilized arguments in favour of their participation that range from scientific authenticity to economic growth. The very enactment of the FCA shifted the powers of decision making around forest diversions from the states to the centre. The next set of negotiations emerged with the Godavarman interim judgement of the Supreme Court. The executive, especially the MoEF, saw the Supreme Court's intervention as interference. We explore this controversy in this section through two long-drawn-out clashes: first, between the Supreme Court and the

MoEF on the issue of composition of the FAC and second between the FAC and ministers of MoEF.

The MoEF–Supreme Court Stand-off on FAC

As mentioned in earlier sections, the MoEF's FAC has a critical role to play in environmental decision-making pertaining to industrial projects. The Forest Conservation Rules prescribed the nature of expertise needed by this committee. The actual constitution of the committee, however, proved more controversial with an interesting debate ensuing from 2006 about the composition of the FAC (Anonymous 2007; Dutta and Kohli 2007; Jain 2007).

The term of the then FAC had just ended, and the committee had to be reconstituted. The Godavarman bench, while trying to address the issue of the implementation of the FCA, suggested the names of specific individuals who could become non-official members. This judicial intervention, on a matter that was under the routine administrative function of the ministry, was not well received. The names suggested by the court and its monitoring body, the CEC, were not acceptable to the MoEF and till April 2007, a stalemate prevailed as to who was eligible to be on the FAC. The first set of substantive orders that impacted the functioning of the FAC came on 15 December 2006 when the Supreme Court stayed the re-constitution of the FAC following the refusal of the MoEF to accept the credibility of the persons suggested. The suggestions had emanated from the members of the Supreme Court's monitoring body on forest matters, that is, the CEC and the *amicus curiae* (friend of the court) for the Godavarman case, Harish Salve. The Court felt no need to arrive at a set of names for this committee through public negotiation or by an open application process (see Dutta and Kohli 2007; Kohli 2008).

By mid-2007 (and articulated through the 24 April 2007 order), a compromise had been arrived at. While the court temporarily agreed to the names suggested by the MoEF, it directed that all decisions of the new FAC would be reviewed by the CEC. The order read:

> We make it clear that pending the decision of the larger question, all clearances [by the FAC] of fresh cases shall be subject to approval by this Court. Before giving approval, we would like to have responses from CEC in respect of each clearance. In order to avoid delay, we direct the

concerned Ministry to give a copy of [each] clearance to CEC so that CEC would give its response expeditiously. We will examine each clearance and decide whether to grant or not to grant the approval thereto. Once the approval is granted by this Court, the matter may be placed before the Central Government for disposal in accordance with law. (Dutta and Kohli 2007)

The Supreme Court clearly recognized the significant powers of the FAC and MoEF and hoped to make those powers subservient to the decisions of the CEC and the highest Court itself. It also made the advisory body answerable to the Court rather than just to the MoEF, directly attempting to break the line of authority the MoEF had over this advisory committee. Clearly, this was because it felt that the FAC–MoEF combination was not really working in the interest of forests. The Supreme Court's move was seen as an abuse of power and a case of over-interference by the Supreme Court in executive decision-making (Rozencranz and Lele 2008).

It was only in May 2008 (as per the 2 May 2008 order of the Supreme Court in the Godavarman case) that the Supreme Court and MoEF agreed upon the names of the three new non-official members and the powers of the FAC with respect to forest diversions were reinstated to the advisory body without any monitoring role of the CEC.

The FAC and MoEF Stand-offs

The law defines the FAC as an advisory body. What this implies is that the decisions of the FAC are not binding on the MoEF but are recommendatory in nature. The final decision to grant or reject forest clearance vests with the ministry. Therefore, there have been several instances of the minister over-ruling the FAC, mostly in favour of diversion but sometimes not. Much of this had remained a point of internal discussion and articulated through dissenting notes. However, between 2009 and 2011, some of the fractures between the FAC and MoEF came to light due to the ministry's more significant media and web presence. During the above-mentioned phase, there were at least three reported instances when the MoEF went ahead and approved forest land diversion despite strong objections raised by the FAC. These included the 140 MW Alaknanda hydro-electric project in Uttarakhand of the GMR group, the Chiria mines of the Steel Authority of India (SAIL) in the dense Saranda forests in Jharkhand and POSCO India Ltd's steel plant and port (Mandal 2011).

The Saranda forests are part of West Singhbhum, known for its rich flora and mega fauna, especially elephants. These forests, mostly classified as Reserved Forests (RFs), are home to the Ho and other adivasi (tribal) communities. The perennial rivers, Karo and Koina, pass through these forested areas supporting diverse floral and faunal resources. The FAC, based on the ecological and human importance of the forests, had recommended that forest diversion not take place. However, in his speaking order of 9 February 2011, the then Minister Jairam Ramesh explicitly stated that while it is not his intention to interfere with the functioning of the FAC, he as a minister had to take a 'broader view' while arriving at decisions related to forest clearance. The broader view he was referring to was the need to sustain economic growth and resurrect public sector undertakings. The minister gave nine reasons in favour of the forest diversion. The first of these was that SAIL deserved special treatment for being a *maharatna* public sector company with a good record for corporate social responsibility. The next two justifications were that SAIL had a Rs 18,000 crore initial public offering on the anvil with 50 per cent of the proceeds meant for the Government of India and that the prime minister of India had written to the chief minister of Jharkhand in 2007 to renew the mining lease for Chiria in the broader national interest. The Minister did acknowledge that Saranda was an ecologically sensitive area, and as a sop, set up a multi-disciplinary expert committee for monitoring compliance with the conditions laid out for carrying out mining in the area (Anonymous 2011a; Kohli 2011; MoEF 2011). The Wildlife Trust of India was deputed to prepare a wildlife and biodiversity management plan for the area in partnership with SAIL. But this was done without making it clear how post facto biodiversity management can substantially mitigate the biodiversity loss caused by open-cast mining.

Examples exist of the FAC–MoEF tussle going the other way. In the case of the Renuka Dam in Himachal Pradesh, which is to supply drinking water to the city of Delhi, the minister, Jairam Ramesh rejected the forest clearance even though the FAC felt that forest diversion for the project should be approved. The go-ahead to the Rs 3,600 crore project was refused on the grounds that 17 lakh trees had to be cut, despite the fact that land acquisition was supposedly almost over and the Himachal

Pradesh Power Corporation Limited (HPPCL) was in the final stage of inviting global tenders to implement the project (Jebraj 2010).

The discussions between the FAC and MoEF took an interesting turn in September 2011, when the three non-official members on the FAC wrote a letter to the new minister, Jayanthi Natarajan, seeking her intervention to end the 'structural deficiencies' of the committee. The non-official members, including Ullas Karanth, Amita Baviskar, and Mahesh Rangarajan said, 'The present style of working is deeply inimical to our long-term goals and may well weaken the Forest Conservation Act.' In fact, their letter stated that the present functioning of the evaluation process places an unrealistic burden on FAC members and did not allow them to do justice to their responsibilities as non-official members. It went on to highlight that decisions are taken based on inadequate and inaccurate information and that there is no guarantee that the conditions imposed would be enforced (Anonymous 2011c). The points raised in the letter were not new, and had, in fact, been raised by civil society groups and former FACs as well. But this was perhaps the first time that expert members of the FAC clearly expressed their complete inability to function in an environment of administrative short-cuts and procedural violations endorsed by the FD and the ministry. There has been no response from the ministry to this letter as yet.

Legalizing Forest Conversions: The Role of Valuation, Compensation, and Zonation

Institutionalizing Compensation

The legal framework for forest diversion and dereservation prescribed in the FCA did not operate as an end in itself. Rather, it had an inbuilt mechanism of compensating the loss of forest. As originally envisaged, the compensatory formula ensured afforestation either on forest or non-forest land. What is interesting, however, are the layers of monetary compensations that have been devised and added to the forest clearance framework, largely again through the intervention of the Supreme Court in the Godavarman case. Today forest clearance operates with a clearly designed and institutionalized system that is based on a centralized, standardized system of valuation and compensation, thus giving a semblance of an efficient and scientific forest regulation qua conservation. In many ways, the dual strategies of valuation and compensation have come to govern the mechanics of the FCA. Together, they have contributed to the conversion of forests into decontextualized, mobile, and tradable commodities (Kohli and Menon 2011).

FCA's Land Compensation Scheme

Under the FCA, the requirement for CA is one of the most important conditions stipulated when an approval is granted. The FCA requires that loss of forest land is compensated nearby or at least within the state where forest land is lost. Such a system of CA requires a project authority to compensate for the loss of forests in physical terms. All proposals seeking FCA approval (with a few listed exceptions) are accompanied by a comprehensive scheme for CA. According to the law, CA is required to be done on an equivalent non-forest area at the cost of the user agency. Whenever non-forest land is not available, which is to be certified by the chief secretary of the state, CA should be undertaken on twice the extent of degraded forest area, again at the cost of the user agency (MoEF 2004). The preference for non-forest land indicates an interest in expanding the FD's estate.

The idea behind CA was presumably to ensure that some of the environmental benefits of the forests lost to conversion are restored or regained elsewhere. But in practice, the implementation of CA proceeds in locally convenient ways that undermines or complicates forest governance. In the northeast, CA seems to be a tool for the FD to increase the area under its control. For instance, the letter of clearance granted in 2002 to the Bairabi Hydro-Electric Project in Kolasib district of Mizoram for the conversion of 9,294 ha of forest land stated that 1,666 ha was actually unclassified *jhum* (shifting cultivation) land. In the case of the Tuivai project, again in Mizoram, the forest loss due to the project was to be compensated in three ways: the declaration of a new protected area, that is, the Lengteng Wildlife Sanctuary; jhum land of old-growth (at least six to seven years) with five times the number of trees as trees being lost in the submergence to be brought under the FD and regular CA in accordance with the FCA. If this were to all be implemented, the area brought under the FD's jurisdiction would be much higher than what would be lost by diversion. Indeed, most of the hydel power projects that are to come up in the Northeast involve CA and then the reservation of lands under shifting cultivation. Thus, jhum cultivation, a practised unnecessarily demonized by the government, is doubly affected. First, jhum areas are included in the forest area to be diverted for the project, as in Bairabi. Second, as in the Tuivai case, jhum areas are to be taken over for compensatory afforestation and thus brought under the control of the FD. (Menon *et al.* 2003).

In contrast, state governments bent on pushing economic growth proposals requiring forest clearance would rather not bring new areas under CA at all. Gujarat is one such example. The intense industrialization on the Kutch coast in the last two decades following the Bhuj earthquake is a direct result of the tax exemptions offered by the state government, purportedly for rebuilding the economy. The Mundra Port and Special Economic Zone Limited (MPSEZL) project involved the use of 2,000 ha forest land under dense mangrove cover and ecologically fragile mudflats in Mundra taluka of Kutch. The mangrove cover was critical to the health of the coastal ecosystem and provided both spawning grounds for marine life as well as small wood for local use. The CA for the loss of these socially useful and ecologically critical forests has been undertaken on defence land at Kaner and Shinapar villages. Movement of civilians in this region is prohibited as it is adjacent to the India–Pakistan border. Hence, the forests developed under CA can neither be used by communities nor monitored by anyone other than government officials (Kohli and Menon 2011).[6]

Depending on their chances of generating investments that may involve diversion, state governments may want to keep land 'forest-free'. The forest clearances granted to projects in Kutch increasingly involve afforestation undertaken within the campuses of project offices and on highways. By doing so, they not only meet the legal requirements but also prevent more lands from being used for it and are therefore relieved of the burden of having to undergo the processes of clearance for those lands. By not tying up degraded or unforested lands to forest regulation, they also subvert the centre's control on their investment decisions.

Assessments carried out by official agencies reveal that even though finances are made available by user agencies to the FD, CA actually never really takes place. The funds made available to the relevant FDs are spent on purchases such as computers and vehicles rather than plantations and conservation efforts. This is discussed in further detail in the following sections.

Supreme Court's System of Valuation

Over the years, discussions on CA in the Godavarman case have led to two crucial institutional reforms relating to CA. The first reform was the setting up of an ad hoc Compensatory Afforestation Planning and Management Authority (CAMPA) located in New Delhi. The second reform was the putting in place of a system of payment of Net Present Value (NPV) for the diversion of forest land for non-forest use. These mechanisms work on the premise that, while forest land diversions would be inevitable, those who are converting it to non-forest use need to compensate for it in both physical and monetary terms: physical by doing or getting CA done, and monetary by paying a sum far in excess of afforestation costs. The justification lies in reverting both the land and financial compensation back into the hands of the FD who would ideally use it for forest conservation.

I) THE GENESIS AND DEBATES AROUND CAMPA: Before the intervention of the Supreme Court (in the Godavarman case) on the issue of CA, the money for the same was being deposited with the respective state governments in whose states forest loss towards a particular non-forest use was taking place. The user agency's role in afforestation ended as soon as they deposited the amount required for CA with the concerned FD that was then entrusted with the task of carrying out the afforestation.

The first sets of debates around the efficacy of this system had begun in the Supreme Court in 1998–9. However, it was by way of its order dated 3 April 2000 that the Supreme Court observed that only 10 per cent of the CA area had actually been afforested. The Court also took note of the status of cases approved for diversion of forest land, and the extent of use of funds realized and yet to be received by the states (Kohli *et al.* 2011). It observed that there was a shortfall of 36 per cent in terms of the utilization of CA funds across all states cumulatively. Even though states had received the money from user agencies, the same had not been utilized.

Subsequently, in its order dated 8 September 2000, the Court observed that the practice of user agencies merely applying for and depositing money for CA may not be in keeping with the FCA and its rules and stated that, prima facie, the responsibility of afforestation should be with the user agency. The Court also took note of the additional solicitor general's recommendations that the guidelines on CA needed to be upgraded. It observed that not only should it be mandatory to follow the rules and guidelines in this regard, but also that CA should be a time-bound exercise. Furthermore, it emphasized the need for an environmental audit to be carried out by user agencies with respect to survival rates of saplings as well as a process to determine whether or not approvals for diversion should be withdrawn on the basis of such audits. Thus, the court clearly placed the onus of CA on the user agencies, as opposed to the states (Kohli *et al.* 2011).

The above points are critical to understand the working of CAMPA over the last decade and to what extent it has been true to the initial concerns and observations discussed in the Supreme Court. The Supreme Court directed the constitution of CAMPA by its order dated 30 October 2002. The body was to manage a Compensatory Afforestation Fund (CAF), the collected NPV (see more details in the next section), and any other amounts recoverable under the FCA from user agencies for non-forest uses of forest land. The CAMPA was also authorized to disburse funds to the concerned states in predetermined instalments as set out in the state level Annual Plan of Operation (APO). This was important as state governments were now made to justify to CAMPA as to why they need the funds and for what purpose. Many state governments took objection to this as they felt that the money which was being generated through forest diversions in their states should not have to be requisitioned in a convoluted and centralized way (Kohli *et al.* 2011).

Since the order of the Supreme Court dated 30 October 2002, money towards CA has been deposited with the CEC by user agencies, with the intention of transferring the same to CAMPA upon its constitution. However, the delay in the constitution of CAMPA resulted in about Rs 79 crore remaining un-utilized with the CEC. It was not until 23 April 2004 that the MoEF issued a notification for the same; even then no functional institution was set up. The issue was intermittently discussed in the Supreme Court but inconclusively, and hence finally, the Court ordered the setting up of an ad hoc CAMPA to collect funds from all the states towards CA until a full set-up was in place. An order dated 15 September 2006 gave the details of the ad hoc committee. The Court took note of the amount of money already collected by the CEC while CAMPA was being set up, which amounted to nearly Rs 5,600 crore of CAF and NPV funds. However, delays continued and the ad hoc body was made functional only in July 2009 soon after Jairam Ramesh took charge as Minister of State (independent charge) for MoEF and disbursement of the funds to the states was initiated. By way of its order dated 10 July 2009, the MoEF ordered the release of funds from the ad hoc CAMPA for utilization, based on a scheme proposed by the CEC. According to the MoEF, by then the funds available amounted to Rs 9,900 crore of principal and Rs 1,300 crore of interest.

The above chronology shows how compensation for forest loss became essentially a task of managing money. Equally important to note is that CA did not allow for the linking of forest loss to loss of livelihoods and local ecology. All revenues from forest diversions were to go to a central fund and then were to be returned to the state FDs based on application for these funds. The activities for which funds could be sought included a wide range of tasks that could broadly be deemed conservation and were not necessarily CA. The dissemination and appropriate utilization of the CAMPA funds continues to be a concern and has been acknowledged by the MoEF as well (Anonymous 2011b; 2012b).

II) DETERMINATION OF NPV: The CEC, in its report dated 9 August 2002, pointed out that plantations raised under the CA scheme can never adequately compensate for loss of natural forests as plantations take much longer to mature and are a poor substitute for natural forests (CEC 2002; Kohli *et al.* 2011). This then formed the basis of a new NPV doctrine. The Supreme Court's Godavarman bench in its order dated 26 September 2005 defined NPV as 'the present value (PV) of net cash flow from a project, discounted by the cost of capital'. The judgement states that the NPV method 'discounts future costs and future benefits by use of appropriate discount rate and brings such costs and benefits to the reference date which in the present case has been assumed to be the year 2005'. When applied to forest land diversion, NPV is understood as the value, in money terms, of the entire set of tangible products (for example, timber) and intangible benefits (flood control, water or soil conservation benefits) flowing from the forest to society. Since diversion leads to loss of all these ecological services, the new user is expected to compensate society for these losses by the payment of NPV. It was also argued in Court that NPV was a means to collect financial resources that are tilled back into conservation efforts.

Central to the method of NPV calculation is the valuation of the forests based on their condition. The NPV was initially estimated to be in the range of Rs 5.80 lakh per hectare to Rs 9.20 lakh per ha depending upon the density and quality of the forest. NPV collections, in addition to CA, had been in place in Madhya Pradesh, Chhattisgarh, and Bihar even before it was debated in Court and the above-mentioned rate was what was in place in these states. In fact, these practices became the basis of discussions in court-related mechanisms for determining and recovery of NPV based on the quality and density of forests.

Subsequently, the Supreme Court set up an expert committee (known as the Kanchan Chopra committee) to advise it on NPV calculation. The recommendations of this committee were further modified by the CEC. Eventually, since 2008, the rate has been fixed between Rs 4.38 lakh per ha to Rs 10.43 lakh per ha (as per the order dated 28 March 2008) based on a detailed chart prepared by the CEC. For calculating the average NPV per ha, the CEC considered seven forest-related goods and services. These include timber, firewood, NTFP, fodder, eco-tourism, bioprospecting, flagship forest species, and carbon sequestration. The nature of goods and services that each class of forests could offer was taken to be proportional to its tree density. Thus, the valuation has become highly tree-centric.

Apart from the numbers, operationalizing of NPV and CA in particular is challenging. Take, for example, the case of Kinnaur district in Himachal Pradesh. A substantial portion of the district is above the treeline and constitutes a high-altitude cold desert area. The forests and alpine pastureland of the region, for the most part, are not ones where high tree density can be found. During a conversation with forest officials of the region in June 2011, we learnt that forest land is continuously being sought for the construction of border roads as well as hydro-power projects, but the district does not have any land where CA can take place. Therefore, if any forest land is diverted in Kinnaur, CA will need to take place in another district of Himachal Pradesh, land for which is yet to be identified (Kalpavriksh 2011; Kohli *et al.* 2011).

Both CA and payment of NPV are now established practices under the FCA. While the first was already envisaged in the legislation, the latter was added as a compensation tool in recognition of the wider range of forest services. In both instances, the controls remain with the central government, much to the discontent of the states. Money from NPV gets collected for all diversions and put in a common pool. The money collected can be used to carry out routine conservation-related operations specified under working plans or for a wide range of other activities that are categorized as forest conservation initiatives. If this fund is being used to finance forest conservation in general, it would have to be maintained forever in order to keep conservation activities going.

Two crucial recommendations of the Kanchan Chopra committee, neither of which were finally accepted by the Supreme Court, are worthy of attention. First, the committee recommended 12 steps that should be followed in order to determine NPV and claims by stakeholders at the forest-range level. Steps 1 to 11 deal with the calculation of NPV and the determination of the claims of all relevant stakeholders to the forest that is to be diverted. These steps include determining the legal status of land involved, its classification, and the kinds of products and services to be valued (such as timber, carbon storage, ecotourism, and NTFP). Step 12 deals with compensation to major stakeholders. According to the committee, the major stakeholders include local communities, the state FDs, and the central government. Compensation to these stakeholders is to be made according to pre-determined norms. Second, the committee also proposed that the different amounts collected as NPV be deposited in funds established at local, state, and national levels. Different mechanisms for administering these funds were also stipulated. However, the Court in its wisdom rejected these recommendations when, in fact, their adoption would have led to a fairer distribution of the compensation (Kohli *et al.* 2011).

The question of monitoring the use of funds becomes more problematic as funds collected from individual projects can be used anywhere in the state. Due to the absence of direct links between projects, their payments under CA and NPV to CAMPA, and their utilization by states, funds will become increasingly difficult to track. This will make it more difficult to assess whether CA and NPV actually compensate for the loss of forest services.

Zoning Out Go–No-Go: It's Anyone's Guess

One of the key issues that has emerged in the discussion around forest trade-offs pertains to the creation of 'go' and 'no-go' forest areas for the purpose of coal-mining. The conversation around 'go' and 'no-go' was initiated by the Ministry of Coal (MoC) and the MoEF in June

2009 in order to identify which blocks in India's existing nine coalfields should be allowed to be mined and which should remain untouched and used as 'strategic energy reserves' in the future. Unfragmented forest landscapes with average crown densities of more than 0.50 were designated as Category A (no-go areas) and areas with less than 0.50 density Category B (go areas).[7]

Using tree-cover density as the main criteria to determine the status of forests was problematic because it made irrelevant all other factors including human dependence on forests. Many areas marked for mining are inhabited by forest-dwelling and dependent communities. Moreover, many of these areas are also valuable ecosystems. For example, open scrub forests or *chhote bade jhaad ka jungle* are not dense but nonetheless ecologically valuable. There is an abundance of such forests in the central Indian forest belt where many coal blocks are located.

The 'go' and 'no-go' distinction must be situated within wider political developments. In March 2010 when the 'go, no-go' list was first released, several forest blocks, including those in the Singrauli region (considered to be India's energy capital) were categorized as 'no-go' zones. Forest areas such as Mahan, Chhattrasal, Amelia, and Dongrital II accounted for 59 per cent of the area (coalfields) that were not allowed to be mined. The full list of areas, available with Greenpeace, an international NGO (procured through RTI), contained 222 coal blocks accounting for 48 per cent of the total coal area that were not to be granted coal mining approvals (Greenpeace 2011).

A year later, after negotiation, the number of 'no-go' coal blocks had come down to 153. On 1 March 2011, the MoEF minister, Jairam Ramesh mentioned in the Rajya Sabha that the forests of Mahan were to be leased to the Essar and Hindalco companies. The forests of Chhatrasal were to be set aside for Reliance's Sassan power plant operations. Dongrital II and Amelia were allocated to Jaypee. Each of these allocations was meant to promote further generation of thermal power.

On 3 February 2011, the Government of India (GoI) constituted a Group of Ministers (GoMs) to look at issues around coal block allocations and streamlining the process of environment and forest clearances. The appointed mediator for this was a special Empowered Group of Ministers (EGoM) headed by Pranab Mukherjee, the current President of India. This EGoM steered the discussions between the two ministries vis-à-vis the future of some large investment projects and in particular coal mines. It also looked at the nature and extent of regulation that the coal sector should be under. The discussions between the EGoM and the two ministries remained inconclusive both with regard to the criteria needed for coal block allocation and specific allocations (Press Trust of India 2012; PIB 2012).

The next environment minister, Jayanthi Natarajan, however, did not favour the 'go' and 'no-go' classification. Instead, she was in favour of a process of identifying inviolate and good forests based on stated scientific parameters, such as protected area status, crop diversity, and biodiversity richness (Narayan and Suneja 2010; Sethi 2012). The then home minister P. Chidambaram was asked to head the GoM to decide on a regulator for the coal sector, with a special coal regulatory authority in the offing. It is not clear what the future criteria for inviolate areas are likely to be, although news reports indicate that deciding upon such criteria are on the back-burner as new individual projects are being decided upon (Anonymous 2012a).

There are a few lessons to be drawn from this discussion around zonation. To begin with, none of the forest lands classified as no-go or inviolate arenas have been permanently designated as such. Moreover, forests not being allowed for coal mining did not guarantee that they could not be diverted for other purposes such as industries and infrastructure projects. Finally, the entire deliberation remained a technocratic and expert-driven exercise with absolutely no space in this discussion for forest-dwelling and dependent communities as well as civil society groups.

In practice, there is no forest land in India that cannot be opened up for industrial and infrastructure purposes. National parks and sanctuaries are considered to be the most sacrosanct areas. However, evidence indicates that even these areas have been opened up for industry, mining, power-generation and other infrastructure purposes by the MoEF in the last few years (Menon *et al.* 2010). There is no guarantee then that areas zoned as 'no-go' will not become 'go' zones if there are counter-rational arguments proposed in the future.

FCA in the Time of FRA: The Making of Forest (Re)Publics

In 2006, another long-pending and crucial layer was added to forest clearance process when the country's parliament brought into force the Scheduled Tribes and Other Forest-Dwellers (Recognition of Forest Rights) Act, 2006 (FRA). The relevance of the FRA goes well beyond the FCA process as it recognizes and vests forest rights and occupation of forest land in forest-dwelling communities (Scheduled Tribes and Other Traditional Forest Dwellers). Rights bestowed include both individual rights and community rights. While there are different sets of criteria for the eligibility of tribal and non-tribal forest-dwelling communities (the former having to show occupation prior to December 2005, and the latter having to show 75 years of residence in the area where the claim is being made), individual and community rights can be claimed by both. However, what it has also meant is that no forest from now on can be diverted for non-forest use till such time these rights are recognized and mechanisms to settle them arrived that (see the chapter by Madhu Sarin in this book). The efficacy and outcomes of this process are being tested in the varied forest contexts (MoEF–MoTA 2010).

The MoEF issued a circular on 30 July 2009 that required claims to be settled prior to any conversion of forest land. Furthermore, the July 2009 order required that written consent or rejection by the gram sabha of the diversion proposal be included in the documents submitted.[8] The circular was forwarded to all chief secretaries on 3 August 2009 (F.No.11.9/1998 – FC [pt]), but the circular has not been implemented consistently, primarily because of its discretionary use and the MoEF's inability to take a position contrary to the state governments. Often state governments insist that no rights exist or that they have already been clarified under FRA. There also continue to be larger unanswered questions concerning the interplay of the FRA and FCA. Is reassurance from the central government or a reliance on documents produced by the state governments adequate to go about the task laid out? Has there been a differential interpretation of the FRA in different projects during the grant of approval? Three cases of forest clearance illustrate the complexities and variations in treatment.

On 24 August 2010, Jairam Ramesh, the then MoEF Minister, rejected the application for the use of 660.749 ha of forest land for bauxite mining in the Niyamgiri Hills of Odisha. Niyamgiri is home to the Dongria Kondh tribal community and is an extremely ecologically and culturally important landscape (MoEF 2010b). The proposal for mining in the area has received international attention from 2005 to 2006 when the issue first came to light. The FAC and then later, the Minister, rejected the proposal for mining in Niyamgiri on the grounds that the FRA process in the area was incomplete and that there would be large ecological and human costs. The FAC in its report of 23 August 2010 indicated that the proposal for bauxite mining in Niyamgiri is a fit case to apply the precautionary principle and recommended temporary withdrawal of the in-principle/Stage I forest clearance that has been granted to the Odisha Mining Corporation. The rejection was partly based on the report of a four-member committee headed by N.C. Saxena (MoEF–MoTA 2010).

Just a month before the Niyamgiri decision, however, the MoEF, in its press release of the 28 July 2010 (MoEF 2010b), indicated its reasons and justifications for the approval of forest land diversion to construct the Indira Sagar (Polavaram) Project in Andhra Pradesh. This approval was based on an assurance by the state government that there are no forest rights that need to be settled in the forest area that was being diverted for non-forest use. The project required 3,731.07 ha of forest land (3,473 ha of notified forest and 258.07 ha of deemed forest). This decision was heavily criticized on the grounds that the Polavaram project threatened to displace more than 270 villages and two lakh people, a majority of whom were tribals. Forest rights groups argued that the majority of tribal communities in this area were entitled to rights under the FRA and, therefore, merely going by what the state government had presented would be unfair.

Finally, the granting of forest clearance to M/s POSCO's steel plant and port in Jagatsinghpur district of Odisha went through an even more tortuous process. As mentioned in earlier sections, POSCO required 1,253.225 ha of forest land for which the MoEF granted conditional forest clearance on 29 December 2009. Due process under the FRA needed to be completed first. This came less than five months after the ministry had itself issued circulars in July and August 2009 contradicting the requirements listed therein, rather then reducing the requirements to a condition. In the meanwhile the MoEF constituted a committee under the chairpersonship of Meena Gupta to review forest and environment clearances granted to the POSCO project.

The final order in the POSCO FRA matter came only on 2 May 2011 (MoEF 2011). This order was issued after the MoEF took into account the submissions of the Posco Pratirodh Sangram Samiti (PPSS), the group leading the local movement against POSCO. However, to make its decision, the MoEF relied on the final word of the state government, who indicated in writing that there are no 'other forest-dwelling communities' in the area and, therefore, approval should be granted. This also went contrary to the observations of the Meena Gupta Committee, which in its report had stated that violations of the July–August 2009 circulars were significant, and that other forest-dwelling communities lived in the project–impacted area (Kohli *et al.* 2012; MoEF–MoTA 2010). The grant of approval in violation of the FRA has been challenged in the Odisha High Court by the people of the area. This, too, is pending a final decision. Further, on 18 October, more than 2,000 people participated in the gram sabha of the Dhinkia panchayat and unanimously passed a resolution asserting their claims under the FRA. The villagers cited a letter (No. TD 11 [FRA]-06/2011) issued by the ST and SC Development Department of the Odisha government on 7 September 2012 instructing district collectors to ensure discussion and planning on the pending claims under the FRA and the consolidation of the list of claims for further monitoring and disposal.

While there is no denying that the recognition of rights is now a legal prerequisite to the diversion of forest land for non-forest use, there continues to be a difference of opinion even within the MoEF on whether the consent of the gram sabha is a mandatory requirement prior to the issuance of forest clearance. It is being stated that while the 2009 circular includes this requirement, there is no explicit requirement prescribed by the parent FRA legislation. While supporters of the FRA push for this liberal interpretation of the law in granting rights and responsibilities to tribal and forest-dwelling communities, the forester members of the FAC have insisted that this is not necessary at the first stage of 'in-principle' clearance and can be obtained before final clearance (Anonymous 2012a). At the same time, the MoEF–MoTA 2010 joint committee has recommended that the best thing to do is to enshrine all the requirements of the FRA into an amendment of FCA and make both laws consistent.

***

Quantitative data pertaining to forest diversion during the three decades after the FCA might not be useful to determine whether the law has achieved its stated goals or outcomes.[9] First, it is not clear what the envisaged outcomes of the FCA were. The Indian Forest Act is primarily concerned with the 'creation' of forests through processes of reservation, settlement of claims, and establishment of privileges of use by communities. The FCA, on the other hand, is burdened with the bigger task of 'conservation'. It is expected to achieve conservation not only on forest land, but also on non-forest land, the distinction between the two becoming more unclear with the Godavarman judgement.

But what conservation means remains unclear. Is it to bring a greater area of land (forest or non-forest) under forest cover? Is it improving the density of existing forests? Is it a way to reconcile economic and ecological uses of forests? Or is it simply a case of facilitating a smoother process of conversion of standing forests to non-forest use for industrialization and infrastructure development? Our argument is that the FCA has as yet to imagine and articulate a statement of collective public intent. Without such a clear statement of purpose that reflects a collective normative position on forests, this Act is merely a set of procedures to be followed for diversion of forest land. Implementors will not know what their decisions must add up to and those who make demands for forest land will have no idea if their demands are reasonable or not. Worse still, procedural integrity and economic progress have by default become the absent values. This only leads to legalized forest clearance that may have greater undesirable effects on forest-dependent communities and ecology. In matters of the environment, decisions have to be tied to clearly stated, unambiguous qualitative outcomes. Otherwise, decision-making around the FCA will be dominated by litigation.

The absence of value statements and outcomes in forest legislations has pushed forest matters into labyrinthine courtroom contests. The popular challenge to decision-making is dominated by litigation where procedural breaches get debated more rather than the complex questions of modern human societies' relationship to forests. The legal debates on 'matters of merit' and the 'application of mind', phrases often referred to in court, help to open up decision-making but this comes so late in the process of governance that it may be impossible to go back on the decision for diversion. Therefore, injecting values and outcomes into the legislation will help create these benchmarks against which the cumulative effects of routine decision-making on forests can be assessed by anyone, including the officials in charge of implementation. If we do not undertake this exercise in a way that is representative of good governance, both governments and the public, run the risk of being instructed by the courtroom on what a forest is, where it exists, who should use it and how.

The second and a more technical reason for not being able to make a confident statement on whether the FCA implementation so far has left forests and forest-dependent people better or worse off is the sloppy classification employed in maintaining data on forests and the difficulty in accessing documentation produced for the purpose of decision-making. All sorts of tree canopy are called forests, including monocultures tree plantations, coffee plantations, and even large parks (Lele 2012). Some of these, such as coffee plantations, are non-forest land uses as per the FCA and could be the result of 'forest clearance' decisions of the FCA. Equally important is that even FSI data on forests does not distinguish between private and public forests. This makes it impossible to attribute the increase or decrease in forest cover to specific actors, particularly the FD. Researchers have pointed to the need for use of better technologies for data-gathering purposes. Available expertise and tools in satellite imagery and spatial mapping now allow for the collection of more nuanced and fine-grained data on forests (Lele 2012; Nagendra 2012).

Documentation produced for the purpose of decision-making is almost entirely treated as classified information available only to the FD, the FAC and the MoEF. It is not made available to the communities who may depend upon the forests in question for their livelihood, or to the researchers or wildlife groups who may be engaged in wildlife conservation. The forest map and estimates of forest density or trees to cut down are not available even to the 'user agency' that seeks to put up the project. Communities, researchers, and NGOs officially get to know about the forest clearance when the list of 'projects pending decision' is put up on the MoEF's website or when the decision has been taken. Moreover, much investigative energy has to be spent by individuals and groups to procure the paper trail for the purpose of challenging these decisions in courts. Without access to such data, it is impossible to understand and meaningfully debate conversions of forests for development purposes.

In spite of several efforts aimed at decentralization and the rainbow of movements for greater public participation in governance, the forest clearance process remains almost bereft of such opportunities. The only positive development of late has been the FRA. Following the enactment of this legislation, forest clearance cannot be sanctioned without first recognizing individual and community rights in the concerned forest areas. This is a significant step towards the democratization of forest governance, but is only a start. Much remains to be done to bring forest-related decision-making into the public sphere where the prioritization of forest uses is arrived at through debate and discussion that pay attention to the history of conservation, local traditions of use, and planetary ecological concerns.

A system of periodic and transparent review of environment law performance needs to be built in to the legal framework so as to prevent decisions based on populist pressures, litigation, and change of government. Decisions related to forest conservation involve long gestation periods during which the effects of the decision may not be evident. Sudden changes in policy or law that result in transfer of decision-making powers, procedures, budgets, and monitoring responsibilities have debilitating impacts on forest-related or forest-based schemes, programmes, and projects. The experience with the FCA illustrates this.

India is often cited as a case of good laws but poor implementation. The closed process of decision-making lends itself to such selective implementation and enforcement of decisions. In most cases, lower-rung FD officials, the FAC and even the MoEF are unable to withstand pressures and end up compromising laws, allowing the illegal use of forests and condoning violations. Environmental laws are considered risks to investment and equated with the licence raj. If decision-making processes are made more inclusive of citizens and transparent, there is a better chance of resolving sticky issues. Whether the compulsions of growth allow this remains to be seen.

Notes

1. In subsequent discussions in this chapter, the word 'diversion' might be used to explain specific instances and examples where de-reservation, diversion for non-forest use, felling of trees, or a combination of these has taken place.

2. Forest diversion for non-forest use can take place either with no de-reservation of land and no felling of trees, either, or both. When either de-reservation or felling occurs, permission to use forest land for a non-forest purpose would have to be obtained.

3. Since the time this chapter was finalized, there have been significant decisions of the National Green Tribunal (NGT) related to the FCA and its jurisdiction. The analysis given here does not include these changes.

4. The details of the above procedure are specified in the FCA 1980 and the Forest Conservation Rules, 1981 (as amended in 2003) and the various forms annexed to the rules.

5. The figures were released in a note titled 'Forest Clearance Approvals granted from 13 July 2011 to 12 July 2012 by Jayanthi Natarajan as Minister for Environment and Forests' uploaded on the website of MoEF in October 2012.

6. We are grateful to Bharat Patel and Vimal Kalavadiya, our colleagues on an ongoing research project on CA in Kutch district, for this information.

7. Comments of the MoEF on a Draft Note for Cabinet Committee on Infrastructure Regarding Need for Making Available More Coal Bearing Areas for Enhancing Coal Production (undated). [Information collected through Right to Information in 2011 by Greenpeace India].

8. There are multiple legal interpretations of this consent clause as the FRA itself does not list a mandatory requirement for the same. However, it has been argued that if one is to interpret clauses of the FRA that give rights under Section 3 and powers under Section 5 to protect and conserve biodiversity/habitats, such consent becomes mandatory. Moreover, the circular does require that consent is mandatory by indicating that consent or rejection is to be noted and reported to the MoEF.

9. We are grateful to our colleague Vivek Maru for numerous discussions on the importance of environmental and social outcomes of environment laws.

References

Anonymous. 2004. 'NHPC continues dumping in contempt, alleges Forest Department'. *Sikkim Now*, 28 August.

———. 2007. 'Govt, SC disagree over forest panel members'. *The Indian Express*, 6 January.

———. 2011a. 'MoEF grants forest clearance for SAIL's Chiria mines'. *The Hindu*, 10 February; http://www.thehindu.com/news/national/moef-grants-forest-clearance-for-sails-chiria-mines/article1200017.ece.

———. 2011b. 'Jairam unhappy with forest scheme implementation'. *The Hindu*, 28 March; http://www.hindu.com/2011/03/27/stories/2011032765021200.htm.

———. 2011c. 'Projects and Forests: Flawed Clearances and Complicit Foresters'. *Economic and Political Weekly*, 8 October, 46 (41): 15–17.

———. 2012a. 'P. Chidambaram to Head Group of Ministers on Coal Regulator'. *The Economic Times*, 6 June; http://articles.economictimes.indiatimes.com/2012-06-06/news/32079086_1_coal-sector-coal-blocks-coal-mining.

———. 2012b. 'Jayanthi Natarajan Turns Down Proposal Seeking Rs 1K crore under CAMPA'. *The Economic Times*, 27 January; http://articles.

economictimes.indiatimes.com/2012-01-27/news/30670322_1_campa-sustainable-forest-management-auditor-general.

Barry, J. 2007. *Environment and Social Theory*, second edition. London: Routledge.

CEC (Central Empowered Committee). 2002. 'Recommendations of the Central Empowered Committee in Interlocutory Application No. 566 of 2000 in Writ Petition (Civil) 202 of 1995'. New Delhi: Supreme Court of India.

Centre for Science and Environment (CSE). 2011. 'Forest and Environment Clearances, Public Watch No.1'. New Delhi: CSE, New Delhi.

Das, S. 2009. 'Pressure of Conversion of Forest land to Non-forest Uses in India'. Paper presented at *Indian Forestry: Key Trends and Challenges*, organized by the Rights and Resources Institute and Indian National Trust For Art and Cultural Heritage (INTACH) at New Delhi on 5–9 March.

Dutta, R. and K. Kohli (eds). 2007. 'Highlights of the Godavarman (Supreme Court) and CEC Hearings in April 2007'. *Forest Case Update*, Issue 34, April 2007; www.forestcaseindia.org, New Delhi.

Greenpeace. 2011. 'Singrauli: The Coal Curse (A Fact-finding Report on the Impact of Coal Mining on the People and Environment of Singrauli)'. Bengaluru: Greenpeace-India.

Jain, S. 2007. 'We Need Experts, Not Activists, Said Govt, Rejecting All 9 Names Proposed by SC Panel'. *The Indian Express*, January 10; http://www.indianexpress.com/news/we-need-experts-not-activists-said-govt-rejecting-all-9-names-proposed-by-sc-panel/20576/.

Jebraj, P. 2010. 'Environment Ministry Blocks Renuka Dam Project'. *The Hindu*, 14 October 14; http://www.hindu.com/2010/10/14/stories/2010101462640400.htm.

Kalpavriksh. 2011. 'Compensatory Afforestation and Net Present Value Payments for Diversion of Forest Land in India'. New Delhi: Kalpavriksh and Heinrich Boell Foundation.

Kohli, K. 2006. 'Mine? What Mine? Ah, Yes, the Mine'; www.indiatogether.org, December 2006.

———. 2008. 'Still Advising the Forest Committee?'; www.indiatogether.org, 11 June 2008.

———. 2010. 'Round and Round the Sacred Hills'; www.indiatogether.org, 5 August 2010.

———. 2011. *Hiatus, Controversies, Collapses*; www.mylaw.net, 24 December 2011.

Kohli, K. and M. Menon. 2009. 'Liberalisation and Environmental Legislation in India,' in A. Perry-Kessaris (ed.), *Law in Pursuit of Development*. Abingdon: Routledge-Cavendish.

———. 2011. 'Banking on Forests: Assets for a Climate Cure?' New Delhi: Kalpavriksh and Heinrich Boll Foundation.

Kohli, K., A. Kothari, and P. Pillai. 2012. 'Counting Coal: A Discussion Paper by Kalpavriksh and Greenpeace'. Delhi/Pune: Kalpavriksh, and Bengaluru: Greenpeace India.

Kohli, K., M. Menon, V. Samdariya, and S. Guptabhaya. 2011. 'Pocketful of Forests: Legal Debates on Valuating and Compensating Forest Loss in India'. New Delhi: Kalpavriksh and WWF-India.

Lele, Sharachchandra. 2012. 'Standalone Agency to Map Greenwealth'. *The Economic Times*, 12 May, p. 9.

Mandal, P. 2011. 'Forest Advisory Committee Turns Toothless'. *Business Standard*, 20 December; http: //www.business-standard.com/article/economy-policy/forest-advisory-committee-turns-toothless-111122000058_1.html.

Menon, M., K. Kohli, and V. Samdariya. 2010. 'Diversion of Protected Areas: Role of the Wildlife Board', *Economic and Political Weekly*, July 26, 45 (26–27): 18–21.

Menon, M. and N. Vagholikar. 2004. 'Violating the Teesta', *The Statesman*, 26 June.

Menon, M., N. Vagholikar, K. Kohli and A. Fernandes. 2003. 'Large Dams in the North East: A Bright Future?'. *The Ecologist Asia*, 11 (1): 3–8.

Merchant, C. 1980. *The Death of Nature: Women, Ecology and the Scientific Revolution*. San Francisco: HarperCollins.

Mining Zone Peoples' Solidarity Group (MZPSG). 2010. *Iron and Steel: Iron and Steal: The POSCO India Story;* http://miningzone.org/wp-content/uploads/2010/10/Iron-and-Steal.pdf.

Ministry of Environment and Forests (MoEF). 2004. 'Handbook of Forest (Conservation) Act, 1980 (with amendments made in 1988), Forest (Conservation) Rules, 2003 (with amendments made in 2004), Guidelines and Clarifications (up to June, 2004). New Delhi: Ministry of Environment and Forests, Government of India.

———. 2009. 'Project for Bauxite Mining in Orissa'. *Press Information Bureau*, 27 November.

———. 2010a. 'Decision on Grant of Forest Clearance in Kalahandi and Rayagada Districts [in] Orissa for the Proposals Submitted by the Orissa Mining Corporation Ltd (OMC) for Bauxite Mining in Lanjigarh Bauxite Mines', Ministry of Environment and Forests, 24 August 2010.

———. 2010b. 'Indira Sagar (Polavaram) Multipurpose Project on Godavari River in Andhra Pradesh given final clearance under FCA, 1980'. *Press Release*, July 28. New Delhi: Ministry of Environment and Forests.

———. 2011. *POSCO*, Ministry of Environment and Forests, 2 May 2011.

———. 2012. 'Forest Clearance Approvals Granted from 13 July 2011 to 12 July 2012 by Smt. Jayanthi Natarajan as Minister for Environment and Forests', Ministry of Environment and Forests; http://moef.nic.in/assets/ec-fc-clearances.pdf.

Ministry of Environment and Forests and Ministry of Tribal Affairs (MoEF–MoTA). 2010. 'Manthan: Report of the Committee on Forest Rights Act'. New Delhi: Ministry of Environment and Forests and Ministry of Tribal Affairs, Government of India.

Nagendra, H. 2012. 'Use Eye in the Sky to Manage Forest Cover', *The Economic Times*, 12 May, p. 9.

Narayan, S. and K. Suneja. 2012. 'After Go, No-go Classification, Inviolate Area Haunts Industry'. *Financial Express*, 1 June.

Press Information Bureau (PIB). 2012. 'Recommendation of Group of Ministers (GoM) Regarding Inclusion of Damage to Crops etc. due to Cold Wave/frost as an Eligible Calamity for Relief under NDRF/SDRF'. Press Information Bureau, 20 July 2012.

Press Trust of India (PTI). 2012. 'GoM to Discuss Clearances to Mahan, Chhatrasal Mines Tomorrow', *The Economic Times*, 29 May 29; http://articles.economictimes.indiatimes.com/2012-05-29/news/31888052_1_coal-blocks-forest-clearances-coal-ministry.

Rajan, R. 2006. *Modernising Nature: Forestry and Imperial Eco-Development 1800–1950. Oxford Historical Monographs.* Oxford: Oxford University Press.

Rajshekhar, M. 2012. 'UPA Fastest in Granting Coal Mining Clearances during its Eight Years', *The Economic Times*, 15 June; http://articles.economictimes.indiatimes.com/2012-06-15/news/32254609_1_environment-clearances-forest-clearances-coal-projects.

Rangarajan, M. 1996. *Fencing the Forest: Conservation and Ecological Change in India's Central Provinces 1860–1914.* New Delhi: Oxford University Press.

Rozencranz, A. and S. Lele. 2008. 'Supreme Court and India's Forests', *Economic and Political Weekly*, 2 February, 43 (5).

Sahu, P.R. 2010. 'Rahul Woos Anti-mine Orissa Tribals', *Hindustan Times*, 26 August; http://www.hindustantimes.com/News-Feed/Orissa/Rahul-woos-anti-mine-Orissa-tribals/Article1-591927.aspx.

Satapathy, R. 2012. 'Vedanta Shuts Down Lanjigarh Refinery', *The Times of India*, 14 October 14; http://articles.timesofindia.indiatimes.com/2012-10-14/bhubaneswar/34448411_1_mtpa-refinery-plant-bauxite-lanjigarh-refinery.

Sethi, N. 2012. 'Forests Off Limits for All Mining?' *The Times of India*, 13 June; http://articles.timesofindia.indiatimes.com/2012-06-13/flora-fauna/32214343_1_catchment-forest-survey-forest-areas.

SOME CROSS-CUTTING ISSUES

KUNDAN KUMAR

Erasing the Swiddens

Shifting Cultivation, Land and Forest Rights in Odisha

Swidden or shifting cultivation is one of the main land-uses in the world's forested areas. In India, shifting cultivation was widely practised in forested areas in the pre-British period and early British period. Shifting cultivation was part of a mixed use 'agroforestry land-scape that included domesticated forests, grasslands and farms nestled amongst each other' (Sivaramakrishnan 1999). Forests and agriculture as discrete, dichotomous land-uses were imposed on these mixed use landscapes by the British practice of categorizing large areas of forested landscapes as 'legal forests'. This approach was continued by the post-colonial Indian state. Currently, almost a quarter of India's land area is categorized as legal forests. By independence, shifting cultivation as a land-use survived only in certain pockets of central India and in the northeastern part of the country. Even in these areas, shifting cultivation lands are being aggressively converted into legal forests and state land, marginalizing adivasi communities practising shifting cultivation (Darlong 2004). This has become an important cause of conflict and contestation in these areas.

The dominant Indian perspective on shifting cultivation is that it destroys forests and leads to ecological degradation (Dhebar 1961; GoI

1988; GoI 1997). The conceptual separation between forests and agriculture, rooted in the Western paradigm of nature versus culture as well as in the origins of modern agricultural systems, has meant that policymakers have been unable to visualize shifting cultivation as part of an integrated livelihood and landscape system. In the context of Northeast India, this perspective is undergoing change based on recent research and increasingly policymakers are talking about 'improving' as opposed to 'replacing' shifting cultivation (GoI 2002). However, in other parts of India, shifting cultivation continues to be perceived negatively and the government continues to pursue the policy of 'replacing' shifting cultivation.

In this chapter, I examine the processes through which hill slopes customarily used for shifting cultivation in Odisha were taken over by the state as forest land or revenue wasteland through a deliberate policy aimed at stopping shifting cultivation. This is a repeat of similar processes that happened in other parts of the country with the important difference that the tenurial transformation in Odisha happened after independence under a democratic dispensation. An additional difference is that in spite of state attempts to stop shifting cultivation, it persists in parts of Odisha. The non-recognition of adivasi rights to hill slopes cultivated by them has had serious implications on their access to land and livelihoods in Odisha. The study uses data and analysis conducted during dissertation research undertaken by the author during 2005–7 in Odisha. The study draws from archival and other government records as well as village-level case studies.

Global Discourses on Shifting Cultivation

Shifting cultivation is estimated to support the livelihoods of 300–500 million people worldwide (Brady 1996). Conklin (1957) defines shifting cultivation as any cultivation system in which the fields are cleared and cultivated for periods shorter than those for which they are fallowed. Though the term shifting cultivation is used for diverse forms of agriculture (Brookfield and Padoch 1994), it is characterized by shifting fields rather than crops, fallow periods longer than cropping periods, recycling biomass generated in the fallow period for nutrients and high dependence on manual labour.

In past policy discourses, shifting cultivation has been seen to be a major ecological and political problem. It has been blamed for causing deforestation (Myers 1993; UNEP 1992) and reducing ecological biodiversity, destroying landscapes, and hydrological regimes. In most developing countries, shifting cultivation is practised by minorities, often indigenous people, and hence disciplining shifting cultivation has also been part of controlling and regulating these minorities (Scott 1999).

However, research on shifting cultivation has highlighted that shifting cultivation's role in deforestation is minuscule as compared to conversion of forests to permanent agriculture, plantations, and pastures. Ecological investigations have revealed the sophistication of many traditional shifting cultivation systems, which ensure sustainable production without any external inputs and the conservation of enormous agro-biodiversity (Kunstadter *et al.* 1978; Ramakrishnan 1992). A lot of the deforested areas attributed to shifting cultivation are actually regenerating forest fallows within the shifting cultivation cycle (Potter *et al.* 1994).

Over the past few centuries, shifting cultivation systems have undergone changes that have reduced their viability and sustainability. For example, in the context of Nepal's hills, Rasul and Thapa (2003) point out that factors such as increasing population, scarcity of land, introduction of new technology especially terracing, changes in land tenure and taxation systems have led to the conversion of shifting cultivation to permanent cultivation. In other areas, intensification of use has led to a reduction in fallowing cycles to such an extent that these cultivation systems have become short fallow agriculture.

While the academic discourse on shifting cultivation has become much more nuanced, at the policy-level, shifting cultivation continues to be stigmatized in many developing countries. In countries like India, governments continue to see shifting cultivation through the lens of 'forest destruction' and 'unproductive agriculture'—a stigma created by the British colonial government and followed by the technocrats in post-British India.

Shifting Cultivation in Mainland India

In India, shifting cultivation, commonly referred to as 'jhum', is associated mainly with the Northeastern part of the country. Yet, just a century back, shifting cultivation was practised in almost all forested tracts in India in different forms and names such as *podu* and *bagada* in

Odisha, *bewar* and *dhya* in the central provinces (Baker 1991; Rangarajan 1996), *khandad* in the Dangs of Gujarat (Skaria 1999), *ponnakadu* in Tamil Nadu (Saravanan 1998), *kumri* in North Kanara (Chandran 1998), and *dalhi* (Saldanha 1998) in Thane district of Bombay Presidency.

As the British expanded their control over forested areas, shifting cultivation was seen as a major threat to valuable forest resources. The difficulties of generating land taxes from often rebellious shifting cultivators and the preference for stable, tax-paying sedentary peasantries, was another reason for strongly discouraging shifting cultivation (Jyotishi 2000; Kumar *et al.* 2005). Forest and land laws were used to restrict shifting cultivation. The major forest laws, namely, the Madras Forest Act, 1882 and the Indian Forest Act, 1927,[1] explicitly criminalized shifting cultivation activities such as felling trees and setting fires. Land laws often withdrew or limited the legitimacy of shifting cultivation as a valid cultivation practice. This was effected through survey and settlement procedures, wherein land under permanent, lowland cultivation was recognized as tenancy land and shifting cultivation on land on hill slopes was designated as either revenue wasteland or forest land.

Francis Buchanan highlighted widespread kumri cultivation (shifting cultivation) in the Western Ghats of Uttara Kannada in 1801 (Chandran 1998: 678). Restrictions on kumri were imposed from 1848 and became progressively stricter. Kumri was also banned in Coorg in 1848 and in Belgaum in 1856 (Stebbing, quoted in Rangarajan 1996). Dalhi cultivation by Warlis, Thakurs, Katkaris, and other adivasis of North Konkan and Thane districts was restricted from as early as the 1850s. Attempts to control shifting cultivation in the Central Provinces started from 1862 with the passing of forest rules that banned dhya by the Gonds, Korkus, Baigas, and Bhumias and other hill tribes in well-timbered areas (Rangarajan 1996). By 1900, shifting cultivation was eradicated in almost all forests directly under British control in the Central Provinces. In the Dangs of Gujarat, restrictions on shifting cultivation (khandad) in teak and sissoo forests were initiated in 1844. By 1909, the claim could be made that khandad in the Dangs had been completely stopped (Skaria 1999).

Similar historical narratives with respect to the erasure of shifting cultivation are available from other parts of the country (see Chandran [1998] and Pouchepadass [1995] for Uttar Kannada, Saravanan [1998] for parts of Tamil Nadu). By independence, shifting cultivation had more or less been stamped out in most of mainland India by the colonial regime and was confined to parts of the Eastern Ghats in Odisha, northern Andhra Pradesh, and the Bastar region of Madhya Pradesh. In the remote and hilly forested terrains of the Eastern Ghats, particularly in Odisha, the British were unable to stop shifting cultivation and it continued as a major land-use practice even after independence.

Shifting Cultivation and the Colonial State in Odisha

Present-day Odisha was constituted through the amalgamation of parts of British-ruled Madras Presidency, the Central Provinces, and Bengal as well as 24 princely states (GoO 1958). Shifting cultivation in Odisha was found in hilly areas that were part of Madras Presidency and few of the princely states.[2] Various adivasi communities in Odisha traditionally practised shifting cultivation, including the Kondhs, Parojas, Gadabas, Bondos, and Saoras in south Odisha and the Juangs, Bhuiyans, and Eranga Kols in northern Odisha. For most of these adivasi groups, shifting cultivation has been an intimate part of their culture and cosmology. For instance, among the Kondhs, a clan group lays claim to 'owning' the earth within a territorially defined area, and the political, social, and cultural structure of Kondh tribes is built around the relationship with land (Bailey 1960). Such territory included not only cultivated paddyland, but also swidden areas, forests, streams, and hills. Anyone who wanted to settle within the territory needed to obtain permission from the clans. Similar systems existed for other adivasi communities.

Historically, adivasi-dominated areas in Odisha have been comparatively autonomous of various rulers and empires, because of both the difficulties of the terrain and the fierceness of the adivasi communities (Padel 1995). The relationship between the various adivasi communities and the local rulers tended to be reciprocal (Bailey 1960). In many princely states, the coronation of the ruler needed to be ratified by the adivasis through rituals and ceremonies.

British attempts to expand their control over these adivasi-dominated areas in Odisha during the nineteenth century were met with fierce resistance. For instance, the Kondhs fought against the British for decades before they were 'pacified' in the 1850s (Padel 1995). Though these wars were fought over the issue of human sacrifice, Bailey points out that the real reason was that the Kondhs feared that their lands would be taken away or taxed (Bailey 1960). In response to the resistance by the Kondhs, the British government issued a proclamation in 1855 that the Kandhs will never be taxed in perpetuity. Special administrative arrangements called 'Agencies' were designed by the British for adivasi areas directly under their control and normal laws were not generally extended to these areas. Part of the shifting cultivation areas also lay in the territories of the nominally sovereign princely states.

The apparatus of land revenue extraction in these remote and hilly adivasi regions remained rudimentary till the 1950s as compared to that in the lowlands. Elaborate survey and settlement of lands and maintenance of tenancy records that formed the basis of the land administration system of the British was only partially extended to these areas. Areas such as the present Kandhamal district were left untouched by land settlements, record maintenance, and land revenue up to independence.[3] In other areas, instead of land tax based on acreage and land category, simple taxation based on the number of ploughs or hoes owned or based on eye estimation of area was levied on shifting cultivation land. Ramdhyani (1947) points out that shifting cultivation was permitted in the Juangpirh of Keonjhar and in tracts of Bamra, Bonai, Ranpur, and Kalahandi princely states. In many of these princely states, shifting cultivation was taxed using the plough tax, hoe (*kodki*) tax or house tax, illustrating that the rulers accepted the legitimacy of shifting cultivation as a land-use.

Shifting cultivation increasingly came to be viewed by the British as leading to destruction of valuable timber forests.[4] Efforts to create legal forests and enforce restriction on shifting cultivation often led to unrest and conflicts in adivasi areas. One of the first attempts to reserve forests in shifting cultivation areas of Odisha was made in 1908 in south Odisha, where 131 sq. miles of forests were declared as Reserved Forests (RFs) using the Madras Forest Act, 1882. The effort to curb shifting cultivation in these reserved forests led to great resentment among the Saora tribals, and after an agitation in 1922, around 26 sq. miles of these forests were de-reserved. Faced with similar possibilities of resistance, there was little serious effort by the British to restrict shifting cultivation in major shifting cultivation areas such as Kandhamal and Baliguda where no legal forests were created till after independence.[5] In other shifting cultivation affected areas, legal forests were created but no serious efforts were made to enforce forest laws seriously.

In princely states such as Kalahandi, Bonai, and Bamra some forest reservation in shifting cultivation areas was carried out under specific laws and rules.[6] In most cases, this meant that a certain area was declared as RF without following any rights settlement process[7]. No serious effort was made by rulers to stop shifting cultivation even inside RFs in these princely states. In Keonjhar princely state, the ruler refrained from notifying any forests in the areas seen as homes of shifting cultivation communities such as the Juangs (Juangpirh) and Bhuiyans (Bhuiyanpirh). At the time of independence, shifting cultivation continued to be practised in hilly, adivasi areas of Kandhamal, Gajapati, Koraput, Rayagada, Malkangiri, and Kalahandi districts in central-south Odisha and Keonjhar and Sundergarh districts in north Odisha.

At the time of independence, shifting cultivation continued to be practised in hilly, adivasi areas of Kandhmal, Gajapati, Koraput, Rayagada, Malkangiri, and Kalahandi districts in central–south Odisha and Keonjhar and Sundergarh districts in north Odisha.

Shifting Cultivation in Odisha: Current Situation

There is no reliable estimate on the extent of shifting cultivation in Odisha based on actual measurement. The existing estimates vary widely. Part of the confusion arises from how to define shifting cultivation and whether it should also include shifting cultivation fallows. When I discuss shifting cultivation areas, I include both current cultivation and fallowed areas within the definition. Also, in certain areas, pressure on land has meant that shifting cultivation has virtually become a system of short fallows on hill slopes. I also include short fallow areas on hill slopes within the ambit of shifting cultivation from the perspective of tenure. Using the above criteria, one can identify two major zones in Odisha where shifting cultivation can still be found (see the map in Figure 8.1).

The first estimate of the extent of shifting cultivation in Odisha was made by H.F. Mooney (ex-conservator of forests), who estimated that approximately 32,681 sq. km (almost 20 per cent of the area of Odisha) was affected by shifting cultivation in 1951 with approximately one million adivasis dependent on shifting cultivation (Dash 2006; GoO 1958, 1959). Over the years, other estimates of areas affected by shifting cultivation in Odisha have been made, ranging from 1,177 sq. km to 30,233 sq. km (see Table 8.1).

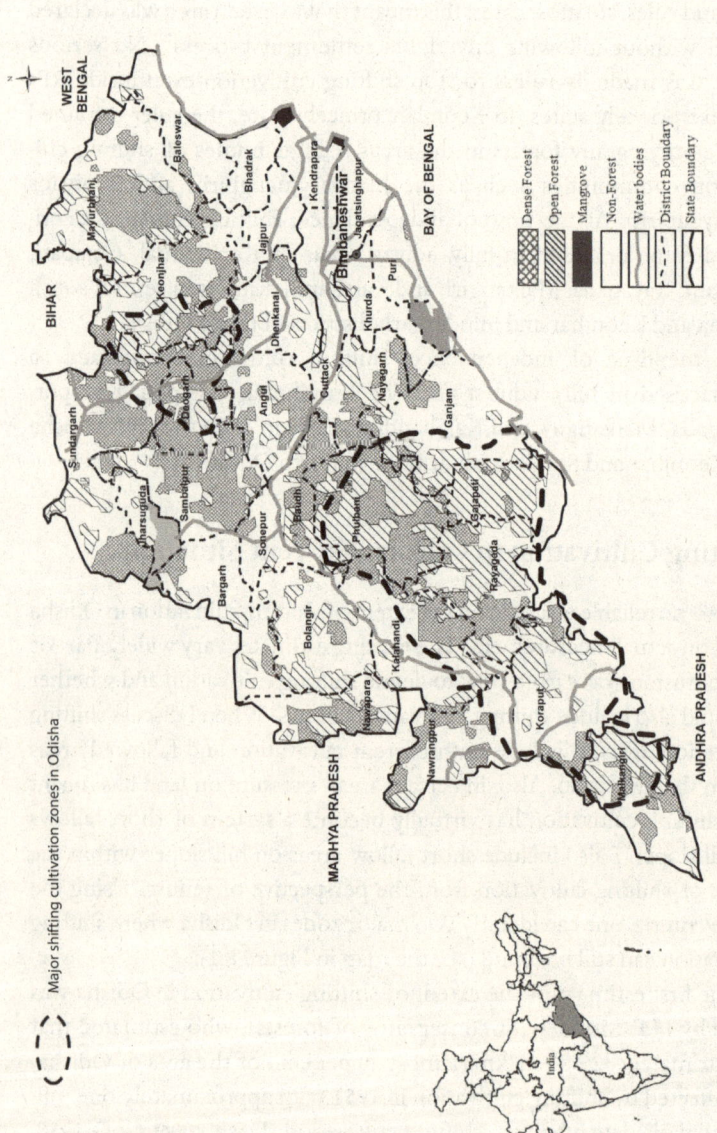

FIGURE 8.1 Major shifting cultivation zones of Odisha

Source: Base map from Forest Survey of India; shifting cultivation area marked approximately by author.

TABLE 8.1 Different estimates of areas affected by shifting cultivation in Odisha

Source	Year	Area in sq. km
H.F. Mooney	1951	32,681
ICAR	1958	8,000
Dhebar Commission	1961	8,333
French Institute, Puducherry, and ICAR	1967	30,233
Task Force on Development of Shifting Cultivation Areas, Ministry of Agriculture, GoI	1983	26,490
Tropical Forest Resources Assessment Project, FAO	1981	16,580
Wastelands Atlas of India	2003	1,177

Source: Adopted from Dash (2006). FAO data from FAO (1981).

None of the estimates seem to have been validated through ground truthing. Mooney's estimates were made through eye estimation. The Indian Council of Agricultural Research (ICAR) and Dhebar Commission estimates also seem to have been based on eye estimation and guesswork. The Tropical Forest Resources Assessment Project by the Food and Agriculture Organization (FAO) in 1981 used coarse resolution Landsat Imagery to estimate that 16,580 sq. km of forest area were affected by shifting cultivation in Odisha[8] (FAO 1981). However, the more recent *Wasteland Atlas of India* estimates approximately 1,177 sq. km as being affected by shifting cultivation (GoI 2003). The widely fluctuating data estimates make it almost impossible to get a clear macro picture of the extent of shifting cultivation in Odisha.

Inferences about the area under shifting cultivation can, therefore, only be drawn through indirect means. The only effort to comprehensively enumerate shifting cultivation on the ground was taken up during the survey and settlement of the Thuamulrampur block of Kalahandi district during the 1950s. The total area under shifting cultivation in this block was estimated to be 88,770 acres (Sunderajan 1963). In comparison, the *Wastelands Atlas* data estimates only 35,960 acres of shifting cultivation for the whole of Kalahandi district.[9] This seems to imply that the *Wasteland Atlas* data might be a gross underestimation. The extraordinary high extent of degraded forest cover and land with scrubs as shown in the *Wasteland Atlas* in traditional shifting cultivation districts arouses suspicion that much of shifting cultivation land has been put in these categories rather than being categorized as shifting cultivation area. High-resolution Google Earth maps make it clear that in most of the shifting cultivation areas, hill slopes continue to be under cultivation (see Annexure I for reference geographical points where shifting cultivation or short fallows in different districts can be seen on Google Earth). In many areas, due to high population pressure and reduction of area available for cultivation, shifting cultivation has become short fallow cultivation on hill slopes. This is especially true for areas in Koraput (Kashipur, Dasmanthpur, Semiliguda, and Pottangi blocks) and Kalahandi districts (Thuamulrampur block). Studies taken up by the author in Pottangi block of Koraput district showed a fallowing of as low as three years which has led to replacement of forests by scrub growth: see Figure 8.2 (Kumar *et al.* 2004b). In districts such as Kandhamal and Gajapati, most of the shifting cultivation areas retain secondary forest growth, and the cycle is often longer, ranging from four to 15 years (see Figure 8.3).

At present, the cultivated areas as well as fallows on the hill slopes in the shifting cultivation areas are categorized either as state-owned forest land or as revenue wasteland. However, the cultivators lay claim to customary rights over these lands. I discuss below the processes through which this transformation from customary land to state land took place.

Land Laws and the Non-recognition of Shifting Cultivation after Independence

In 1940, an enquiry committee set up by the government (the Partially Excluded Area Enquiry Committee) came down heavily on shifting cultivation and suggested that podu (shifting cultivation) should be abolished as early as practicable (GoO 1940). This committee report seems to have formed the basis of government policies on shifting cultivation in Odisha. The official documents of the decades immediately after independence exemplify the deep bias against both shifting cultivation and the adivasi communities. Shifting cultivation was held to be primitive, destructive, leading to floods, drying up of perennial springs, soil erosion, and so on, an 'evil' and 'pernicious' practice, which would shortly lead to the denudation of hill tracts which will no longer yield

FIGURE 8.2   Short fallows on hill slopes in Koraput district
Source: Author.

food crops (GoO 1958). Shifting cultivating adivasis were seen to be
'indolent and carefree in spirit' who lack all desire for self-improvement
(GoO 1958) and their primitiveness required that they be educated and
resettled in colonies.[10] The state of Odisha used two main strategies to
delegitimize shifting cultivation: the non-settlement of land rights on
shifting cultivation lands, and conversion of shifting cultivation land to
legal forests.

FIGURE 8.3   Long fallows shifting cultivation by Kutia Kondhs, Balliguda block, Kandhamal district
Source: Author.

In most adivasi parts of Odisha, the first land survey and settlements[11] were carried out only after independence.[12] Given the principles of formalization of land occupancy as laid down in various land laws at the time,[13] the survey and settlements should have recognized occupancy rights on both the valley land and the hill slope land under cultivation. This did not happen, primarily because of the strong bias against shifting cultivation in policy discourse.

The first major step taken to curtail legal recognition of shifting cultivation in the post-colonial period occurred in the shifting cultivation areas of erstwhile Kalahandi princely state, during the survey and settlements of the Dongarlas.[14] The government of Odisha issued orders in 1953 to deny adivasi occupancy rights over shifting cultivation land while preparing a special category of records called *Dongar Khasras* for these lands.[15] The board explicitly instructed that these Dongar Khasras were not occupancy rights (Sundarajan 1963). Dongar Khasras were prepared for 86,000 acres. The legal basis of denial of occupancy rights over shifting cultivation land was never made explicit by the government.

The next major decision taken regarding shifting cultivation areas was in undivided Koraput district (current Koraput, Rayagada, Malkangiri, and Nowrangpur districts). The first ever land survey and settlement was completed in this area in 1964. During the Koraput survey and settlement, the government of Odisha took an initial position that even though shifting cultivation was harmful to forest growth and soil conservation, given the livelihood dependence of people on shifting cultivation, it should be regulated rather than completely stopped (Behuria 1965). It was decided that shifting cultivation land on hill slopes of one in ten gradients would be settled with cultivators and the areas on higher slopes were to be treated as encroachment liable for eviction in due course. The legal and procedural basis for this decision, especially eviction, was not made explicit but the process of demarcation of areas for shifting cultivation up to one in ten gradients began in 1954–5 (Behuria 1965).

However, in 1960, the Board of Revenue (BoR) took a decision that rights on any shifting cultivation land were not to be recognized in any form. The BoR used a provision in the land laws,[16] which says that the person who had occupied ryoti land for a continuous period of 12 years shall automatically be considered a ryot with occupancy rights. The BoR decided that 'shifting cultivation couldn't be construed as continuous possession for a period of twelve years' and therefore, should not be recorded as ryoti land with occupancy rights. Rather, such land should be classified as government land.[17] The basis of such a view, given that the same family or community possessed these lands for generations in accordance with adivasi customary law, was not made clear.[18] The clever interpretation of the law carried out by the state government ensured that adivasi rights over shifting cultivation land was not recognized during the Koraput survey and settlement.

Once the precedent was set in the Koraput survey and settlement, it became easier for the government to follow the same principle of non-recognition of rights over shifting cultivation land in other parts of the state. The main shifting cultivation zones affected in this manner were Kandhamal district, the present Gajapati district and Juangpirh and Bhuyanpirh in Keonjhar district. In the Kondhmal survey and settlement, almost half the area within the village boundaries were classified as state-owned revenue forests, even though they were commonly used for shifting cultivation.

The non-recognition of adivasi rights over shifting cultivation land and their categorization into government land also became a major factor in the increasing conversion of these lands into legal forests, a process which had started well before independence and accelerated in the post-independence period.

Legal Forests and Shifting Cultivation

A major proportion of shifting cultivation land in Odisha has been converted into legal forests. This happened through two legal pathways. One was the application of forest laws such as the Madras Forest Act, 1882, the Indian Forest Act, 1927, or the Odisha Forest Act, 1972 to declare RFs or Protected Forests (PFs). The second was through the land survey and settlement processes, which also provided for creation of legal forests within village boundaries. Notification of forests in the remote adivasi tracts of Odisha was started in earnest in the 1960s. Large areas of land in the major shifting cultivation affected areas were notified as legal forests after independence: the details are given in Table 8.2.

The modes of change of the legal status of shifting cultivation land through forest and land laws can be depicted diagrammatically as in Figure 8.4.

Creation of Forests in British-ruled Areas before Independence (Typology I)

Small areas of forests were reserved using the Madras Forest Act, 1882 during the British period in parts of the Ganjam Agency (current Gajapati district) despite resistance from shifting cultivators. After independence, these areas continued to be treated as RFs.

In the Jeypore zamindari, large areas of reserved and protected land, were specially created under Section 26 of Chapter III of the Madras Forest Act, 1882 (also see Table 8.2, row 3). The main reason for notifying these areas as reserved and protected land was to avoid the rights' settlement process that creating RFs would have entailed (Behuria 1965). Many of these forest areas had ongoing shifting cultivation. In the

TABLE 8.2 Creation of legal forests after independence in areas where shifting cultivation is prevalent

Districts/areas with shifting cultivation	Pre-independence status	Legal forests at time of independence	Legal forests at present	Legal forest area created after independence
Juangpirh and Bhuiyanpirh (Telkoi and Banspal blocks), Keonjhar district	Part of Keonjhar princely state	No legal forests	One RF (875 hectare [ha]). Approximately 600 sq. km* more classified as revenue forest	608 sq. km
Kandhmal and Gajapati district	Kandhmal, Baliguda, and other Ganjam Agency areas under direct British rule and in zamindaries	Approximately 520 sq. km as RFs	Reserved Forests: 2426 sq. km DPFs, reserved lands, UDPFs: 2010 sq. km forests created through revenue survey and settlements: 2708 sq. km Total legal forests: 7144 sq. km	6,624 sq. km
Rayagada, Koraput, and, Malkangiri districts	Jeypore Zamindari	4140 sq. km as Reserved land or Protected land	RF: 2,136 sq. km Proposed RFs, Reserved land, UDPFs: 4,417 sq. km Forests created through revenue survey and settlements: 3,948 sq. km **Total legal forests: 10,501 sq. km**	6,461 sq. km

Source: Calculations and mapping carried out by the author from diverse data from multiple sources, including Forest Working Plans, Census Data, Survey and Settlement Reports and other archival sources.

Notes: [*] Census of India, 2001, provides the forest area included within village boundaries. In two blocks of Telkoi and Banspal, which overlap with the Juangpirh and Bhuiyapirh talukas, the total forest area inside the village boundaries adds up to 62,491 ha, that is, 624 sq.km.

DPF = demarcated Protected Forest; UDPF = undemarcated Protected Forest

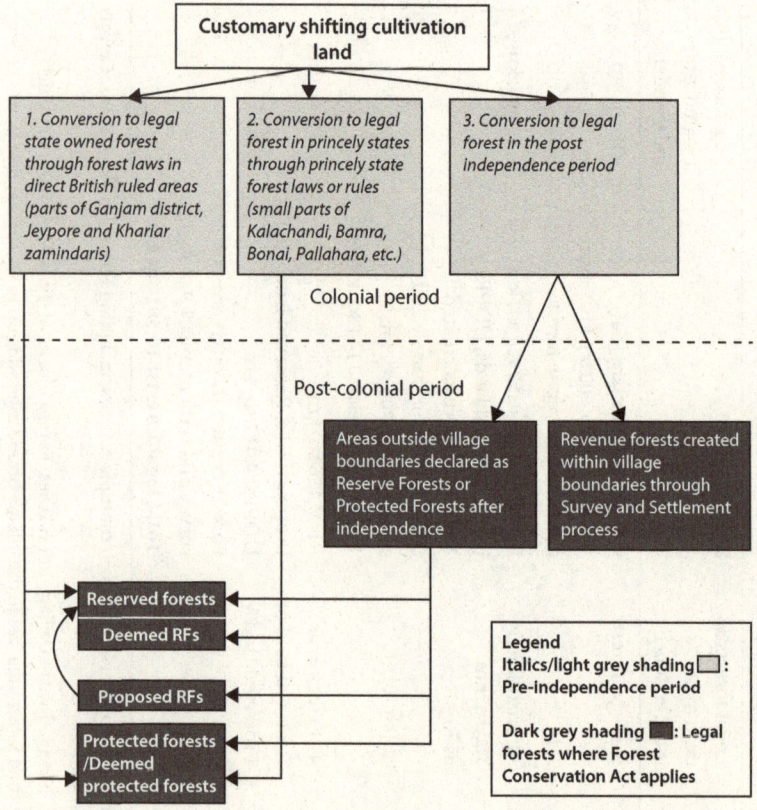

FIGURE 8.4   Changes in legal status of shifting cultivation lands over time
Source: Author.

post-independent era, these areas were 'deemed' to be PFs under the Odisha Forest Act, 1972 (Kumar *et al.* 2005).[19] No effective rights' settlement has been taken up in these areas.

Forests in Princely States (Typology II)

Through an amendment, the Government deemed all RFs and other legal categories of forests in the ex-princely states as Reserved Forests or Protected Forests under the Indian Forest Act, 1927. (Kumar *et al.* 2005).[20] This implied that all these areas were to be treated as RFs or

PFs, even though no proper rights' survey and settlement had been carried out while originally notifying these forests. The areas where shifting cultivation land was affected by this process were in parts of Kalahandi, Bamra, and Bonai states.

Forests Notified after Independence (Typology III)

This is by far the most important category as can be seen from Table 8.2. Large areas of forests were notified as RFs in shifting cultivation areas of Kandhmal, Gajapati, and the undivided Koraput district. In principle, the Government of Odisha (GoO) had already accepted that shifting cultivators had no rights over the shifting cultivation land on the hill slopes, making it easier for the FD to include large stretches of such land within the reserved forests. Other stretches of forest have been brought under Section 4 of the IFA in preparation of reservation or are yet to be demarcated. No rights settlement of any kind has taken place in these lands.

Apart from the forest laws, the GoO also used the survey and settlement processes to create forests within village boundaries. Table 8.2 shows that almost 6,600 sq. km of land have been categorized as revenue forests through survey and settlements in Kandhmal, Gajapati, and the undivided Koraput district. Since these areas were next to the settlements within village boundaries, they were often being used for shifting cultivation or short fallows.

All the forests as notified through the above typologies were not necessarily areas affected by shifting cultivation. However, empirical case studies, the limited secondary data, and anecdotal evidence tell us that large areas of shifting cultivation land were included within legal forests. For instance, in the Kirkicha watershed, almost 150 acres of shifting cultivation area was included within the recently notified Mohangiri Proposed RF. Another 32 acres of shifting cultivation area was found to be on land categorized as revenue forests in 1980s.

In Mangara village of Koraput district, the author documented a few existing patches of shifting cultivation deep inside the RFs, and villagers mapped out the erstwhile shifting cultivation areas in the RFs, which they have been forced to leave by the FDs since the late 1990s (Kumar *et al.* 2004b). In Gaurigaon village of Kondhmal district, around 61 acres of shifting cultivation land (by 29 families) was included within revenue forests in 1982 during the survey and settlements.

Similar situations have been observed in many traditional shifting cultivation areas but there is no systematic study of how much shifting cultivation area was converted to legal forests through forest laws. Perusal of forest notification proceedings also throws little light on this issue as the forest settlement officers (FSOs) have almost always ignored claims of shifting cultivation. A perusal of forest reservation proceedings for Kandhmal district by the author turned up just one case of where the FSO acknowledged that there was an existing practice of shifting cultivation in the proposed reserved forest, and then promptly proceeded to reserve the area.

That the legal 'forests' created through forest laws included large areas of shifting cultivation areas can also be inferred from the information in forest working plans. Extracts from the Rayagada Forest Division's Working Plan that describe the constitution of different working circles illustrate this (see Table 8.3).

TABLE 8.3 Shifting cultivation and Rayagada Division Working Plan prescriptions

Working circles	Total area	Shifting cultivation affected area	Remarks in the Working Plan
Improvement working circle	41,526 ha	Part of the circle	All the blocks with congested crops at pole stage mainly of podu origin were included in this circle
Rehabilitation-cum-soil conservation working circle	27,768 ha	Part of the circle	It included podu areas thoroughly degraded due to repeated hacking
Teak plantation working circle	2,507 ha	2288 ha podu-affected area	Podu areas and existing teak plantations
Protection working circle	987 ha	Half of Rafukona RF (647 ha) under shifting cultivation	

Source: Rayagada Division Working Plan (GoO 2006).

Similarly, the Balliguda Working Plan (covering part of Kandhmal district) states:

> Large forest areas in the division have been destroyed since time immemorial due to the pernicious practice of podu and this has changed the complexion of some of the forest areas completely. Some of the podu areas which once supported lofty trees are now bereft of vegetation. Extensive forest areas now bear a crop of poles and young saplings due to recent podu". (GoO 1989)

De Jure Tenure and De Facto Land-use in Shifting Cultivation Landscapes

Customary shifting cultivation land has been either converted into legal forests or into government-owned revenue wasteland. In shifting cultivation areas of Kandhmal and Gajapati districts, where the shifting cultivation cycle was longer and the fallows generally had secondary forests, most of the hill slopes were converted into legal forests, both through forest laws and through survey and settlements. In Koraput and Rayagada districts, where the cycles were shorter and the hill slopes degraded into scrub, most of the hill slopes used for cultivation were categorized as revenue wasteland.

Four-fifths of the total land area in the shifting cultivation affected districts has been categorized as government land through forest notifications and survey and settlements. In order to understand the legal categorization of land in shifting cultivation districts, we used Census 2001 data (which along with population data, also provides the total area of the villages, the total patta[21] (private land) and the total government land[22] in the village) to find villages in adivasi areas where an average household owned less than 0.1 ha of land. In five districts we found 727 villages where an average household owned less than 0.1 ha of land. Scheduled tribes (STs) formed three-fourths of the population of these villages. While comparing the area of privately held land with government-owned land within the boundaries of these villages, it became clear that there is large extent of state-owned land in these villages (see Table 8.4).

Two inferences can be drawn from these data. One is that there is a substantial number of villages in the shifting cultivation areas where

TABLE 8.4   Analysis of land availability per household in villages where an average household held less than 0.1 ha

Districts	No. of villages[23] with less than 0.1 ha of land/hh	Area of patta land/hh	Area of government land/hh	Ratio of patta land/hh to government land/hh
Koraput	157	0.01	1.91	1: 191
Rayagada	101	0.03	5.65	1: 155
Kandhmal	116	0.02	2.61	1: 130
Malkangiri	205	0.02	1.69	1:84
Gajapati	139	0.04	1.89	1: 47

Source: Based on calculations by author using Census of India data.

households hold none or a very minimal amount of land on a legal basis. Second, in these villages, relatively large areas are categorized as government land. The secondary data indicates that only a small part of the actual land under customary use by adivasi communities has been legally settled with them. The difference between the de jure legal ownership of land in shifting cultivation areas and the de facto land-use based on customary land rights is obvious to anyone who visits these areas. Normally, the bottom valleys are used for terraced rice cultivation, the lower slopes for various vegetables and dryland crops and the hill slopes for shifting cultivation or short fallow cultivation. The people have rights on the land in valley bottoms and lower slopes. Large parts of the de facto cultivated areas lying on the steeper slopes are categorized as government land, mostly forest land.

More systematic and direct data to substantiate the above argument is available from case studies done by us in a number of villages. A detailed case study undertaken in the Kirkicha watershed is discussed to illustrate the micro-level manifestation of government policies on land settlement.

Case Study: Kirkicha Watershed, Kalahandi District

The Kirkicha watershed is located in the Thuamulrampur block of Kalahandi district. The watershed consists of four villages (five settlements) and large patches of proposed RFs and has a total area of 2,652 acres (Orissa Remote Sensing Agency [ORSAC]). Out of this, 1,275 acres come within the boundaries of the four villages, whereas the rest of the area is categorized as proposed RF. The total number of households in these five settlements is 129, with a majority being Kandh adivasis.

In the pre-independence period, the area was under the jurisdiction of the Karlapat zamindari of the Kalahandi princely estate. Two of the four villages, Mohangiri and Khandala, went through survey and settlements in 1963 and 1988–9. During the 1963 settlement, lowland was settled in the name of the cultivators. As the Kalahandi princely state had levied taxes on shifting cultivation land based on seed capacity, as discussed earlier, kodki pattas were given to the shifting cultivators. These were not treated as ownership rights. The other two villages, Usamaska and Kirkicha, were left unsurveyed.

In the next survey and settlement in 1988–9, a revisional survey of Mohangiri and Khandala were taken up. During this survey and settlement, the kodki pattas for shifting cultivation were withdrawn and the land became unencumbered government land. The other two unsurveyed villages were surveyed in the 1990s and the final record of rights published in 2002. Thus by 2002, the whole watershed was officially surveyed and formalization of land rights completed.

Table 8.5 summarizes the results of an exhaustive plot-to-plot survey of the watershed. The survey tried to reconcile the legal status of land with the actual land-use. All plots of land under cultivation were mapped and measured, their land-use compiled and ownership details collected through interviews. This was compared with the legal land record and the cadastral map. The first row of data pertains to the total area under each legal category of land according to government records. The second row gives details of the area under permanent cultivation as found by the plot survey. The third row details the areas under shifting cultivation. The second, third, and fourth columns give the area of government land under cultivation.

Based on these data, the following inferences were made:

1. Only 8 per cent of the total land in the watershed was settled with the villagers, that is, the villagers have rights over only 8 per cent of the land in the watershed.
2. Within the village boundaries, that is, the area which has been subjected to survey and settlements, only 16 per cent of the land has

TABLE 8.5 Legal land categories and actual land-use in the Kirkicha watershed*

	Legal private land (acres)	Land legally categorized as government land (acres)				Total (acres)
		Forest land (within village boundary)	Forest land (in PRF)	Non-forest revenue land	Total govt. land	
Total extent (legal) **	214	290	1376	772	2438	2652
Permanent cultivation including short fallows	194	80	0	188	268	462
Shifting cultivation	0	46	141	114	301	301
Total land cultivated	194	126	141	303	570	764

Source: Based on calculated data which draws from actual plot-to-plot analysis and Record of Rights.

Notes: [*] Numbers may not sum accurately due to rounding.

[**] The legal area of private land, forest land inside the watershed, and non-forest government land was obtained from the Record of Rights for the four villages in the watershed. The area of forest land outside the village boundaries was determined by subtracting the total area of the four villages from the total area of the Kirkicha watershed as determined by the Orissa Remote Sensing Agency (ORSAC). All the land outside the village boundaries are categorized as reserved or proposed reserved forests.

been settled with cultivators, and 84 per cent categorized as government land.

3. In spite of having legal rights over only 214 acres, the 129 watershed families were actually cultivating 764 acres of land, including 570 acres of government land.

4. More than half of the 570 acres of cultivated government land was put under shifting cultivation.

5. Out of 301 acres of land under shifting cultivation, 187 acres is categorized as legal forests.

Each shifting cultivation family claims rights over two to three patches that they farm rotationally. Adivasis claim customary ownership over all the land cultivated by them. Customary rights over the land were passed from generation to generation.

The plot-wise analysis and its comparison with the legal status of land reveal the disjuncture between the legal status of land and the actual ground-level situation. This difference is notwithstanding the fact that survey and settlement took place in Kirkicha in the 1990s. A major reason for the discrepancy between the actual area cultivated and the area officially settled as private land with cultivators is the non-recognition of shifting cultivation land.

Other Examples

I found a similar gap between de facto and de jure land tenure in another case study in the Gaurigaon watershed of Kandhmal district. In this village, the total area is 675.5 acres. The survey and settlement process in 1982 earmarked 574.5 acres as government land and settled 100 acres of land in the name of cultivators. A majority of the land settled in the name of cultivators was settled with non-adivasis living in neighbouring villages. Only 44 acres was settled with the 29 adivasi households of Gaurigaon. In this village also, both permanent (including short fallow) and shifting cultivation is taken up in land categorized as state-owned as shown in Table 8.6.

In the village of Dekapar, Koraput district, villagers estimate that approximately 430 acres of land was under occupation by their forefathers. Out of this, only 152 acres at the valley bottom was settled as patta land during the survey and settlement. The rest of the land on the hill slopes,

TABLE 8.6   Cultivation on land categorized as government land (area in acres) in Gaurigaon village

Legal status	Permanent cultivation, including short fallows	Current shifting cultivation	Total
Forest land	71.3	60.65	131.95
Non-forest govt land	22.87	0.00	21.87
Total	**94.17**	**60.65**	**153.82**

Source: Based on calculated data that draws from actual plot-to-plot analysis and Record of Rights of the village.

customarily used for shifting cultivation, was either classified as government-owned revenue wasteland or as forest land (Kumar *et al.* 2005).

Similarly, in Kadalibadi village of Juangpirh (Banspal block), Keonjhar district, the highly vulnerable Juang adivasis have been given rights to only 25 ha of land out of the total of 283 ha in the village. The customary communally held shifting cultivation land on the hill slopes was categorized as government land in the survey and settlement of the 1980s. In Bangusahi village of Gajapati district, the residents legally own only 70 acres of land but occupy 160 acres of customary shifting cultivation land categorized as forest land.

I have chosen individual examples from five different districts to illustrate that shifting cultivation or short fallows on hill slopes still continues in Odisha in diverse, widespread locations and to highlight that most of the land cultivated by the adivasis in these areas are categorized as state land. These are samples that were aimed at actually estimating the physical discrepancies between de jure and de facto land tenure.

Consequences of Non-recognition of Rights

Special Problems of Forest Land

Forest laws criminalize shifting cultivation through multiple modes. Cultivating land categorized as forests is an offence, so also the felling of trees and setting fires. Even trespassing in RFs is supposed to be an offence. Thus as soon as land is categorized as forests of any category, shifting cultivation on these lands is criminalized. The FD has been trying very hard to 'eradicate' shifting cultivation on RFs, and has used all means at its disposal including threats, fines, arrests, and court cases; this has been a continuous source of conflict and means to harass adivasis. For example, in the G. Udaygiri Forest Range in Kandhmal district, out of 87 cases forwarded to the courts during 2001–6, 47 are related to shifting cultivation (Kumar *et al.* 2008). Court cases are catastrophic to adivasis, who are completely unfamiliar with the complexities of the legal system.

The FD, through coercion and filing of cases, has been able to stop shifting cultivation in the RFs in certain areas. An example is the eastern part of Kandhmal district where due to changing livelihoods patterns and continuous imposition of legal penalties by the FD, shifting cultivation has almost ceased (Kumar *et al.* 2008). In other areas, the FD strategy has not been so successful because of remoteness of terrains, lack of alternatives for cultivators, and increasingly, the influx of left-wing insurgents.

There is high resentment amongst adivasis against the FD for its attempts to stop shifting cultivation. The issue of cultivation on forest land has become a running, low-level battle between the adivasi peasantry and the FD. Another strategy adopted by the FD is that of taking up plantations of forestry species on shifting cultivation land, including through joint forest management (JFM) programmes. In our research, we have come across conflicts over such plantations because the FD collaborates with dominant local interests, often non-adivasis, to take up JFM plantations on shifting cultivation land. This has been the case in Mangara and Podagarha villages, Koraput districts (Kumar 2005; Kumar *et al.* 2006). Conflicts between JFM (funded under the Japanese Bank for International Cooperation [JBIC]) and shifting cultivation has led to the arrest of a number of adivasis who resisted plantations (Bidyut Mohanty, personal communication, 24 August 2008). Similarly, in the Kirikicha watershed we found that a large patch of sloping land, which was being used for short fallow cultivation by villagers, was brought under JFM plantations after evicting the cultivators.

Loss of Access to Shifting Cultivation Land

In general, in spite of its criminalization, shifting cultivation or hill slope cultivation continues, especially in the revenue wasteland and revenue forests. However, when the government requires these types of land for any other purpose, conflicts arise. Plantations for different purposes are the most common reason for a de facto takeover of shifting cultivation or short fallow lands. Other causes are land diversion for development, industrial or mining projects, and leasing of land to private parties.

The first major plantations of cashew and eucalyptus were taken up in the shifting cultivation affected watersheds of the Machkund reservoir in Koraput district through the soil conservation department. There was resistance against this takeover of customary land. B.K. Roy Burman has documented the burning of plantations by adivasi shifting cultivators (GoI 1986). Despite resistance, cashew plantations on shifting cultivation land continued. In 1981, the government transferred large areas of cashew plantations to the Cashew Development Corporation, a state-owned corporation, which manages them commercially and auctions the plantations every year to the highest bidders. This effectively privatized customary shifting cultivation land and led to conflict (Mishra 2007).

Plantations continue to be taken up in shifting cultivation land, often in the name of adivasi development under watershed programmes and employment schemes. A study in Dekapar village of Koraput district revealed how the same patch of shifting cultivation land has been taken over four times by the FD over the last three decades for plantations despite the fact that each time local people have destroyed the plantation. Nonetheless, each new take over has resulted in out-migration to Andhra Pradesh or neighbouring Nowrangpur (Kumar *et al.* 2005). In the case of Kadalibadi village in Keonjhar village (described earlier), 116 ha of the shifting cultivation land in the village used by extremely vulnerable Juang adivasis has been converted to plantations by the FD (Rath 2005).

The government has often leased out shifting cultivation land to coffee growers (generally rich non-adivasis) and in one case in Keonjhar, to a tea-growing company. All these lands were customarily claimed by one adivasi community or other. With an increasing emphasis on bio-fuels and carbon-sequestration, many companies are eyeing the vast revenue wasteland and forest land in the shifting cultivation areas. This again may lead to the de facto loss of these lands.

Displacement due to dams, industries, and mining has also led to loss of access to shifting cultivation land without any compensation.

Koraput district has four major reservoirs and their construction led to the loss of both valley land and shifting cultivation land. Adivasis obtained some compensation for the valley bottom land, but none for the customarily held land on hill slopes as they had no legal rights over these areas. Major industrial and mining projects such as National Aluminium Company Limited (NALCO) have also been allotted vast areas of land, including hills that were used for short fallow cultivation by adivasis.

Ecological Degradation

There are different ways in which the non-recognition of rights can lead to ecological degradation and loss of productivity for the cultivator. First, stopping shifting cultivation in RFs and in plantations seems to be an important factor in the reduction of fallow cycles and overuse of certain hill slopes. In areas of Koraput district, degradation has reached a stage where secondary forests have been replaced by shrubs and grasses and the ability of the land to regenerate is deeply compromised.

Second, it is quite clear that the ambiguity in the ownership of the hill slopes has meant that the cultivators have had no tenurial security in making major investments in reducing degradation of the hillsides. This is significant in view of intensification of shifting cultivation and reduction of fallow periods. Adivasis in the hill areas are often highly skilled in soil and water management, including soil conservation. The Saoras of Gajapati district are adept at creating paddy terraces on steep slopes; they could easily transfer their soil and moisture conservation knowledge to shifting cultivation slopes. Other adivasi groups have similar skills. In some areas, in spite of not having legal tenure to land, adivasis have tried to improve the land through indigenous stone-bunding and vegetative bunds. However, without secure rights on these lands and the risk of losing the land, it does not make sense to invest and maintain such conservation measures.

Third, lack of security over the land also makes it difficult for adivasis to plant trees, especially horticultural species. In many parts of Gajapati and Koraput districts, adivasis have started planting cashew on their lands (Kumar *et al.* 2004b). They could plant cashews on degraded hill slopes too if their rights are recognized.

At the same time, the government has spent hundreds of crores over the last four decades on soil and moisture conservation work in shifting cultivation areas. A long list of programmes and projects for watershed development and forestry, taken up on shifting cultivation land, have come and gone with little impact. Large investments in ecological regeneration have been negatively affected because of the difference between the de jure and de facto land tenure.

State Interventions for Shifting Cultivation Amelioration

Initial efforts to address shifting cultivation included schemes for resettling shifting cultivators in adivasi colonies and allocation of agricultural land for settled cultivation. A scheme for resettlement of shifting cultivators in the plains was initiated in the First Five-Year plan itself. The scheme provided land and basic agricultural implements to shifting cultivating communities. Most of these colonies did not do well (Dash 2006).

In the 1980s, programmes such as the Economic Rehabilitation of Rural Poor (ERRP) were used to take up plantations on shifting cultivation land with usufruct rights[24] to cultivators. The scheme was dogged by a number of problems, which included conflicts between the customary owner of the land and the new allottees, non-allotment of the plantation to beneficiaries, and poor survival of plantations. Another initiative was a small-scale programme funded by the central government named 'Plantation by Podu Cultivators'.[25] The scheme[26] envisaged plantation of commercial fruit trees such as mango, tamarind, jackfruit, guava, and orange by the podu cultivator. The cultivator was to be given usufruct rights of trees and intercrops were to be raised. The scheme was on a small scale.

An important area-based initiative to address hill slope cultivation and shifting cultivation on a watershed basis was taken up in Kashipur block in the early 1990s through the International Fund for Agriculture Development (IFAD)-sponsored Odisha Tribal Development Project (1988–97). It covered 12,500 adivasi families and another 4,000 local non-adivasi households of Kashipur block in Rayagada district. In Kashipur, most shifting cultivation land had been converted to short rotation fallows on hill slopes and were highly degraded. For the first time, GoO recognized the need to settle hill slopes up to 30 degrees slope with cultivators.[27] During the project, a record of rights for *dongar* (hills) land was issued to 6,837 adivasi households in 236 villages covering a total area of 17,175 acres. This order was not applicable to forest lands. According to the evaluation report of the Orissa Tribal Development Project (OTDP), the allocation of land rights has helped in restoring agro-ecological balance in this area (IFAD 1999). However, during my research, I found out that the rights' settlement process did not recognize the traditional tenure system. As a result, land customarily owned by one household was often settled with another household, causing conflicts (Kumar *et al.* 2004a).

The Kashipur order was extended to all adivasi areas of the state in 2000 due to pressure from civil society.[28] However, the actual implementation of this circular has been tardy due to lack of political will and problems in surveying the hill slopes. A later government initiative that tries to address the issue of land tenure in shifting cultivation areas is the Odisha Tribal Empowerment and Livelihood Project, funded by IFAD and the Department for International Development (DFID) that has been initiated in 30 adivasi blocks of Odisha and that is supposed to be implemented on a watershed basis. One of the main components of this project is to facilitate settling of land in accordance with the Kashipur circular. However, in most watersheds covered under this programme, the shifting cultivation lands have been categorized as forest land.[29]

Responses of the Shifting Cultivators: Taking Back the Land

As adivasis have become more aware, they have re-asserted their claims to their customary landscapes. Many adivasi communities have taken up community protection of degraded hill slopes to regenerate forests as a re-assertion of their control on these lands. NGOs and grassroots organizations working for adivasi development have encouraged adivasi communities to start regenerating forests. As per a report prepared for FAO using satellite data and ground truthing, it was found that, in just six blocks of Kandhamal district, almost a thousand communities were involved in community forest protection over 37,000 ha of land (Singh *et al.* 2005). Most of these areas were shifting cultivation land, which have regenerated into secondary forests due to protection by communities. Hundreds of villages in Koraput and Rayagada districts are involved in protection of forests in their erstwhile shifting cultivation land. In Dekapar village of Koraput district, the villages are protecting hill slopes that surround a major water spring and proudly list out the name of persons who 'donated' their short fallow land[30] for protecting the slopes (Kumar *et al.* 2005). Similarly, in Malkarbandh village in the same area, the adivasi villagers donated paddy land to landless swiddeners so that they could protect the catchments of a water-harvesting structure that was being used for short fallow cultivation (Kumar *et al.* 2004a). Such examples of protection of hill slopes categorized as government land abound across the adivasi districts.

In many cases, these assertions of control have also extended to plantations taken up in shifting cultivation land. In 22 villages in Koraput district, adivasi communities forcefully took over the cashew plantations leased out to the Cashew Development Corporation by the state government (DTE 2007). The Cashew Development Corporation used to auction these plantations to outside contractors; the local villagers, claiming that the plantations were taken up in their ancestral land, have forcefully re-occupied them. This has led to conflict and filing of court cases against the adivasis. The struggle over cashew plantations has expanded across the landscape, resulting finally in a government order in 2008[31] that declared that these cashew plantations will be allocated to adivasis only. It is clear that as the adivasi communities become more assertive, such efforts to reassert rights on land under government control will increase.

The Possibilities of FRA, 2006

The enactment of the Scheduled Tribes and Other Forest Dwellers (Right to Forest) Act, 2006 (henceforth FRA), which aims to redress the historical injustices done to adivasis and other forest-dwellers by the reservation of forests, opens new possibilities for resolving some of the contradictions regarding shifting cultivation land. Till now, forest land could not be settled with cultivators without getting the permission of the central government. The FRA eased the situation by providing for recognition of rights over forest land occupied by adivasis and other forest-dwellers up to a maximum of 4 ha. This provision can also be applied to forest land under shifting cultivation. Land claimed as shifting cultivation land is held by households and can be settled in the name of households under this law. Our studies show that in most cases households do not lay claim to more than 10 acres of land over the total shifting cultivation cycle.

Unfortunately, the FRA does not mention shifting cultivation as one of the rights to be recognized. This means that in shifting cultivation slopes with secondary forests, even if settled with individual families, the clearing of fallows for the next cycle could imply violating the Indian Forest Act, 1927 which forbids felling of trees, and thereby allows the FD to charge those who shifting cultivate. Only forest land which has no effective forest vegetation and is cultivated as short fallows might actually end up being actually allocated to shifting cultivators.

Many communities who have regenerated part of their shifting cultivation land into secondary forests did not have any formal rights over these regenerated forests. The government's JFM programme can provide space for community control over these forests. But JFM does not recognize any legal rights of communities over this forest land and leaves the control of the forest firmly in the hands of the FD. Many adivasi communities are not interested in an arrangement which leaves them at the mercy of the department. In Mandaguda village of Kandhmal district, the villagers refused to agree to a proposal to include community protected forest into JFM. One villager had this to say, 'Whether it is reserved forest or village forest we the people are protecting it to meet our future needs. We did not want to involve the forest department. Hence, we did not involve the department in the process' (Kumar *et al.* 2008).

The provision of 'community forest resources' in the FRA opens the possibility of shifting cultivation communities asserting effective control over all the forest land in their traditional boundaries. Rights on such forest areas also could result in accessing funds for development of such forests, including possible carbon sequestration funds. If they are given rights, many shifting cultivation communities, if provided incentives to protect and regenerate forests, may bring in additional shifting cultivation areas under forest cover.

Rethinking Forests and Land Rights in Shifting Cultivation Landscapes

Shifting cultivation landscapes in the Eastern Ghats, including in Odisha, are in an environmental and social crisis. Part of the reason for the crisis is the gap between the de jure and de facto tenure over land and forests. The state has taken over the responsibility of over three-fourths of these forested landscapes even though it has no effective capacity to govern them sustainably so that the needs of the millions of people who live in these landscapes are addressed.

Adivasi societies treat the landscape as a unity and their agro-ecological and livelihood systems incorporate all elements of the landscape. The notion of 'owning the earth' within a territory was deeply embedded in the culture and economy of many of these communities and their political and social structures developed around these relationships with land. The imposition of artificial legal categories such as forests and non-forest land and private and state land was a major disruption for these communities. The imposition of legally defined forests on the pre-existing systems of rotational agro-forestry that were closely linked to other economic activities such as wetland cultivation and forest product gathering has led to dispossession of adivasis, conflict, and degradation of the hillsides.

In the absence of alternatives, adivasi communities continue to eke out a living from the same hill slopes, albeit illegally. The constructed 'illegality' of shifting cultivation has placed tremendous powers in the hands of petty officials, especially forest officials, and has led to a regime of coercion, exploitation, and marginalization. Marginalization through criminalization is an integral part of the larger, vicious cycle of exploitation and surplus extraction from adivasis by a nexus of local elites, moneylenders, and the officialdom. 'Illegal' land-use also means that cultivators have little incentive to invest on the land, or capacity to access government incentives and to improve these lands, thus leading to further land degradation.

Both the Odisha and Indian governments are investing large amounts of money and effort in various development programmes, including watershed development, in shifting cultivation areas. These programmes have become more participatory over time. Yet the lack of de jure rights means that adivasi communities are reluctant to take up state-sponsored natural resource-based interventions. Thus even on pragmatic grounds, it is urgent to resolve the issue of rights and tenure over land.

In the spirit of Schedule V of the Constitution, the rights of STs should extend to all the land that they depend upon for livelihoods, including the hill slopes should be protected. The special protections provided in the Constitution must become the key reference point for any strategy which tries to untangle the mess created by faulty land and forest laws so as to ensure justice and development for adivasi communities. The solution in these areas lies in providing the adivasi communities more autonomy and control over the landscapes, and allowing their community governance systems to manage the resources as a whole. The potential of rationalizing the land and forest tenure landscapes is visible in the cases of large-scale forest protection and soil and moisture conservation efforts taken up by communities in these landscapes. These grassroots initiatives indicate the direction that tenure reforms should take in these areas. The main principles should be to ensure control of communities over their landscapes with support and incentives provided for better landscape governance.

The settlement of non-forest land with adivasi households and communities can be carried out within the existing framework of laws given some policy changes. Both the Panchayati Raj Extension to Scheduled Areas (PESA), 1996 and the Forest Rights Act provide the space for a transformation of local governance of resources by recognizing the autonomy of adivasi communities. However, PESA's provisions have been subverted by the state governments. Even in case of FRA, there seems to be little interest in ensuring that the law is implemented properly in the context of community rights.

The rationalization of legal land tenure in shifting cultivation areas needs to be supplemented with support of different types that will allow local adivasi communities to stabilize both their livelihood base and access to land-based resources. This would need a revamping of the programmes for land-based and natural resource development. These programmes should be reoriented to ensure that communities use these investments for stabilizing shifting cultivation and improving landscape management. Also these programmes should be accompanied by participatory research aimed at studying shifting cultivation landscapes in the Eastern Ghats as integrated agro-ecological systems and to devise participatory processes that can help reverse and stabilize the degradation of landscapes and improve their productivity.

Providing rights over the hill slopes to the communities of shifting cultivators can also enable them to participate in activities such as carbon sequestration through planting trees and regenerating forests, growing commercial crops like cashew on barren slopes and undertaking agro-forestry, including crops such as coffee and other plantation crops. These have the potential of augmenting household and community incomes and reducing poverty in these areas. This will also create incentives for local communities to regenerate forests and hillsides and move towards sustainable governance of these landscapes.

Annexure I

Examples of shifting cultivation in Odisha on Google Earth		
District	Location (latitude/longitude)	Comments
Kalahandi, Trampur block	19 35 28.47 N 83 08 33.16 E	Shifting cultivation. Mainly Kondhs tribe
Kalahandi, Trampur block	19 30 48.78 N 83 00 52.74 E	Degraded short fallows landscapes
Rayagada, Kashipur block	19 15 04.76 N 83 00 36.80 E	Short fallows landscape with soil conservation measures under OTDP
Koraput district	18 31 51.40 N 82 46 01.77 E	Degraded short fallows landscape
Kondhmal district	19 50 52.22 N 83 56 42.17 E	Shifting cultivation inside Kotgarh Wildlife Sanctuary, Kandhs
Gajapati district	19 33 48.15 N 83 57 10.09 E	Shifting cultivation, Saoras
Keonjhar district	21 46 12.67 N 85 12 01.29	Shifting cultivation, Juangs

Source: Author.

Notes

1. The Indian Forest Act, 1927, in Section 10, directs that during the reservation of forests, claims for shifting cultivation should be submitted to the state government for approval. In case the state government approves that shifting cultivation could take place, some area could be excluded from reservation or certain areas could be separately demarcated where shifting cultivation is con-

ditionally allowed as a privilege. It also mentions that 'Shifting cultivation shall in all cases be deemed a privilege subject to control, restriction and abolition by the State Government'.

2. The princely states, called Garhjat, often were a continuation of military fiefdoms ruled by subordinate chiefs who had earlier owned allegiance to the Mughals and the Marathas.

3. In Kandhamals, a special survey and settlement was taken up in 1921–5 only to assess the land of non-adivasis plains people who had moved into the area to take advantage of not having to pay tax revenue (Kumar *et al.* 2008).

4. In 1872, the superintendent of the Tributary Mahals, advised the Garhjat rulers (princely states) that *toila* (shifting cultivation) was one of the chief causes of destruction of timber and suggested that it should be brought under the ruler's supervision and regulated (Rath 2005).

5. Demarcation of forests was initiated in the Balliguda area before independence. One of the major concerns of forest officials was that the reservation of forests would lead to agitations and trouble in the hill tracts which needed to be faced 'boldly' (Nicholson 1939: 35).

6. These laws and rules were generally based on the Indian Forest Act, 1927.

7. For instance, Ramdhyani writes about Kalahandi princely state, 'Reserve forests are simply those which may be reserved, and the rules do not prescribe any matter to be taken into account while making reservations' (Ramdhyani 1947).

8. According to the report,

A detailed and extensive field visit of Odisha state was undertaken to supplement and correct the office interpretation. Interpretation carried out in the office was checked in the field and areas marked as closed forest, degraded forest, water bodies, non-forest etc... Data regarding areas affected by shifting cultivation were obtained from related working plan and local forest officers for the entire state and reported on to 1:250 000 scale, Survey of India toposheets. The data collected on toposheets were used in delineating the areas affected by shifting cultivation in closed forest and degraded forest. (FAO 1981)

Even this exercise depended on secondary data rather than ground truthing.

9. Shifting cultivation is also practised, albeit on a smaller scale, in Lanjigarh block and Madanpur Rampur blocks of Kalahandi district.

10. These discursive memes are to be found in almost all the official documents which sought to deal with shifting cultivators in Odisha.

11. Survey and settlements consist of the process of mapping and categorizing land-use and landownership through drawing up of cadastral maps and creating formal records of rights. Thus survey and settlements basically became the governmental instrument to formalize the property rights system in land.

12. In most princely states, hilly adivasi areas were left out of survey and settlements processes. For instance, in the Keonjhar princely state, areas inhabited by Juangs (Juangpirh) and Paudi Bhuiyans (Bhuinaypirh) were left out of the settlements of 1899 and 1911 (Ramdhyani 1947).

13. The main principle sought to be followed was land to the tiller.

14. The mountainous southeastern part of Kalahandi district was known as the Dongarlas.

15. These special category of records were called Dongar Khasras, which contained the name of the cultivator with the parentage, caste, name of the Dongar, cultivated crops grown, seed capacity of the slopes under cultivation, number of kodkis (spades) possessed by the cultivator and rent settled for cultivation.

16. The Madras Estate Land Act, 1908, Section 15, subsection 3.

17. Letter no. 434-XXI-57/58-L-LRS dated 12 March 1959.

18. In Odisha, most shifting cultivation tribes also practise irrigated paddy cultivation in valley bottoms. Shifting cultivation supplements the terraced rice cultivation; and the investment of labour in creation of paddy fields precludes regular movement of settlements (Bailey 1960).

19. Section 33 (4) of the Odisha Forest Act, 1972 provides that all lands notified as forests under Chapter III of the Magras Forest Act, 1882, were deemed to be protected forests.

20. Section 20-A of the Indian Forest Act, 1927 (Odisha Amendment).

21. The Census of India, 2001 VD table contains village-wise data on irrigated land (no. 138) and unirrigated land (no. 139) which were added to get the total private land in each village.

22. Similarly, the Village Details in Census Data table contains village-wise data on Culturable Waste (no. 140), areas not available for cultivation (no. 141) and forest land within village boundaries (no. 126), which were summed up to get the total government land in the village.

23. In the census data, the spatial area details, including the patta land and government land is sourced from the revenue Record of Rights and, therefore, reflect the legal or de jure position of land tenure. The average patta land per household in a village was calculated by dividing the sum of total patta land by the number of households. This allowed us to identify villages in a district in which only negligible amount of land (less than a 1,000 sq. m. or one-fourth of an acre per average family) is legally owned by the inhabitants.

24. Circular no GE(GL)-S-*/81-37565/REV, Revenue Department, GoO, laid down the operation guidelines for provision of usufruct rights on plantations taken up under the programme on government land (except RFs). Each beneficiary was to be given usufruct rights (Dafayati rights).

25. Circular no. GE (GL)– S–69/79–3755/Rev, Revenue Department, GoO, dated 18 January 1980.

26. 'Measures against shifting cultivation allotment of podu-affected government land for plantation by podu cultivators—conferring of usufruct rights on temporary basis' as per GoO, Revenue Department Letter No. GE (GL) – S – 69/79 – 3755/Rev dated 18 January 1980.

27. GoO letter no. TD-I(IFAD)-18/91/2628/HTW dated 10 April 1992 issued following a review meeting on the Odisha Adivasi Development Project sponsored by IFAD under the chairmanship of the chief minister of Odisha.

28. GoO letter no. 14643-R-S-60/2000 dated 23 March 2000.

29. The situation has changed with the passage of the Scheduled Tribes and Other Traditional Forest Dwellers (Recognition of Forest Rights) Act, 2006 and now the OTELP is involved in trying to get the forest land settled with cultivators.

30. It is interesting to see that even land under short fallow in Dekapar, which consists only of shrubs and weeds, came back as a good secondary forest after the fallowing was stopped. This is a testimony to the potential of regeneration of these slopes.

31. Letter no. GU (Cashew) 54/08_32240/Ag Dated 11/11/2008: Minute of the meeting chaired by Hon'ble Chief Minister, Orissa on 31 July 2008 regarding usufruct rights over existing cashew plantations on government land.

References

Bailey, F. 1960. *Tribe, Caste and Nation. A Study of Political Activity and Political Change in Highland Odisha.* Manchester: Manchester University Press.

Baker, D. 1991. 'State Policy, the Market Economy, and Adivasi Decline: The Central Provinces, 1861–1920', *Indian Economic Social History Review*, 28 (4): 341–70.

Behuria, N.C. 1965. 'Final Report on the Major Settlement Operation in Koraput District, 1938–64'. Cuttack: Government of Odisha Press.

Brady, N. 1996. 'Alternatives to Slash-and-burn: A Global Imperative', *Agriculture, Ecosystems and Environment*, 58 (1): 3–11.

Brookfield, H. and C. Padoch 1994. 'Appreciating Agrodiversity: A Look at the Dynamism and Diversity of Indigenous Farming Practices', *Environment (Washington DC)*, 36 (5): 6–11.

Chandran, M. 1998. 'Shifting Cultivation, Sacred Groves and Conflicts in the Colonial Forest Policy in the Western Ghats', in R.H. Grove, V. Damodaran, and S. Sangwan (eds), *Nature and the Orient: The Environmental History of South and Southeast Asia*, pp. 674–707. London: Oxford University Press.

Conklin, H. 1957. 'Hanunoo Agriculture'. A report on an integral system of shifting cultivation in the Philippines. *FAO Forestry Development Paper*, 12. Rome: FAO.

Darlong, V. 2004. 'To Jhum or Not to Jhum: Policy Perspectives on Shifting Cultivation. Guwahati: The Missing Link (TML)'. Guwahati: Society for Environment and Communication.

Dash, B. 2006. 'Shifting Cultivation Amongst the Tribes of Odisha', *Odisha Review*; http://orissa.gov.in/e-magazine/Orissareview/july2006/engpdf/76-84.pdf.

Dhebar, U. 1961. 'Report of the Scheduled Areas and Scheduled Tribes Commission'. New Delhi, Government of India, 21.

FAO. 1981. 'Tropical Forests Resource Assessment Project: Forest Resources of Tropical Asia (India brief)'. Rome, FAO; http://www.fao.org/docrep/007/ad908e/AD908E13.htm.

Government of Odisha (GoO). 1940. 'The Partially Excluded Area Enquiry Report'. Cuttack: Government of Odisha Press.

———. 1958. 'Report of the Administration Enquiry Committee'. Cuttack: Government of Odisha Press.

———. 1959. 'Forest Enquiry Report'. Cuttack: Government of Odisha Press.

———. 1989. 'Working Plan for Baliguda Division'. Cuttack: Forest Department.

———. 2006. 'Revised Working Plan for the Proposed Reserved Forests, Demarcated Protected Forests and Compensatory Afforestation Area of Rayagada Forest Division for the period 2006–07 to 2015–16. Cuttack: Forest Department.

Government of India (GoI). 'Planning Commission. 1986. Report of the Study Group on Land Holding Systems in Adivasi Areas'. New Delhi: Planning Commission.

———. 1988. 'National Forest Policy 1988'. New Delhi, Ministry of Forest and Environment, Government of India; http://envfor.nic.in/divisions/fp/nfp.pdf.

———. 1997. 'Report of the Expert Committee on Conservation and Management of Forests in the North-East (Rajamani Commission Report)'. New Delhi: Ministry of Forest and Environment, Government of India.

———. 2002. 'Report of the Task Force on Strategy for Management of Shifting Cultivation'. New Delhi: Ministry of Forest and Environment, Government of India.

International Fund for Agriculture Development (IFAD). 1999. 'India: Completion Evaluation of Odisha Development Project. Seven Lessons Learned'; http://www.ifad.org/evaluation/public_html/eksyst/doc/agreement/pi/Odisha.htm.

Jyotishi, A. 2000. *Swidden Cultivation: A Review of Concepts and Issues*. Bangalore: Institute for Social and Economic Change.

Kumar, K., P.R. Choudhary, and J. Kerr. 2004a. 'Tenure and Access Rights as Constraints to Community Watershed Development in Odisha, India'. Paper presented at 'Tenth Conference of the International Association for the

Study of Common Property', 9–13 August. Mexico: Oaxaca; http://dlc.dlib. indiana.edu/dlc/bitstream/handle/10535/1780/Kumar_Tenure_040804. pdf?sequence=1.

Kumar, K., Y. Giri Rao, and S. Sarangi. 2004b. 'Odisha (India) Case Study Report: IUCN's Sustainable Livelihoods, Environmental Security and Conflict Mitigation Project'. Bhubaneswar: Vasundhara.

Kumar, K. and Y. Giri Rao. 2004. 'Perpetuating Injustices: Adivasi Rights and Forest land Cultivation in Odisha (Mimeo)'. Bhubaneswar: Vasundhara Online; http://www.vasundharaOdisha.org/Discussion%20paper/Perpetuating-injustice-Land.pdf.

Kumar, K., P.R. Choudhary, S. Sarangi, P. Mishra, and S. Behera. 2005. 'A Socio-Economic and Legal Study of Scheduled Tribes' Land in Odisha'. Bhubaneswar: Vasundhara; http://vasundharaodisha.org/download22/ World%20Bank%20Study%20Report.pdf.

Kumar, K., S. Behera, S. Sarangi, and O. Springate-Baginski. 2008. '"Historical Injustice": The Creation of Poverty through Forest Tenure Deprivation in Orissa'. University of East Anglia Working Paper. Norwich: UEA.

Kunstadter, P., E. Chapman, S. Sabhasri, and M. Mai. 1978. *Farmers in the Forest: Economic Development and Marginal Agriculture in Northern Thailand*. Honolulu, USA: University Press of Hawaii, published for the East-West Center.

Mishra, A. 2007. 'Odisha Adivasis Struggle to Retain Cashew Rights', *Down to Earth*, 15 August; http://www.downtoearth.org.in/full6.asp?foldername=2 0070815&filename=news&sec_id=50&sid=20.

Myers, N. 1992. 'Tropical Forests: The Policy Challenge', *The Environmentalist*, 12 (1): 15–27.

Nicholson, J.W. 1939. 'Annual Progress Report on Forest Administration in the Province of Orissa for the Year 1937–38'. Cuttack: Government Press.

Padel, F. 1995. *The Sacrifice of Human Being: British Rule and the Konds of Odisha*. Delhi/New York: Oxford University Press.

Potter, L., H. Brookfield, and Y. Byron. 1994. 'The Sundaland Region of Southeast Asia', in J. Kasperson, R. Kasperson, and B. Turner (eds), *Regions at Risk: Comparisons on Threatened Environments*, pp. 460–518. Tokyo: United Nations University Press.

Pouchepadass, J. 1995. 'British Attitudes towards Shifting Cultivation in Colonial South India: A Case Study of South Canara District 1800–1920', in D. Arnold and R. Guha (eds), *Nature, Culture, Imperialism: Essays on the Environmental History of South Asia*, pp. 152–84. New Delhi: Oxford University Press.

Ramakrishnan, P. 1992. *Shifting Agriculture and Sustainable Development: An Interdisciplinary Study from North-Eastern India*. Paris: UNESCO; New Delhi: Oxford University Press.

Ramdhyani, R.K. 1947. 'Report on Land Tenures and Revenue Systems or Odisha and Chattisgarh State'. Berhampur: ILP Press.

Rangarajan, M. 1996. *Fencing the Forest: Conservation and Ecological Change in India's Central Provinces, 1860–1914*. New Delhi/New York: Oxford University Press.

Rasul, G. and G. Thapa. 2003. 'Shifting Cultivation in the Mountains of South and Southeast Asia: Regional Patterns and Factors Influencing the Change', *Land Degradation & Development*, 14: 495–508.

Rath, B. 2005. *Vulnerable Adivasi Livelihoods and Shifting Cultivation: The Situation in Odisha*. Bhubaneswar: Vasundhara.

Saravanan, V. 1998. 'Commercialisation of Forests, Environmental Negligence and Alienation of Adivasi Rights in Madras Presidency: 1792–1882', *Indian Economic and Social History Review*, 35 (2): 125.

Saldanha, I. 1998. 'Colonial Forest Regulations and Collective Resistance: Nineteenth-century Thana District', in R.H. Grove, V. Damodaran, and S. Sangwan (eds), *Nature and the Orient: The Environmental History of South and Southeast Asia*, pp. 708–33. London: Oxford University Press.

Scott, J.C. 1999. *Seeing Like a State: How Certain Schemes to Improve the Human Condition Have Failed*. New Haven: Yale University Press.

Singh, K., B. Sinha, and S. Mukherji 2005. 'Exploring Options for Joint Forest Management in India. Forest Policy and Institutions Working Paper. A World Bank/WWF Alliance Project. Rome: Food and Agriculture Organisation.

Sivaramakrishnan, K. 1999. 'Transition Zones: Changing Landscapes and Local Authority in South-west Bengal, 1880s-1920s', *Indian Economic Social History Review*, 36 (1): 1–34.

Skaria, A. 1999. *Hybrid Histories: Forests, Frontiers and Wildness in Western India*. New Delhi: Oxford University Press.

Sundarajan, S. 1963. 'Final Report on the Survey and Settlement of the Kashipur, Karlapat, Mahulpatna and Madanpur-Rampur Ex-Zemindaries in the District of Kalahandi'. Cuttack: Odisha Government Press.

9

DHRUPAD CHOUDHURY

The Forest Rights Act, Northeast India, and Shifting Cultivators

A Commentary

Reactions to the promulgation of the Scheduled Tribes and Other Forest Dwellers (Recognition of Forest Rights) Act (FRA), 2006 in Northeast India have been somewhat muted—at best, lukewarm, suggesting a sense of disappointment, dejection, and perhaps, even disillusionment and a sense of alienation. Unlike the rest of the country, no euphoria or protests have marked the notification of the Act, and to observers outside the region, this has been quite baffling. The reasons, however, are not difficult to seek. The FRA offers no rights, privileges, or concessions other than those already enjoyed under existing provisions of different state and district council laws or constitutional provisions; in fact, the Act does little but reiterate the privileges enjoyed under existing provisions. The Act makes no attempt to address ambiguities or redress shortcomings under existing laws and hence fails to address concerns that have arisen subsequent to the formulation of such laws. On concerns arising out of the Supreme Court rulings in regard to the *Godavarman* case, the Act remains completely silent. The muted reaction, therefore, is not surprising. What, then, were the expectations from the Act? What are the issues and concerns of Northeastern upland communities that need to be addressed but remain unresolved? To appreciate the expectations of the Northeast communities, it is necessary to first understand the rights enjoyed under customary laws, particularly in regard to shifting cultivation and access to forest resources as well as the restrictions on these rights brought in by state and district council legislations and more recently, the Forest Conservation Act (FCA) and the Supreme Court's rulings.

Traditional Rights of Access and Use: Shifting Cultivation and Forest Resources[1]

For members of ethnic communities of the uplands of Northeastern India, customary laws have governed their rights of access and use of natural resources for all activities concerning their sustenance, be it food security, shelter, or energy needs. The fundamental principle underlying all customary laws governing natural resource management, irrespective of the nature of the traditional institutions upholding such laws, is that of sustenance and access to natural resources for livelihood security. Members of upland ethnic communities, thus, usually have access to land for the purpose of shifting cultivation or to forest products from forests and regenerating fallows for collection of wild edibles for consumption, timber and thatch for house building, or to fuelwood for household energy requirements. Customary laws permit members to hunt wild animals for food, but strict regulations restrict overhunting and ensure conservation and management of each species. Most communities also maintain community forests with clear categorization of use ranging from sustenance to ecological services as well as for pure preservation.

Despite variations in the traditional institutional framework and access regimes found among different ethnic communities of Northeast India, two fundamental attributes that can be gleaned among all are tenurial security and the underlying philosophy of equitable access to resources; the few exceptions to this general framework are situations where a highly autocratic chieftainship prevails and access to resources for members depends solely on the whims and subjective judgment of the former. Among the majority, however, the twin principles form the basis of governance of natural resources and are critical constituents contributing to the strong social capital characteristic of most Northeastern indigenous communities. This is expressed most succinctly in the access regime applicable in shifting cultivation. The framework for access to land for shifting cultivation is based on the principles of common property regimes, irrespective of whether the traditional institution governing the access regime is a village council composed of elders with universal representation of clans and households, or a chieftain aided by a council of clan elders (a few exceptions to this arrangement do exist, as pointed out earlier).

The Access Framework in Shifting Cultivation Systems: Tenurial Security

On establishment of a village, the traditional institution in consultation with the villagers earmarks areas for the purpose of shifting cultivation and selects the locations for each year's cultivation. Within such sites, allocation of clan patches is done on the basis of clan hierarchy based on the seniority of founding clans (determined on the basis of initiatives and efforts taken to establish the new village). Within the clan patches, choice of plots is done according to seniority of members (by age) and this includes widowed and female-headed households. This principle is followed for all cultivation sites till the village decides that they have adequate land parcels for shifting cultivation that allow a sufficiently long fallow cycle for complete recuperation and recovery of plots in each cultivation parcel. The rest of the land in the village is earmarked for homesteads, burial grounds, and forests of different usage categories, including hydrological services and conservation such as catchment and sacred forests.

Once shifting cultivation sites are determined, clan patches earmarked, and plots allocated to households, the tenurial rights of individual households in each of these parcels are guaranteed by the clan and the traditional institution. These rights are heritable and hence passed down through generations; however, such rights are inalienable and cannot be transferred to anyone, except in exceptional situations and that too, only with the approval and ratification of the clan council as well as the traditional institution. Thus although no formal documentation is done, tenurial security within the customary system is as secure as with a formal land titling, with the exception of the right to transfer ownership. This security is not confined to members of the founding clans only, but extends to later migrants as well. While it is obvious that later migrants to the village will not have plots allocated to them in earlier parcels, the village accommodates their access rights (and tenurial security) in parcels that are opened up subsequent to their entry into the village community, thus ensuring that even the migrants have a right to productive land for shifting cultivation.

The Access Framework in Shifting Cultivation Systems: Equitable Access to Productive Resources

In addition to ensuring tenurial rights to all members of a village community, traditional institutional arrangements also ensure equitable access to productive resources, particularly in regard to land for shifting cultivation. Although tenurial rights are guaranteed for all households within a particular shifting cultivation parcel, access frameworks introduce the concept of plot size flexibility to operate the principle of equitable access to productive land. Unlike private property regimes, plot sizes in shifting cultivation are not fixed and remain flexible, operating on the principle of 'mouths to feed within the household' for the particular year in question. In effect, plot sizes are determined on the basis of members of an individual household present in the village for the year of cultivation. Plot sizes, therefore, would vary according to the number of individuals required to be supported. This also links up to labour availability and hence optimizes returns to labour. With surplus land thus generated due to plot size rationalization with household size, households that do not have tenurial rights within the particular parcel (usually later migrants and new entrants) are accommodated to ensure their access to land for cultivation. This access, however, is restricted to only that particular year and does not impart tenurial security that can be inherited, thus safeguarding the tenurial security of the original household. The instrument of plot size rationalization to current household size is followed strictly, however, and not relaxed under any circumstances, thus allowing institutions to implement the principle of equitable access to productive land.

Equitable access to resources is also ensured in the context of resource use in forests and regenerating fallows. Rights to forest produce for consumption, shelter, or sustenance is ensured to all members of the village community, but under strict regulations that ensure sustainable harvest and management. These rights are also applicable in shifting cultivation fallows. As a family completes the cultivation phase and moves on to a new plot, the plots gradually regenerate into secondary forests. Access to forest resources within such regenerating fallows is not restricted solely to the family that has tenurial rights on the plot for cultivation, but is open to any member of the village. The only exceptions are with regard to use of any trees or other resources that the household may have planted or introduced and nurtured (such as a bee colony) in their plot during the cultivation phase.

Legislations, Curbs on Traditional Rights, and Ambiguities

State and District Council Legislations and Curbs on Traditional Rights

Although rights of access and use enjoyed by the upland communities are protected by provisions of the Constitution,[2] several legislations enacted by the state and various autonomous district councils in the region govern the practice of shifting cultivation.[3] A detailed discussion on the provisions of each of these legislations and their implications on rights enjoyed under traditional customary laws is beyond the scope of this commentary; however, the salient points of the implications of these legislations will be highlighted in the subsequent sections to underline the main argument made.

All the legislations enacted for the management of shifting cultivation recognize the right of community members to practise shifting cultivation but grant this as a privilege or concession provided the practice does not degrade the land or cause undue degradation of forests. The role of traditional institutions in allocation of land is taken over by the executive committee of the district council, or the district commissioner (or the land conservator in the case of the Balipara/Tirap/Sadiya Frontier Tract Jhumland Regulation) and these authorities attain appellate and final powers of decisions. In all these legislations, however, this power can be vested by the concerned authority on the traditional institution—thus, under the Garo Hills regulation, the Nokmas are fully empowered to allocate land for jhumming (Section 3); similarly, Section 15 of the Balipara/Tirap/Sadiya Frontier Tract Jhumland Regulation provides for any or all of the powers granted to the land conservator to be vested to the tribal council. Under Section 6 of both the Lushai Hills District (Jhumming) Regulation, 1954 as well as the Pawi Autonomous District (Jhum Regulation) Act, 1983 allocation of jhum plots under traditional norms (privilege to select) is replaced by one of drawing lots. These legislations, therefore, dilute the absolute control hitherto enjoyed by traditional institutions under customary norms and in the case of the Lushai Hills and Pawi Regulations, completely dissolve the tenurial security hitherto enjoyed under customary laws.

Although all legislations respect the principle of inalienability of the land and restrict transfer of land, exceptions are made in case the authority permits it. The principle, however, is totally ignored with regard to government efforts to promote permanent cultivation (and probably to facilitate replacement of shifting cultivation). In fact, in so far as introduction of terraces for wet rice cultivation is concerned, some of the legislations empower the executive committee of the district council to take unilateral decisions. Thus, the Garo Hills Regulation empowers the executive committee to:

> ... settle culturable wasteland in any Akhing amongst the permanent residents thereof and on the merits of each case and taking the recommendations of the Nokma concerned into consideration. Provided that no one of the same Akhing be interested in bringing permanent cultivation to any culturable plain and make it available for settlement, the Executive Committee shall give one month's notice through the Nokma of the Akhing of the intention of the council to settle such culturable wasteland with any suitable person permanently resident in the District who undertakes to accept settlement of any particular portion of such land to bring it under permanent cultivation. (Garo Hills District Jhum Regulation, 1954; Section 8)

Section 9 (1) of the Garo Hills Regulation further goes on to empower the executive committee to similarly enforce the introduction of horticulture 'notwithstanding any objection of the Nokma of the concerned Akhing'. Section 9 (2) goes even further and empowers the executive committee to grant permanent settlement to the person willing to start such permanent cultivation even if the person is not a permanent resident of the village and despite objections raised by the Nokma. This section of the regulation not only severely dilutes the traditional control enjoyed by the Nokma in regulating settlement, but also introduces an external authority to grant ownership to individuals who may not be even permanent residents of the village. This raises three important issues—the first concerns the compromise and dilution with regard to the authority of traditional institutions and the establishment of a dual authority in the guise of the executive committee; the second concerns the dilution of common property regimes and the enforced introduction of private property against the will and consent of the villagers. The third aspect that underlines such enforcement, particularly among Garo society, is the potential dispossessment of land by women who are the traditional owners of land in a matrilineal society.

As mentioned earlier, the privilege of shifting cultivation is given provided the practice does not degrade the land and cause undue degradation of forests, erosion, or potential reduction of water resources. All the legislations retain the power to suspend shifting cultivation in case the authorities judge that the practice is responsible for undue erosion or degradation, possible reduction of hydrological flow, and damage to forests or environment. The authorities can force the traditional institutions to ensure that fast-growing tree species are introduced to facilitate afforestation and prevention of erosion. The council is legally bound to provide financial resources to the cultivator to do so and the legislations allow the cultivator rights to the trees planted thereof. Interestingly, in all cases, cultivation is suspended 'for a period not exceeding ten years'. Although seemingly insipid, this provision provides the first evidence to suggest that authorities acknowledge that shifting cultivation fallows are arable land and not forests and acknowledge that the cultivator will return to the plot for the purpose of cultivation at the end of the fallow period. Explanation 1 of Section 2 (b) of the Balipara/Tirap/Sadiya Frontier Tract Jhumland Regulation also recognizes that what is considered jhumland 'shall be deemed to be so notwithstanding the fact that a part or the whole may have been planted with fruit trees, bamboos, or tung, or reserved for growing firewood'. These provisions clearly establish the nature of fallows and the recognition that they are, ultimately, arable land and hence should not attract the provisions of any legislation governing forest conservation and management.

Unclassed State Forests and the Ambiguity Concerning Shifting Cultivation Fallows

Shifting cultivation fallows regenerating into secondary forests, however, are classified as 'Unclassed State Forests' in administrative land-use classifications. This category of forests includes all forests that are not under reserves, protected areas, or somehow under direct control of the forest bureaucracy. In practice, although communities continue shifting cultivation and enjoy access to forest products from such forests, an ambiguity continues in respect of the application of the provisions of the FCA and the Supreme Court's rulings. While the community claims traditional rights of control and use (and in practice, enforces them), the forest bureaucracy disallows any extraction or logging from such forests, except in cases with an approved work plan or scheme. This restricts the rights of the community to forest produce and in case of extraction of timber and other produce even for the purpose of domestic use, potentially makes such acts unlawful and criminal under the Hon'ble Court rulings. In fact, in some cases, the Net Present Value (NPV) conducted for conversion of such forests for development purposes and the resultant revenue collected has directly been claimed by the forest department (FD) with no knowledge or acknowledgement of the community owning such forests. This ambiguity is the prime concern of Northeast communities that requires immediate redressal.

Reaching Out to Communities in the Northeast

It is clear from the previous section that state and district council legislations have restricted the role and control of traditional institutions as well as introduced restrictions to the rights enjoyed under customary laws. In all the cases, the authority of traditional institutions has been compromised. In the case of the Garo Hills, the legislation has introduced provisions that override the traditional rights of the Nokmas to regulate settlement and has provided an instrument to confer permanent ownership on individuals engaged in permanent cultivation, thereby enforcing private property regimes within a common property framework contradicting traditional property regimes. Such transformations in property regimes not only contradict traditional frameworks, but they also dilute traditional institutions and, more importantly, challenge the integrity and continuation of the institution of equitable access to productive resources. In the process, they weaken and threaten the continuity of social capital characteristics among such communities. Lastly, the categorization of community forests and regenerating fallow forests as 'Unclassed State Forest' creates an ambiguity in regard to the rights enjoyed by communities in such forests for sustenance and unduly creates tensions between the forest bureaucracy and communities.

The FRA, in Chapter II Section 3 (j), describes rights that are recognized under any state law or laws of any autonomous district council or autonomous regional council or that are accepted as rights of tribals under traditional or customary law of the concerned tribes of any state as the forest rights of forest-dwelling Scheduled Tribes (STs) and other traditional forest-dwellers on all forest land. This provision restricts the rights to those described under the relevant state or district council legislation and thus fails to address the limitations enforced by the latter. A reading of the rules raises a further concern—Chapter III Section 11 (2) '... and vested with the community up to a maximum of four hectares'—restricting community-holding for cultivation to a mere four hectares is preposterous in the case of shifting cultivation. The rules, however, make no mention of the special situation of shifting cultivation and hence, prescribe no exceptions.

As outlined in the commentary, existing legislations have thrown up several fundamental concerns that require to be redressed—and at the earliest. For the ethnic communities of Northeast India, the most pressing concern is that of categorizing all community forests and regenerating fallows as 'Unclassed State Forests'. Without this ambiguity being addressed, historical injustices also cannot be rectified. The FRA, in so far as the Northeast is concerned, and particularly with regard to shifting cultivators, requires suitable amendments. It is only with such amendments that communities in the Northeast will welcome the Act as a true instrument to remove injustices.

Notes

1. See Choudhury (1998 and 2006).

2. Most states in the region are governed by special constitutional provisions: Nagaland (Article 371A), Assam (Article 371B), Manipur (Article 371C), Mizoram (Article 371G), Arunachal (Article 371H). In addition, Articles 244 and 244A makes provisions of the Vth and VIth Schedules of the Constitution applicable to Scheduled (Tribal) Areas and Tribes.

3. These are: The Balipara/Tirap/Sadiya Frontier Tract Jhum Land Regulation, 1947; The Garo Hills District (Jhum) Regulation, 1954; Lushai Hills District (Jhumming) Regulation, 1954; Pawi Autonomous District (Jhum

Regulation) Act, 1983; The Mikir Hills District (Jhuming) Regulation, 1954; The Nagaland Jhumland Act, 1970; The Sylhet Jhum Regulation, 1891. For a full list of legislations with relevance for shifting cultivation, see Darlong (2004).

References

Choudhury, D. 1998. 'Conservation and Local Communities in North-East India', in A. Kothari, N. Pathak, R.V. Anuradha, and B. Taneja (eds). 1998. *Communities and Conservation: Natural Resource Management in South and Central Asia*, pp. 435–48. New Delhi: Sage Publications and UNESCO,

———. 2006. 'Equitable Resource Access under Shifting Cultivation Systems in North East India', in ANGOC and ILC *Enhancing Access of the Poor to Land and Common Property Resources: A Resource Book*. Rome: Asian NGO Coalition for Agrarian Reform and Rural Development (ANGOC), Quezon City, and International Land Coalition (ILC).

Darlong, V.T. 2004. *To Jhum or Not to Jhum: Policy Perspectives on Shifting Cultivation*. Guwahati: The Missing Link, Society for Environmental Communication.

10

PRAKASH KASHWAN
VIREN LOBO

Of Rights and Regeneration

The Politics of Governing Forest and Non-forest Commons[1]

India is in the process of revisiting the property rights question vis-à-vis its forests. The Forest Rights Act of 2006 (FRA), discussed in Chapter 3 of this volume, has reinvigorated the debate about forest rights and forest conservation issues. The Act has refocused attention on the 'historical injustices' caused by the drawing of forest boundaries without adequate investigation of pre-existing adivasi settlements, cultivation, and forest use. The Act stipulates that where 'forest'land was historically used for subsistence-oriented non-forestry purposes, such use should be recognized and the 'forests' should be redesignated as lands with restricted private property rights and/or community rights, as the case may be.

The debate on historical injustices and traditional rights has involved an overly simplified treatment of 'communities' and 'resources': the debate has treated local communities as a homogenous block (Lele 2004; Li 2002). A number of manifestations of such a simplified reading of community are apparent. The Ministry of Tribal Affairs (MoTA) pushing 'speedy implementation' of the Act assumes that it is possible to *implement* (that is, dispense) traditional rights merely by virtue of the formal encoding of 'community participation' in the statute. On the other hand, activists seem to put uncritical faith in the 'corporate community', which is able to develop a consensus over the character and content of these rights. Short-sighted analysis also exists amongst those opposing the FRA. Detractors of the FRA assume that it is justified and possible to conserve forests by continuing to ignore the problem of contested tenures, and the longstanding conflicts engendered by that (Saberwal 2000). They have also tended to see forests as something apart from other common land and hence ignored the fact that other common land face similar problems of contested tenure. The literature on the politics of the commons (Agrawal 2001; Peluso and Vandergeest 2001; Pérez-Cirera and Lovett 2006) and commons-dependent livelihoods suggest that none of these simplified images both of proponents and detractors of the FRA are true in terms of the practical politics of day-to-day struggles around the commons (Brara 2006).

In this chapter, we problematize simplifications: We argue that nongovernmental organizations' (NGOs') approaches to resource conservation fail to address the question of inter-and intra-community equity. We also argue that social movement approaches advocate community rights to forests without attending to the challenges of effectively conserving the forests that different sections within local communities depend on, albeit to different degrees. In turn, we grapple with the complexities of delineating and dispensing rights to historically marginalized sections within 'local communities'.

The next section lays out the broad framework in which we analyse access to resources and rights given longstanding intra-community inequities that in part have contributed to the marginalization of a section within the local community. We dissect the monolithic idea of 'encroachment' in order to clarify the conceptual terrain and focus on questions of legitimacy, the drivers behind 'encroachment' on the commons, and the implications of different types of forest to non-forest land conversions. This is followed by an empirical analysis of the complexity of causal drivers that are rooted in the political economy of nature–society interactions and an analysis of a much ignored phenomenon—that of decimation of non-forest common land and the impact it is having on the forest-dependence of local communities. Following this, we briefly present the typical NGO-type approach to resource conservation and contrast this to a social movement approach to land rights' mobilization. The analysis points out how both of these approaches suffer from a homogeneous view of the community and fail to address the challenges rooted in intra-community inequities. Finally, the chapter concludes with an illustrative discussion of interventions that have successfully integrated aspects of conservation and equity, and draws succinct lessons from the analyses presented. The intention is not necessarily to suggest models to be replicated but to highlight selective approaches to addressing some of the key issues raised in the chapter.

Framework for Looking at the Commons Reforms

In this chapter, we build on the premise 'that framing and implementing forest policy has always been a political process (though informed by ecology, economics and sociology)'. It is also our belief that, given the socio-political and economic environment, any attempts to balance and prioritize different benefits from forests are likely to produce winners and losers amongst different sections of society. Moreover, some of the drivers that lead to differential intra-community distribution of benefits are related to the historical inequities within local communities. The strategies that locally influential actors employ to capture a disproportionate share of development and conservation programmes are also part of the 'politics' around resource conservation and resource rights. In order to reveal local politics around resource claims, we begin by looking into the simplified imageries of 'community' employed by NGOs working on forest conservation, as well as by social movements fighting for forest rights. We also compare and contrast the array of resource claims and propose a more nuanced vocabulary for discussing transformation of land claims and land-use.

Forest Conservation and Intra-community Inequities

Community-based natural resource management (CBNRM) has long been criticized for its somewhat simplistic understanding of the community and by implication, its limited success at achieving political and social empowerment of marginalized sections (Agrawal 1997; Leach *et al.* 1999). In a recent study on CBNRM in South Asia (Menon *et al.* 2007), the authors conclude that most CBNRM interventions that operate at the level of community have failed to address inequities within the communities. Even interventions targeted at the household level have made limited strides towards a more egalitarian approach to development. In fact, the study concludes that most NGOs end up working with those that are 'willing to cooperate'. Factoring in the time and labour opportunity costs faced by the poorest among the community explains why NGO interventions are often coordinated by local elites. The authors point out that 'given the centrality of "community formation" to the discourse of CBNRM... little attempt has been made to actively engage with "particular interests"...'. They go on to suggest that 'particularly "tricky" socio-economic concerns that may be central to some actors have not been raised. The most obvious of these is access to land...' (see also Leach *et al.* 1999). This critique of CBNRM becomes extremely relevant given that the interventions around regeneration of the commons often interfere with the land and livelihood rights of the marginalized within local communities.

We leverage these insights to scrutinize the debate around land rights in forests and other commons on the one hand, and the regeneration of the commons on the other. Prior experiences of land reforms, and also land allotment as a result of grassroots mobilization (such as in Madhya Pradesh, partly as a result of a sustained campaign run by Ekta Parishad), has shown that if meticulous strategies are not developed to account for local socio-political inequities, implementation of justified reforms is invariably impaired (Brara 2006; Srivastava 2006). We believe that understanding intra-community distributional issues is likely to be important in proper implementation of reforms. It must be noted that implementation failures attributed to bureaucrats are often partly a result of the local elites vying to corner benefits of entitlements meant for ordinary citizens (Bardhan 2002).

Developing a Nuanced Vocabulary of Rights and Violations

The fact that adivasis and other forest-dwellers lived in forested regions much before governments demarcated vast areas as public forest land is now widely recognized (Sarin 2008). Stakes for these communities in forest land may be referred to as 'traditional rights' and are within the boundaries of what is now public forest land. The Forest Rights Act, 2006 (FRA) currently under implementation recognizes these rights in spirit. We do not consider these to be cases of forest *conversion* because the term 'conversion' assumes that the previous land-use was physical forest. Therefore, we propose that historical mistakes related to forest settlement must be identified and settled once and for all. We can then pay attention to deliberating over rights over management of forests (Lele and Srinidhi 1998). At the same time, it is precisely for the sake of conceptual clarity that we urge caution on extending rights without a careful scrutiny of the eligibility of the potential rights-holders.

Instances of local elites and those living in large villages/towns in forest fringe areas converting forest areas into agriculture areas or grazing lands to support or subsidize their household consumption, or even for commercial activities such as milk production, should be considered as cases of 'forest conversion'. In some ways, these groups contribute most to the confounding of forest use by traditional rights-holders. Commercial use of forests by some households may contribute to forest degradation, which in turn may adversely affect the livelihoods of those deriving subsistence from forests (Nagendra *et al.* 2006). The better-off within historically marginalized communities will continue to frustrate attempts at empowering the most marginalized. Therefore, a more differentiated and nuanced view of the 'community' should inform contemporary strategies at rectifying historical injustices. In recognizing and privileging bona fide livelihood needs, the FRA supports such a view of preference in dispensing rights. However, very little discussion has gone into interpreting this provision of FRA, leaving this effectively to the discretion of local officials.

Another type of muddled situation relates to the *non-forest* commons, such as community pastureland and revenue wasteland. The cases of the existing private holdings of local community members erroneously included within the boundaries of common pastures may be rare. To the extent this holds true, most private encroachments of village pastures should not be classified at par with the traditional forest land rights recognized under the FRA.[2] Even so, given that rural and forest communities depend on a portfolio of commons, whatever happens in one type of commons is likely to affect others. For instance, if the traditional grazing land is brought under private cultivation, the grazing is likely to shift to forests, leading under certain circumstances to increased grazing pressure on the remaining non-forest common land or on the forest areas.

On the other hand cash crop plantations promoted by state governments, we argue, should be defined as 'forest diversions'. The term 'diversion' emphasizes the non-forestry goals related to these activities, even though the land cover may still remain in the form of a 'forest' of some sorts ('eucalyptus forest' or a 'cardamom forest'). In many cases, because of their superior bargaining powers in the political arena, the planters may have gained full legal backing for their occupation of forest land: 99-year leases and so forth. However, we argue, such 'legal backing' is *conceptually* no more sacrosanct than is the 'illegality' of occupation of forest land by those now recognized to hold traditional land rights within the boundaries of what is presently public forest land. The plantation leases, or the terms and conditions thereof, must be rethought in the changed social and environmental goals of our times. In sum, a more comprehensive view of forest tenure and non-forest common regimes working in tandem with a careful analysis of local disparities is the guiding theme through the rest of this chapter.

We utilize the framework discussed above, a twin focus on intra-community heterogeneity and landscapes (as opposed to a single resource such as forests), to offer a broad critique of both the forest conservation and forest rights discourses that are commonplace in NGO/social movements' parlance. The following section discusses the broader canvass of the local commons and problems with focusing solely on forests as opposed to the wider commons, particularly in the context of intra-community heterogeneity.

Bullock Capitalists and the Commons

In the villages that E. Somanathan visited in the Central Himalayan region, he found a 'pattern of the more well-to-do and powerful in a village being the most assertive (in taking over common land)' (Somanathan 1991: PE-43). In one particular instance he was told that *maaldaar*s, as some people occupying forest land were called, keep occupied civil forest land as their 'private grass preserve' (Somanathan 1991). In the framework proposed above, it is these maaldaars, who are the source of forest conversions at the local level. Indeed, civil forests 'diverted' to private occupation are the source of multiple livelihood benefits to the poor in the villages. But, concern over private diversion of public land goes beyond forest land. This necessitates looking at the issue of the village commons in their entirety. The linkages between forest and non-forest commons are historical as well as contemporary. Brara discusses how the inadequacies of the process of settlement led to decimation of the village commons:

> ... pasture-land... wrongly classified as 'unoccupied' land often went unchallenged because ordinary users were unaware of these legal provisions (meant to) protect their customary grazing right.... The state could reserve its 'unoccupied' lands for the constitution of forests under the Rajasthan Forest Act, 1953. Subsequently, forest produce was presumed to be property of the state government until the contrary was proved. Prior rights of pasturage were thereby subject to controls imposed by the forest department. (Brara 1989: 2247–8, 51)

Villagers treat all available public land as a pool available for free-ranging grazing, gathering of fuelwood and other non-timber forest produce. Reduction in the size of any one category of public land affects the availability of overall livelihood benefits. Ironically, the 'social forestry' programmes implemented in the name of aiding supply of fuelwood and fodder led to the decimation of the village commons. As part of working arrangements in social forestry, 'village wastelands' were turned over to Forest Departments (FDs) for raising social forestry plantations. Subsequently, these village 'wastes' were declared as Protected Forests (PFs) or Reserved Forests (RFs) (Saxena 1992). Thus, both from a historical as well as contemporary perspective, it is important to consider the intricacy of linkages between private land and non-forest commons, such as pastureland. The next section traces contemporary linkages between the forest and non-forest commons and how socio-political dynamics fuelled by intra-community heterogeneity animate conversion and diversion of various categories of the commons.

Dependence and Inter-CPR Linkages

Early studies by Jodha (in Iyengar 2001) revealed that wasteland contributed significantly towards rural livelihoods, mainly in the form of subsistence consumption, employment, and income of the poor. The data collected by Jodha showed that CPRs contributed between 11 and 17 per cent to the total income of the poorest households in the communities living in arid and semi-arid regions. In subsequent writings,

Jodha suggested that the contribution of CPRs to income was between 14 and 23 per cent (Jodha 1989). Given the apparent decline in national poverty levels, increasing reliance on markets, and patterns of seasonal and circular migration, one would have expected the proportion of commons-related income to go down. However, revisiting a subsample of villages from his seminal studies, Jodha (2006) found that in the 24 villages revisited, the share of CPR related income exceeded 50 per cent. One reason for this high percentage could be attributed to methodological differences between his earlier studies and the more recent 'revisits', as Jodha acknowledges. Even so, it is safe to assume that the poorest peasants continue to depend critically on public land for their livelihoods as his and other studies show (see also, Menon and Vadivelu 2006).

Studying CPRs in the state of West Bengal, Beck and Ghosh (2000) concluded that CPRs accounted for about 12 per cent of the income of the rural poor. In addition, they argued that at a macro-level CPRs contribute an estimated US $5 billion a year to the incomes of poor rural households in India. Despite excellent insights that we get from CPR studies in India, many studies, including the surveys conducted by the National Sample Survey Organisation (NSSO), define CPRs based on their de jure status, which result in many types of de facto CPR uses being left out (Beck and Ghosh 2000; Menon and Vadivelu 2006). Equally important is the fact that studies related to forest dependence rarely account for non-forest CPRs (Conroy et al. 2001).

Forest-CPR linkages are important for several reasons. First, considering the diversified livelihood portfolios that rural communities, the poor in particular, rely on, it is unthinkable that strategies for forest rights could be treated in isolation of non-forest CPRs (Baumann and Farrington 2003; Gupta 1991). Often, different categories of common land offer different products and services. While forests may offer valuable food supplements and non-timber forest produce, pastureland may be better suited for grazing purposes. Second, after decades of diversion of CPRs for 'developmental purposes', we are left with an inadequate area of common land, which means that the remaining commons offers valuable subsistence benefits for the most marginalized among the local communities. Such dependence is not going to go away even if every poor household is allotted a piece of private agriculture land (Menon and Vadivelu 2006). Third, efforts by community groups to manage different types of commons often tap into and rely on a variety of institutional niches. Robbins' (1998: 410) work in western Rajasthan suggests that 'multiple, contending institutions govern village resources in a state of legal pluralism'. Multiplicity of institutional forms helps build resilience, that is, the ability of a system to overcome sudden shocks (Adger 2000). Finally, whatever happens in the non-forest commons affects the health of forests and vice-versa. Bon (2000) cites the case of privatization of village pastures in Himachal Pradesh that led to the intensive lopping and heavy grazing of village forests. This resulted in degradation and disappearance of an important species of the *Quercus* family.

Diversion and Distortions

While Jodha's insights on CPR dependence of the rural poor received significant attention, a second insight about the depletion of CPRs through allotments of various sorts, particularly the political economy of such allotments, has been less seriously internalized within the CPR discussions. Ironically, diversion of CPRs has been justified in the name of the poor and the marginalized (and thus the 'silence' on such diversions) even though it has ended up benefiting the non-poor more often than not. Although Bokil (1996) found that privatization of commons improved food security for the poor households, he concluded that the rights to these often degraded and non-irrigated land were not sufficient to pull a household out of poverty. The poor who lacked complementary resources to develop and use the newly received lands were forced to sell their land, often at unfavorable terms of exchange (Bokil 1996). Accordingly, Bokil (1996: 2260) suggests that any attempt at improving livelihoods of the poor should simultaneously consider aspects of equity, productivity, and sustainability.

Hiremath (2001) finds that 49–86 per cent of the privatized CPRs ended up in the hands of the non-poor in different areas. Similarly, Nadkarni (2000: 1184) concludes that the more powerful among local communities encroach on CPRs to a greater extent. It is important to recognize that this would not have been possible without the ability of powerful people to exploit the institutions of democracy, in particular those related to electoral politics. In states such as Rajasthan, the authority of allotting 'government wasteland' used to be vested in the head of elected village councils. Brara (1989) finds that the better informed village elites were able to exploit the processes of settlement directed at conferring legal security to the de facto status of grazing lands. Village elites managed to falsely claim prior occupation of grazing lands by exploiting a clause that allowed the government to give public land to the poor.

Bon (2000) depicts a similar situation. He recounts how 'influential and well-informed' community members were able to exploit a loophole in the Himachal Pradesh Village Common Land Vesting and Utilization Act of 1974 to illicitly occupy and subsequently get legal titles by exploiting their political connections. His study showed that class hierarchies explained the 'allotment and the size of encroachment' better than caste hierarchy (Bon 2000: 2570). The Act itself was meant to 'apply scientific methods of management on open wastelands which were previously used as common pastures' (Bon 2000: 2569). By depleting the size of common pastures, it eventually ended up degrading the existing village forests. Similarly, an overly zealous Karnataka FD's devotion to planting village grazing land with eucalyptus under the social forestry scheme led to a serious fodder crisis during the 1986 drought, even though the programme itself was aimed at easing the fuelwood and fodder scarcity. Thus, a number of examples exist of how the rural poor collectively have lost a significant part of the source of their subsistence through the decline of CPRs.

The intent of distributing the commons to the poor also needs to be critiqued. Does decimation of the village commons, on which the rural poor depend critically for parts of their subsistence, look like meaningful land reforms? What appears necessary is a much more nuanced exposition of how communities use the commons, including the forest commons. Social movements have focused more or less exclusively on demanding rights without undertaking a serious analysis of how individual rights over cultivable lands relate to the overall livelihood security. Similarly, very little thinking seem to have gone into the substantive aspects of how 'resource rights' translate into actual benefits of managing forests for local livelihoods from sustainable harvesting of forest produce for subsistence and cash income.

NGOs, too, have focused almost exclusively on collective management and regeneration of forests without significant success at translating regenerated resources into tangible income gains for the 'local community'. They have continued to work with the FD despite the continued lack of accountability and responsiveness of forest officials. Where do collective rights of communities over the forests managed under JFM, stand vis-à-vis the powers of the FD over the same land? Despite the lack of progress being made in this regard since the first JFM resolution of 1990, recent reports continue to reflect optimism about JFM (Pai and Datta 2006). Given the early warnings that naive enthusiasm might undermine concerted thinking and efforts on making real progress in JFM, such simplistic optimism may be self-defeating (Lele 1999; Saxena 1992; 2005). Finally, while some NGOs that earlier worked on 'forest conservation' have stayed away from the discussion of rights of individual households over cultivated forest land, and even collective rights over forests, others have actively persuaded the households to give up individual cultivation rights.

Moreover, the struggle within civil society about the best way of approaching agendas of rights, reforms, and development, has led to a confused medley of interventions that may be contradictory in nature unless implemented in a coordinated way. Very little work has been done to compare divergent interventions aimed at regeneration of and rights in forest and non-forest commons. The following section is an initial attempt at such an analysis from the vantage point of the theoretical framework adopted in this chapter.

Civil Society Divided (and Confused?)

One of the defining features of the debate around the forest question has been the involvement of civil society organizations. From an analytical perspective, at least two types of civil society groups may be identified. One broadly defined group can be termed reformist organizations that believe in working from within the existing state-dominated institutional set-up to achieve outcomes that will hopefully be more pro-poor and pro-environment as compared to the status quo. We consider Seva Mandir, Udaipur, to be a representative case. It has attempted a nuanced understanding of the processes involved in 'restoration' and regeneration of the commons through sustained engagement with 'local communities'.

Several social movements, influenced by Gandhian ideology, such as the Ekta Parishad, or leftist ideology, such as the Jungle Jameen Jan

Andolan (JJJA), have attempted to problematize the reformist approach and the tendencies therein to ignore the political roots of the status quo. These organizations, which we term as social movements, believe in confronting 'the state' in their quest to achieve social change. Since forests and other commons are believed to be under government ownership, these resources become synonymous with state control and, hence, the focus is demanding private tenure rights. Seva Mandir and JJJA make for a particularly interesting comparison because they are active in the same region, and attend to the same problems that are relevant to the discussion here, that is, societal processes of contestation and claims around natural resources such as forests, grazing land, watersheds, and water resources in general.

JJJA leads the campaign for recognition of private claims to forests by adivasis while Seva Mandir advocates reinforcing the commons character of forests (standing on the 'forest land') by strengthening community-based governance of forests. Obviously, the two groups disagree on fundamental issues of how best to organize nature–society interactions. For JJJA, the issue of livelihoods and survival with dignity can only be solved through securing legal rights and a proper recognition of existing livelihood strategies. Livelihoods depend on clearing forests for agriculture and having access to non-timber forest products (NTFPs), timber, and grazing grounds (Sharma 2001). This, notwithstanding, JJJA does not intervene when local elites privatize non-forest CPRs, perhaps because it depends on the support of local elites in mobilizing adivasis for its rallies and dharnas. On the other hand, Seva Mandir's proposition is that the commons are the means of survival for the poor, and 'commons' encroachments' have been undertaken primarily by the more powerful within local communities who have the tacit support of the state machinery (Mehta 2001). However, studies by Seva Mandir and other agencies working in the same region illustrate that both the poor and non-poor have staked claims to cultivable land within the boundaries of public forest land and other commons (Bhise 2004; Conroy *et al.* 2001).

Interestingly, both JJJA and Seva Mandir see community as an organic whole, although each has a different view of the composition of the whole. Seva Mandir believes that most households (irrespective of being adivasi or non-adivasi) within the local community benefit equally from interventions aimed at restoration and regeneration of common land. On the other hand, JJJA recruits only adivasi households to join its mass movement. To the extent JJJA focuses exclusively on adivasi households, their strategy is more equity-oriented at the outset. However, JJJA's strategy of mass mobilization is dependent on mobilizing large numbers of local community members to attend their rallies and demonstrations. This necessitates JJJA to depend on local leaders to coordinate the process of mobilization. For the poorest within the community, it may not be easy to forsake a day's wages to attend rallies.[3] Second, it is extremely difficult for the poorest among these communities to take on a leadership role in such mobilization strategies. Given intra-community heterogeneities, and elite-poor relations, which are antagonistic at worst and patronizing at best, it is difficult to see how JJJA's pro-poor vision may materialize.

A recent study by Seva Mandir (Bhise 2004) highlights some of the challenges of implementing the pro-poor vision of land rights' movements. The case of Viramgam, one of the 25 villages studied, is instructive in many respects. Some influential families from neighbouring villages occupied plots within the Viramgam forest. One of these families is believed to have had 'connections' in the police department. Another family has a member who is a government employee.Both families had significant irrigated farmland in their own village. More importantly, a local communist leader, who also heads a local NGO fighting for people's rights, occupied a parcel of forest land. The forest officials turned a blind eye (they even sought their own transfers), which may be an indication that this case involved very influential people. Community members as well as Seva Mandir staff have attempted pressurizing FD officials to take cognizance of forest land conversions by influential people. Interestingly, even though the assistant conservator of forests (ACF) has issued eviction notices in his judicial capacity, the local forest officials have not actually served those notices to the offenders. The local forest officials are either scared of the consequences of acting against the group or have been bribed.

Another case relates to the village of Khema Dungar, where a local leader Bhimashankar worked as Seva Mandir's point person in the village.[4] Bhimashankar had been apparently instrumental in the successful implementation of the JFM programme in Khema Dungar. However, around 2000–1, Bhimashankar started misusing his position and contacts with Seva Mandir by allegedly deploying some of the community funds to support his private business in the village. Seva Mandir responded by initiating a broad-based discussion within the community about democratization of the local leadership, attempts that were forcefully resisted and actively contested by Bhimashankar and his supporters. However, when Seva Mandir persisted with its efforts, Bhimashankar split ranks and joined the JJJA with whose support he started demanding private land rights for himself and his supporters within the same forest that he had earlier worked to protect as a community resource under JFM. As it turned out, JJJA had then been looking for a point person in the region and Bhimashankar soon became one of the important local leaders and spokesperson for JJJA in the region, often being one of the main adivasi speakers at large rallies.

These two cases of Viramgam and Khema Dungar provide two interesting insights. First, social movements organized in the name of the poor can easily be subjected to rent-seeking by its local leadership, as in the case of Viramgam. The local Communist leader in Viramgam not only advocated legalization of forest land occupied by 65 other families (many of who may not be 'marginalized'), but he himself encroached upon forest land. On the other hand, local leaders can also exploit the mobilization needs of social movement interventions to further their own vested interests, as in the case of Khema Dungar. Second, it is clear that the state–society dichotomy often projected by NGOs, as well as social movements, does not hold given the manner in which they are mutually constituted.

The latter point requires more explanation. Local government officials often act in connivance with local elites. The local forest officials in charge of Viramgam refused to act, even as sections within the local community mobilized against the powerful families. We recognize the possibility that in some cases such mobilization may be driven by other elites within the community. However, even if that were the case, the attempt by a local leader to convert forest land into other uses for his personal gains, as was the case in Viramgam, is justification enough for such counter-mobilization. None of these mobilizations may, however, end up benefiting the most marginalized within the community. This is clearly the case because intra-community heterogeneities interlock with a top-down forest management framework that has been in place for generations (Robbins 1998). Similarly, without putting in place appropriate checks and balances, local leaders representing larger social movements may exploit their positional advantage to further vested interests as opposed to working on egalitarian principles as envisaged by the movement. Thus, JJJA and other local social movements falter by portraying a homogeneous 'local adivasi community' without confronting the divisions within the adivasi communities and how such divisions may affect their ability to reach the most marginalized community members. Moreover, these failures contribute significantly to the destruction of the local commons (including forests), which may end up hurting the marginalized groups that the movement intends to benefit.

Seva Mandir views the 'community' as even more homogeneous. Its interventions and institutions are designed to treat everyone equally and involve the entire community (adivasis as well as non-adivasis) with the assumption that improved productivity of the forest commons will benefit everyone. There is an underlying assumption that *by default* the poor benefit more from the commons, and hence there is a need to preserve and maximize common spaces. Such expectations are belied by growing evidence that local inequalities invariably translate into the poorer sections getting a lower proportion of livelihood benefits from successful regeneration of the commons (Baviskar 2004; Sundar 2001).[5] More importantly, the flow of benefits to different groups depends greatly on the nature of resource-regeneration strategies put in place.

Other studies by Seva Mandir staff and visiting researchers have also produced similar evidence. A research project on silvipasture arrangements underlined the manner in which costs and benefits of JFM (and other enclosure-based strategies for regenerating the commons) were being shared inequitably by different segments within the community. JFM and other silvipasture systems[6] were found to work against the owners of small ruminants such as goats and sheep (Kashwan 2000) and camel-herders of another community (Vardhan in Conroy 2001). Ahluwalia (1997) offers a particularly nuanced view of different kinds of effects that Seva Mandir interventions had for the community as a whole, and different groups within the community. She acknowledges that group-formation activities of Seva Mandir helped communities unite in their fight against a violent landlord. However, she also points out that such success does not necessarily signify the existence of a corporate community that can come together, cutting across internal differences. Communities are instead infused with conflicts grounded in different livelihood strategies adopted by different groups within local communities. In Ahluwalia's (1997) case study area, large landowners were able to afford and benefit disproportionately from the enclosure of village grazing land, while those dependent critically on livestock-based livelihoods opposed such interventions because they did not perceive of any benefits arising.

Seva Mandir's work also offers more clues into their approach to the rights of forest-dwellers. In 2001–3, long before the ratification of the FRA, Seva Mandir organized an effort to actively pursue 'encroachers' to give up their claims on common land in lieu of benefits for an entire community or specific families involved. This action research project, 'Decolonizing the Commons', was based on a normative stand that all 'encroached' forest and pasture land should be reverted back to the commons. The project findings highlighted that all strata of people encroached from the poorest to the wealthiest to most powerful 'local community' households (Bhise 2004).

The process of community-based negotiation aimed at giving up claims to common land also revealed some interesting insights. Villagers in Harinagar, Udaipur, refused to give up their occupation of a community pasture until other households occupying parts of revenue wasteland gave up their claims (Kashwan 2004). However, Seva Mandir refused to intervene on the parcels of revenue wasteland occupied by a section within the community (including some relatively better of families). This decision was taken in line with Seva Mandir's policy of not intervening in the lands categorized as revenue wasteland. Despite the fact that 'revenue wasteland' constitutes a major portion (25 per cent) of about 73 per cent of the land categorized as 'commons' by Seva Mandir, it has deliberately stayed away from addressing issues related to these lands. Therefore, not responding to the demands made by Harinagar villagers to take a holistic look at the 'commons question' was probably a missed opportunity for addressing complex issues of village-level politics around claims over resources.

Revenue wasteland was originally kept for the provision of allotment to the landless families in a village. In practice though, such allotments were made to those who could bribe revenue officials. The remaining revenue wasteland has been occupied by individual households, mostly better-off, but by the poor as well. The first author's surveys in a number of villages selected for watershed interventions revealed that almost 100 per cent of the revenue wastelands were occupied by the villagers. It is not surprising, therefore, that legal administrative categories have little meaning for villagers. As in the case of Harinagar village, community members often look at the sum total of landholdings and claims (by way of occupation) of others within the village in articulating their own claims.

In an extensive analysis of the NSSO 59th round data, Menon and Vadivelu (2006) find that landless households are significantly dependent on CPRs (particularly in regions with a stronger market articulation). However, it is likely that in these regions, the commons have been converted and diverted to private property. In such cases, the landless poor often rely on access to private fallows and crop residues for supporting their livestock. Continued access to arable and the non-arable lands should be central, therefore, to discussions around the commons (Menon and Vadivelu 2006). Thus, we argue that even private landholdings should be the subject of broader discussions around the commons. In other words, any discussion of the 'forest question' or the question of the commons in general cannot be resolved without accounting for the failure to implement meaningful land reforms by successive governments since independence.

Seva Mandir's (and that of other NGOs working on the 'commons' along similar lines) efforts are focused specifically on discouraging and reverting household claims on forest land and community pastures. Moreover, this policy is applied uniformly to anyone in occupation of common land irrespective of a family's level of poverty or prosperity. As Kerr *et al.* (2007) suggest, such a policy risks being anti-poor in a social-context where the poorest community members dependent mostly on cultivable land may be perceived to be the source of problems.

Mehta (2001) summarizes Seva Mandir's position well by arguing that 'the custodians of public lands have systematically vitiated the ability of the poor to properly manage these lands. They have informally privatized these resources in order to derive rents and gain power and control over villagers' (Mehta 2001). Robbins (2000: 439; emphasis added) validates Mehta's analysis but goes further: 'Corruption persists because it is rooted firmly in existing institutions suffused with patriarchy, *class privilege, and caste power* that locally *hold sway in the social whole.* Indeed corruption is the logical extension of those systems of power, extending them into the control of nature.' This is precisely the point—given the highly feudal and inequitable socio-political set-up, civil society's persuasion of the poor to give up participation in insti-tutionalized corruption may neither reform the 'rotten institutions', as Robbins calls them, nor empower the poor.

Moreover, since Seva Mandir has focused almost exclusively on reforming local communities, the onus of setting right these 'land and social relations that discourage community action and emasculate institutions of local self-governance' (Mehta 2001), has fallen dispropor-tionately on local community members. The poorest among the 'local community', with their limited resources and limited bargaining power, find it particularly challenging to renegotiate these relationships in a manner that strengthens 'institutions of local self-governance' without hurting their short-term livelihood needs. The already marginalized poor may suffer silently without being able to withstand the turbulence created by an over-zealous civil society organization.

Thus, paradoxically, both kinds of civil society organizations—of the reformist kind, and the more activist kind—end up either weakening the position of the poor or achieving much less than envisioned originally.

A Way Forward: Acting on Multiple Fronts

We have made the following points in the chapter so far. First, the for-est and non-forest commons cannot be seen in isolation. While these insights have been available for a long time (Herring 1990a, 1990b; Sinha and Herring 1993), the rights discourse, as well as NGO interventions, have not adequately accounted for these linkages. Second, inter- and intra-community inequities are at the very heart of the debate about forest rights (Baviskar 1994), or the 'commons rights' as we would call them. We think that both NGOs and social movements in general have been unable to systematically account for intra-community inequities and the implications, thereof, for their agenda of community develop-ment (in case of NGOs) or social justice (in case of social movements). We (especially, the second author) have ourselves struggled with the challenges of thoroughly investigating, devising, and trying to imple-ment interventions around issues of resource rights and resource management in the commons. Given this, we cannot be content with merely offering a critique of the ongoing approaches without offering alternatives that may help address the issues of intra-community inequity, and CPR inter-linkages that we point out. In this concluding section, we attempt to offer theoretical as well as empirical answers to the questions we raise in this chapter.

Based on intensive CBNRM case studies across many countries, Leach *et al.* (1999: 225) proposed the environmental entitlements framework, which explores 'how differently positioned social actors command environmental goods and services that are instrumental to their well-being'. The authors recognize the central role of institutions, defined as 'regularized patterns of behavior between individuals and groups in society' (Leach *et al.* 1999: 225). We cite this definition here because more often than not institutional interventions are conceptualized in terms of building 'organizational infrastructure' (such as village development committees, forest protection committees, watershed associations). Such approaches fail to appreciate that institutions are socially embedded patterns of relationships and expectations (Hodgson 2006; March and Olsen 1984; Ostrom 2007), which are only partly open to social engineering, and are manifest in good measure in the form of constituted authority relations (Peters 2005).

Leach *et al.* (1999: 241) suggest that a successful intervention aimed at addressing equity gaps through CBRNM 'strategically supports subordinate groups to enhance access to and control over resources by taking "operational clues" from ongoing struggles, knowledge and strategies'. Alternatively, they suggest the need to strengthen ongoing CBRNM interventions by facilitating negotiations between 'actors at local and state level, or between smaller groups of resource users' (Leach *et al.* 1999: 241). We argue that the policies and programmes related to the commons are central to redressing local inequities, as well as inter-regional and national inequities (Shah *et al.* 1998). While the former requires facilitating local negotiations, the latter requires attempts at holding the state accountable to citizens as a whole. We believe that the two must be combined so as to bring far reaching changes to the existing settlement of rights vis-à-vis natural resources (Baumann and Sinha 2001). In what follows, we consider two empirical examples of interventions that have tried to operationalize concerns for equity and the interconnectedness of commons: the work of Deccan Development Society (DDS), Medak, and that of the Kisan Adivasi Sanghatan (KAS), Kesla.

Our sense is that DDS' interventions are conceptually close to the environmental entitlement approach proposed by Leach *et al.* (1999). Their approach recognizes the central role of institutions but contextualizes actor–institution linkages in the context of intra-community differences. DSS mobilizes and organizes groups of single, divorced, or unmarried women, particularly Dalit women, into Sangams. Mobilization and organization is then channelled into helping Sangam members access state facilities meant to serve the most marginalized (Kumbamu 2009). For instance, the Sangams and its members lease private cultivable land, and so help improve land entitlements over time. These interventions have reportedly made inroads into challenging the prevailing gender and social relations (Agarwal 2003). DDS's steadfast focus on Dalits, we argue, has been most productively harnessed in rethinking the watershed development approach (WASSAN 2004: 1).

> DDS started wondering how (watershed development) would affect the poor, especially dalits.... An outcome of this thinking was *Dalit* Watersheds... a new model of watersheds that converted *dalit* lands into ecological food farms working on biodiversity based production system. Farming on these lands was totally organic, crops raised were diverse and the focus was on food production. They were managed completely by dalit women, turning these watersheds into *women's watersheds*. And finally, focusing all development effort on the lands of the poor, these watersheds became intrinsically equity-based.

An external review of the DDS's Dalit Watersheds suggests that these micro/mini watersheds, focusing on specific groups of marginalized households, succeeded as much in livelihoods and ecological gains as the conventional large-scale ridge to valley watershed approaches would (WASSAN 2004). DDS initiatives also ensured that the most marginalized benefited the most.

KAS, Kesla, organized displaced adivasi community members to take on management of reservoir fisheries in Madhya Pradesh (Sunil and Smita 1996). In a nutshell, the project involved struggling against the state to demand justified rights of the adivasi households displaced because of a dam, and excluded from access to the newly built canal infrastructure. This struggle led to the Tawa Matsya Sangh (adivasi fishing cooperative federation) being awarded the harvesting and marketing rights for fisheries in the Tawa reservoir. Not content with winning these battles, the cooperatives and KAS implemented fish harvesting, processing, and marketing at a level of quality that would make it competitive in open metropolitan markets. This required tremendous technical and market-related expertise, the kind of skills that social movements are usually not inclined to master (Kashwan 2006). More importantly, KAS, as well as members of the Tawa Matsya Sangh, are actively engaged with the political process, most notably, but not limited to, their participation in democratic institutions of local governance. The Tawa Matsya Sangh was also able to ensure that fisheries were harvested and maintained in a sustainable manner (Kashwan 2005). Even though the KAS struggle faced tough challenges at the hands of an insensitive state machinery, we believe they have achieved a fair deal of balance between demanding resource rights, managing resources equitably, and continuing to work towards deepening democratic processes from within the existing institutional set-up.

The most important lesson is that livelihoods rights and resource conservation are inextricably linked across the form of property (private, common, public) and the type of resource (forests, pasture, cultivable land, and water resources). Moreover, the cases of DDS and KAS demonstrate that it is possible to judiciously and carefully combine demands for rights and technical interventions aimed at resource-generation with equal zeal and expertise. Indeed, more research is required to reconstruct the processes that may have gone into building these approaches over the years. This is crucial for dealing with the challenges of common land regeneration even while making sure that particular 'interest groups' of marginalized communities are organized, supported, and trained. Doing so ensures that they are able to demand their entitlements and can make good on them despite inequitable local social and political relations which require addressing the finer inter-locking of political economy relations at all levels. Since such interlocking has been in effect for decades, even generations, in the case of social institutions, it is important to address the inner working of such institutionalized local relations. This we think will be a fruitful departure from either the CBNRM approaches that have ignored the rights of marginalized groups within their predominately resource conservation focus or the social movement strategies that have paid virtually no attention to the maintenance of the commons for livelihood benefits of marginalized groups who depend upon these commons.

Notes

1. The authors acknowledge the comments and critiques offered by participants at the workshop on 'Beyond JFM' organized by Centre for Interdisciplinary Studies in Environment and Development (CISED) and Winrock International India (WII). Use of the facilities at the Workshop in Political Theory and Policy Analysis, Bloomington and helpful inputs by Daniel Taghioff and Chantalle LaFontant are gratefully acknowledged.

2. Often, village elites are able to encroach on 'revenue wasteland' that are most likely to be legalized by government provisions of distributing land to the landless (Brara 1989). In those cases, the common villagers are left with no scope of similarly occupying revenue wasteland, and they are forced to occupy grazing land which is less likely to be regularized.

3. For an excellent exposition of different visions of development between social movement activists and local tribal leaders and community members, see Baviskar (1997).

4. This case is based on the authors' own experience of working as the in-charge of NRM activities in Girwa block (within which Khema Dungar is located) during 1999–2003. Names of the villages and individuals have been changed to protect the identity of community members and leaders.

5. While a critical portion of poor forest-dwellers' livelihoods are dependent on forests and other public land, in absolute terms comparatively better-off forest-dwellers gain more from forest-related incomes (Narain *et al.* 2008).

6. Silvipasture interventions refer to the interventions aimed at twin objectives of reforestation and pasture development. Indeed, silvi-pasture approaches practised under JFM are constrained somewhat by what the FD may approve. However Seva Mandir, in its pastureland interventions, where the FD is not involved, has significant autonomy to shape its interventions as it would like, are designed more or less along similar lines.

References

Adger, N.W. 2000. 'Social and Ecological Resilience: Are They Related?' *Progress in Human Geography*, 24 (3): 347–64.

Agrawal, A. 1997. *Community in Conservation: Beyond Enchantment and Disenchantment*. Gainesville, Florida: Conservation & Development Forum.

———. 2001. 'Common Property Institutions and Sustainable Governance of Resources', *World Development*, 29 (10): 1649–72.

Ahluwalia, M. 1997. 'Representing Communities. The Case of a Community-based Watershed Management Project in Rajasthan, India', *IDS Bulletin*, 28 (4): 23–34.

Bardhan, P. 2002. 'Decentralization of Governance and Development', *Journal of Economic Perspectives*, 16 (4): 185–205.

Baumann, P. and J. Farrington. 2003. 'Decentralising Natural Resource Management: Lessons from Local Government Reform in India'. *Natural Resource Perspectives*, Working Paper 86. London: Overseas Development Institute.

Baumann, P. and S. Sinha. 2001. 'Linking Development with Democratic Processes in India: Political Capital and Sustainable Livelihoods Analysis'. *Natural Resource Perspectives*, Working Paper 68. London: Overseas Development Institute.

Baviskar, A. 1994. 'Fate of the Forest: Conservation and Tribal Rights', *Economic and Political Weekly*, 39 (38).

———. 1997. 'Tribal Politics and Discourses of Environmentalism', *Contributions to Indian Sociology*, 31 (2): 195–223.

———. 2004. 'Between Micro-Politics and Administrative Imperatives: Decentralisation and the Watershed Mission in Madhya Pradesh, India', *European Journal of Development Research*, 16 (1): 26–40.

Beck, T. and M.G. Ghosh. 2000. 'Common Property Resources and the Poor: Findings from West Bengal', *Economic and Political Weekly*, 35 (3): 147–53.

Bhise, S.N. (ed.) 2004. *Decolonizing the Commons*. New Delhi: National Foundation for India.

Bokil, M.S. 1996. 'Privatisation of Commons for the Poor: Emergence of New Agrarian Issues', *Economic and Political Weekly*, 31 (33): 2254–61.

Bon, E. 2000. 'Common Property Resources: Two Case Studies', *Economic and Political Weekly*, 35 (28/29): 2569–73.

Brara, R. 1989. 'Commons' Policy as Process: The Case of Rajasthan, 1955–1985'. *Economic and Political Weekly*, 24 (40): 2247–54.

———. 2006. *Shifting Landscapes: The Making and Remaking of Village Commons in India*. New Delhi: Oxford University Press.

Conroy, C. 2001. 'Forest Management in Semi-Arid India: Systems, Constraints and Future Options'. NRI Report no. 2656. Kent, UK: Natural Resources Institute.

Conroy, C., S. Iyengar, V. Lobo, and G.B. Rao. 2001. 'Household Livelihood and Coping Strategies in Semi-arid India: Adapting to Long-term Changes'. New Delhi: Society for Promotion of Wastelands Development.

Gupta, A.K. 1991. 'Household Survival through Commons: Performance in an Uncertain World'. IIMA Working Paper no. WP1991-06-01-01016. Ahmedabad: Indian Institute of Management.

Herring, R.J. 1990a. 'Editor's Introduction: Forests, Peasants, and State: Values and Policy', *Agriculture and Human Values (Historical Archive)*, 7 (2): 2–5.

———. 1990b. 'Rethinking the Commons', *Agriculture and Human Values*, 7: 88–104.

Hiremath, S.R. 2001. 'Community Control', *Seminar*, 499; http://www.india-seminar.com/2001/499/499%20s.r.%20hiremath.htm.

Hodgson, G.M. 2006. 'What Are Institutions?' *Journal of Economic Issues*, 40 (1): 1–25.

Iyengar, S. 2001. 'The Gujarat Experience', *Seminar*, 499; http://www.india-seminar.com/2001/499/499%20sudarshan%20iyengar.htm.

Jodha, N.S. 1989. 'Depletion of Common Property Resources in India: Micro-Level Evidence', *Population and Development Review*, 15: 261–83; http://www.jstor.org.ezproxy.lib.uconn.edu/stable/i330206.

———. 2006. 'Some Places Again: A "Restricted" Visit to Dry Regions of India', in R. Ghate, N. Jodha, and P. Mukhopadhyay (eds), *Promise, Trust and Evolution: Managing the Commons of South Asia*. New Delhi: Oxford University Press.

Kashwan, P. 2000. 'Bada Bhilwada Joint Forest Management, A Case Study'. Silvipasture Management Case Studies by Seva Mandir. N. Jain, R. Jindal, N. Negi, P. Kashwan, and M. Vardhan. Kent, UK: Natural Resource Institute, University of Greenwich.

———. 2004. 'Decolonizing the Commons: A Case Study of Kaliwali Village', in S.N. Bhise (ed.), *Decolonizing the Commons*. New Delhi: National Foundation for India.

———. 2005. *Environmental Governance in India*. New Delhi: Lead India.

———. 2006. *Tawa Matsya Sangh: A Case of Local Governance of Water Resources for Livelihood Security. Lead Global Training Session: Stakeholder Participation in Environmental Governance*. Bhopal: Lead India.

Kerr, J., G. Milne, V. Chhotray, P. Baumann, and A.J. James. 2007. 'Managing Watershed Externalities in India: Theory and Practice', *Environment, Development and Sustainability*, 9 (3): 263–81.

Kumbamu, A. 2009. 'Subaltern Strategies and Autonomous Community Building: A Critical Analysis of the Network Organization of Sustainable Agriculture Initiatives in Andhra Pradesh', *Community Development Journal*, 44 (3): 336–50.

Leach, M., R. Mearns, and I. Scoones. 1999. 'Environmental Entitlements: Dynamics and Institutions in Community-Based Natural Resource Management', *World Development*, 27 (2): 225–47.

Lele, S. 1999. 'Institutional issues in (J)FM(& R)', in Anonymous (ed.), *National Workshop on Joint Forest Management, VIKSAT*, pp. 19–29. Ahmedabad: Gujarat Forest Department and Aga Khan Foundation.

———. 2004. 'Beyond State-community and Bogus "Joint"ness: Crafting Institutional Solutions for Resource Management', in M. Spoor (ed.), *Globalisation, Poverty and Conflict: A Critical Development*, pp. 283–303. Dordrecht and Boston: Reader, Kluwer Academic Publishers.

Lele, S. and A.S. Srinidhi. 1998. 'Indian Forest Policy, Forest Law, and Forest Rights Settlement'. International Workshop on Capacity Building in Environmental Governance for Sustainable Development. Mumbai: Indira Gandhi Institute of Development Research.

Li, T.M. 2002. 'Engaging Simplifications: Community-Based Resource Management, Market Processes and State Agendas in Upland Southeast Asia', *World Development*, 30 (2): 265–83.

March, J.G. and J.P. Olsen. 1984. 'The New Institutionalism: Organizational Factors in Political Life', *The American Political Science Review*, 78 (3): 734–49.

Mehta, A. 2001. 'Empowering Agrarian Society', *Seminar*, 499; http://www.india-seminar.com/2001/499/499%20ajay%20s.%20mehta.htm.

Menon, A., P. Singh, E. Shah, S. Lele, P.S. and K.J. Joy. 2007. *Community-based Natural Resource Management: Issues and Cases from South Asia*. New Delhi: Sage Publications.

Menon, A. and G.A. Vadivelu. 2006. 'Common Property Resources in Different Agro-Climatic Landscapes in India', *Conservation and Society*, 4 (1): 132.

Nadkarni, M.V. 2000. 'Poverty, Environment, Development: A Many-Patterned Nexus', *Economic and Political Weekly*, 35 (14): 1184–90.

Nagendra, H., S. Pareeth, and R. Ghate. 2006. 'People within Parks—Forest Villages, Land-cover Change and Landscape Fragmentation in the Tadoba Andhari Tiger Reserve, India', *Applied Geography*, 26 (2): 96–112.

Narain, U., S. Gupta, and K. van't Veld. 2008. 'Poverty and resource dependence in rural India', *Ecological Economics*, 66 (1): 161–76.

Ostrom, E. 2007. 'Institutional Rational Choice: An Assessment of the Institutional Analysis and Development Framework', in P.A. Sabatier (ed.), *Theories of the Policy Process*, pp. 21–64. Boulder, Colorado: Westview Press.

Pai, R. and S. Datta (eds). 2006. *Measuring Milestones: Proceedings of the National Workshop on Joint Forest Management (JFM)*. National Workshop on Joint Forest Management (JFM). New Delhi: Ministry of Environment and Forests, Government of India and Winrock International India.

Pathak, A. 1994. *Contested Domains: The State, Peasants and Forests in Contemporary India*. New Delhi: Sage Publications.

Peluso, N.L. and P. Vandergeest. 2001. 'Genealogies of the Political Forest and Customary Rights in Indonesia, Malaysia, and Thailand', *The Journal of Asian Studies*, 60 (3): 761–812.

Pérez-Cirera, V. and J.C. Lovett. 2006. 'Power Distribution, the External Environment and Common Property Forest Governance: A Local User Groups Model', *Ecological Economics*, 59 (3): 341–52.

Peters, B.G., J. Pierre, and D.S. King. 2005. 'The Politics of Path Dependency: Political Conflict in Historical Institutionalism', *Journal of Politics*, 67 (4): 1275–1300.

Robbins, P. 1998. 'Authority and Environment: Institutional Landscapes in Rajasthan, India', *Annals of the Association of American Geographers*, 88 (3): 410–35.

———. 2000. 'The Rotten Institution: Corruption in Natural Resource Management', *Political Geography*, 19 (4): 423–43.

Robinson, E.J.Z. 2008. 'India's Disappearing Common Lands: Fuzzy Boundaries, Encroachment, and Evolving Property Rights', *Land Economics*, 84 (3): 409–22.

Saberwal, V.K. 2000. 'Conservation as Politics: Wildlife Conservation and Resource Management in India (1)', *Journal of International Wildlife Law and Policy*, 3 (2): 166.

Sarin, M. 2005. 'Laws, Lore and Logjams: Critical Issues in Indian Conservation', *The Gatekeeper Series 116*. London: International Institute of Environment and Development.

———. 2008. 'Righting the Wrongs Done to India's Forest Dwellers', in S. Maginnis and J. Sayer (eds), *Arborvitae: Rights-based Approaches to Forest Conservation*. The IUCN Forest Conservation Programme newsletter, 36.

Saxena, N.C. 1992. 'Eucalyptus on Farmlands in India: What Went Wrong?' *Unasylva*, 43 (170): 53–8.

———. 2005. 'Bridging Research and Policy in India', *Journal of International Development*, 17 (6): 737–46.

Schlager, E. and E. Ostrom. 1992. 'Property-Rights Regimes and Natural Resources: A Conceptual Analysis', *Land Economics*, 68 (3): 249–62.

Shah, M., D. Banerji, P.S. Vijayashankar, and P. Ambasta. 1998. *India's Drylands—Tribal Societies and Development through Environmental Regeneration*. New Delhi: Oxford University Press.

Sharma, M. 2001. *Landscapes and Lives: Environmental Dispatches on Rural India*. New Delhi: Oxford University Press.

Sinha, S. and R. Herring. 1993. 'Common Property, Collective Action and Ecology', *Economic and Political Weekly*, 28 (27/28): 1425–32.

Somanathan, E. 1991. 'Deforestation, Property Rights and Incentives in Central Himalaya', *Economic and Political Weekly*, 26 (4): PE37–PE46.

Srivastava, R.S. 2006. 'Land Reforms, Employment and Poverty in India'. Paper presented at the International Conference on Land, Poverty, Social Justice and Development by Institute of Social Studies at The Hague on 12–14 January.

Sundar, N. 2001. 'Is Devolution Democratization?' *World Development*, 29 (12): 2007–23.

Sunil and Smita. 1996. 'Alternative Forms of Management in Reservoir Fisheries: Comparative Case Studies from Madhya Pradesh', in R. Rajagopalan (ed.) *Rediscovering Cooperation*. Anand: Institute of Rural Management.

WASSAN. 2004. *On the Margin, Poor and their Lands: A Livelihoods Analysis of the Dalit Watershed Development Program*. Hyderabad: Deccan Development Society.

11

AJIT MENON
VIREN LOBO
SHARACHCHANDRA LELE

The Commons and Rural Livelihoods

Shifting Dependencies and Supra-local Pressures

A core premise of the protests against colonial forestry and the post-independence re-emergence of forest rights movements is that rural communities are strongly dependent upon forests for meeting their subsistence and income needs. Given this strong dependence and deep cultural ties, it is argued that depriving them of access to forests will harm rural livelihoods and lives very substantially. This idea of dependence was recognized to a small extent even by the British, when they yielded some concessions to farmers in certain pockets as far apart as Kumaon in the Himalayas and Uttara Kannada in the Western Ghats. And it was re-iterated time and again in the post-Chipko discourse (see for example, Fernandes 1988), and made one of the pillars of the argument for granting rights to local communities to manage forests.[1] While there is much debate about the adequacy of joint forest management (JFM) as a response to this demand (see Chapter 1 in this volume) and the merits of the Forest Rights Act, 2006 (FRA) as an alternative (see Chapter 3 in this volume), there has been relatively little examination of the core assumption of forest-dependence on which all such approaches rest.

In recent years, there has been a small but significant global literature questioning the potential for forest-based livelihoods. The arguments come from two angles. On the one hand, it is argued that there are inherent limits to the magnitude of economic benefits that can be realized from forests, especially in the context of a growing population and shrinking forests. For instance, Sunderlin *et al.* (2005) argue that while forest resources are crucial to the rural poor to prevent them from slipping further into poverty, there is little evidence to suggest that forest resources are a way out of poverty. This might be the case because products that have to be harvested in large quantities and over large areas have limited markets (Belcher and Schreckenberg 2007). Second, it is suggested that local interest in forests is declining and will continue to do so as alternative sources of livelihood (agricultural or altogether non-land-based) become available, which is the conventional trajectory of development. As Byron and Arnold (1999) argue, low-value products are likely to be abandoned if other income-generating activities are made available. Similarly, firewood may be seen as an 'inferior good' (to use the parlance of economics) that is abandoned the moment superior cooking fuels such as LPG are made affordable for cooking (DeFries and Pandey 2010).

In the wake of this literature, we take a fresh look at the assumption of forest-dependence. We examine this in both empirical and normative terms. Empirically, we argue that the scope of the discussion needs to be expanded to include all common land rather than only forests, because forest and non-forest common land are complementary resources integrated into rural livelihoods. We then examine the extent of dependence on common land as well as the variation in dependence within local communities, and particularly whether there has been a secular decline in dependence over time. We assess both possibilities: that this decline is a result of a 'pull' effect where other livelihoods are increasingly attractive due to developmental processes. At another level, we examine supra-local factors that have progressively limited community access to such lands and 'pushed' them out of dependence, making non-dependence a self-fulfilling prophecy. We show that these factors are not only present in forest land but even stronger in non-forest commons, strengthening the argument for an enquiry into agrarian environments.

Normatively, we examine whether the argument for local control over local public lands, forested or otherwise, must necessarily rest on high economic dependence of the local community, or whether there are other bases for the argument as well.

Defining the Commons

Two important definitional issues confront us when we use the term 'commons'. First, what kinds of physical resources are we referring to, and second, what property regimes are we invoking? Common property resources or the 'commons' has been used in the literature to include all kinds of resources: forests, pastures, water, fisheries, and so on (see www. iasc.org). In this chapter, we focus on land resources, hence the phrase common property land resources (CPLRs: first coined by Iyengar 1989) is more appropriate. CPLRs include both forested and non-forested land, the latter including grassland, long-term fallows and even private land that may be seasonally left open for common grazing (Beck 1994).

Second, although strictly speaking 'common property' refers to a tenurial regime where the resource is legally owned or substantially controlled by a clearly defined group of people, the situation in India requires special treatment. Not only is statutory support for community control of forest land almost non-existent (except in Uttaranchal and some states in the Northeast), but even grazing land (*gauchars*, *charnois*, *charagahs*, *ghasnis*, *gomaals*, or other terms) are not always legally managed by local communities. At the same time, local people *use* these lands extensively, and often the right to use is legally recognized, even though it may be abridged by the state without reference to the user. In other words, common 'property' regimes are mostly absent, what is prevalent is common access.[2] Thus, when we use the term CPLR, we are referring to all lands that are *used* in common by a group with or (mostly) without legal sanction to use, to exclude others or to manage.

Dependence on CPLRs: Magnitude, Nature and Geographic Variation

The extent of such broadly defined CPLRs varies quite significantly across the country, and given that it is defined in terms of de facto access, it is hard to estimate the actual extent. Estimates range from 23 per cent (Kapur 2010) to 29 per cent (Qureshi and Kumar 1998) of the country's land area. Even if approximate, these figures indicate that CPLRs are not an insignificant part of the landscape.

Large-sample research on the dependence of rural communities on CPLRs in India began with a path-breaking set of studies carried out by Narpat S. Jodha, using panel data from across rural India (Jodha 1985, 1986, 1987, 1989, 1990), wherein he showed how the decline of these commons through various pathways had negatively affected rural livelihoods, especially those of the rural poor. Following these studies, a large literature emerged on forest-dependence in eastern, central, and western India (Fernandes *et al.* 1988; Nadkarni *et al.* 1989; Reddy and Chakravarty 1999) and wider CPLR dependence as well (Beck and Ghosh 2000; Iyengar 1989; Nadkarni 1990; Shanmugaratnam 1996; Singh *et al.* 1996). The focus of this research was on material dependence, measured in terms of the imputed value of products obtained from CPLRs in absolute terms and as a fraction of the total imputed income of the household. The common message of these studies was that collection of produce from CPLRs, including forest land, contributes significantly to the imputed incomes of rural households, the contribution ranging from 10 per cent to 40 per cent.

The significant variations in dependence have both an agro-ecological and regional dimension. CPLR dependence is highest in forested landscapes (Arnold and Stewart 1991)[3] and in semi-arid regions[4] in terms of the per cent of total income derived from CPLRs and in terms of the per cent of families dependent on CPLRs. These findings are substantiated both by regional studies (Qureshi and Kumar 1998) and by an analysis of the country-wide NSSO 54th round data (Menon and Vadivelu 2006). Dependence is generally lower in the Gangetic plains and other areas of irrigated agriculture. Having said that, individual case studies such as Beck's (1994) and Beck and Ghosh's (2000) studies of villages in the alluvial belt of West Bengal highlight a high level of dependence in intensive agricultural zones as well, most notably in terms of the collection of gleaned grains.

What Menon and Vadivelu (2006) also highlight from the NSSO data set is the variety of products collected and utilized. The NSS data classified CPLR products as fuelwood, fodder, and 'others' that included manure, fruits, roots and tubers, vegetables, gums and resins, honey and wax, medicinal plants, fish, and leaves and weeds. The NSS estimates highlight that approximately 58 per cent of total CPR product collections are of fuelwood.[5] Fodder constitutes 25 per cent of collections and 17 per cent of collections pertain to 'other' products. A further disaggregated analysis across different agro-climatic zones and agrarian systems shows that in the forested tracts of the country, fuelwood is the most important CPR whereas in the alluvial belt of the Ganges, fodder is more important. Non-timber forest products (NTFPs), not surprisingly, are important mostly in the forested regions of central India and the Northeast.

In parallel, a more anthropological and historical literature has helped elucidate CPLR dependence, the multiple CPLR resources used by communities, and the fluid boundaries between forests and non-forest CPLRs. Examples of such studies include Skaria's study of the Bhil tribals of the Dangs, Gujarat (Skaria 1999), Baviskar's examination of the Bhilala tribals of the Narmada Valley (Baviskar 1999), Satya's enquiry of Berar agriculturalists in central India (Satya 2004) and studies of Gaddi pastoralists of Himachal Pradesh (Saberwal 1999), and Raika pastoralists of Rajasthan (Agrawal 1998). Firstly, what is apparent from these studies is that livelihoods often straddle across different land resources, including agricultural land for cultivation, grazing land in multiple locations for fodder and forests for fuelwood, timber, and non-timber forest produce—be they de jure or de facto commons. Singh, in his study of Himachal villages, speaks of 'intermediate zones', located between cultivated land and forests, often termed waste, that are central to hill livelihoods (Singh 1999). Philip's study of the Nilgiris makes the important point that forests were often the site for shifting cultivation (Philip 2004), again highlighting that legal categories that define CPLRs often are of no consequence for rural communities. And studies done by the Society for Promotion of Wastelands Development (SPWD) highlight that private land as commons are often much more important than land classified officially as pastures.[6] In other words, CPLRs are integrated into wider and complex livelihood patterns in ways that are difficult to capture in property rights classifications.

Secondly, the use of CPLRs complements agriculture in terms of seasonality: most NTFP collection, for instance, takes place during the dry season. Forests, being less sensitive to variations in rainfall as compared to rainfed agriculture, also act as safety nets during droughts, providing critical food resources, and generally providing a back-up source of income. A number of studies have highlighted specifically how access to forest resources prevent communities from sinking into poverty. For example, Reddy and Chakravarty (1999) argue that forestry in the Kumaon Himalayas makes a significant contribution in terms of poverty alleviation.

Moreover, dependence should not be reduced to economic dependence alone. Forests have meanings beyond the economic goods they provide. For instance, Skaria (1999) has highlighted that Dangi self-identity is inextricably linked to the forests within which they live. The same can be said of central Indian adivasis such as the Gonds, Bhils, Oraon, Munda, or Dongria Kondh (Guha 1996). In southern India the Kattunayakans, living in Mudumalai National Park, want to stay there because it is their home, despite the significant rehabilitation package on offer for them to relocate and the fact that development work is not permitted in the newly formed tiger reserve.[7]

To summarize, what the combination of the quantitative and qualitative literature endorses is that CPLRs are critical livelihood resources. It also emphasizes that the non-forest component of CPLRs is as important as the forest component, that the benefits from CPLRs should not be measured in simple economic terms as they may provide livelihood support during the lean season that complements agriculture during the rainy season as well as being inextricably linked with local culture and identity.

Dependence on CPLRs—Uneven, Inadequate, and Waning?

The empirical 'fact' that local communities are heavily dependent upon forests has been the basis for virtually all calls for an increased local role in forest management or common land management, (for example, Agarwal and Narain 1989) whether coming from activists or eventually being echoed by the state. And state policy has accepted this claim, even if its response has been less than adequate. For instance, the National Forest Policy 1988 states that 'the life of tribals and other poor living within and near forests revolves around forests. The rights and concessions enjoyed by them should be fully protected' (Government of India, 1988, Section 4.3.4.3) The same sentiment, namely the importance of meeting the local community's needs, is echoed repeatedly in various state-level orders on JFM starting the early 1990s. The assumption of dependence is thus crucial to the argument for local governance, as it provides both the normative justification and the analytical guarantee: local governance will lead to regeneration that leads to both environmentally beneficial and poverty-alleviating outcomes, and the level of dependence will ensure a strong interest in local governance.

This assumption has rarely been questioned in Indian forestry debates. Even those who reject the idea of participatory management, such as many foresters, assume that rural households are uniformly and heavily dependent on forests. Their recommendation is weaning off these communities from this (destructive) dependence using a combination of technology, economic development, and 'education', (see, for example, debates quoted in Guha 1983). While one element of this assumption, namely, uniform dependence, has been examined by several Indian scholars, the other element, namely, high and unchanged dependence is now under scrutiny, especially from researchers working in other forested regions. We examine both aspects in the following two sections.

Variation in Dependence within Communities

Beginning from Jodha's work itself, most studies have included an analysis of the variation in dependence on CPLRs *within* the local community. The literature has generally found that the percentage contribution of CPLRs to the imputed income is higher among the poor. There is no consensus, however, with regard to the absolute contribution of CPLRs. Jodha (1986) found that across a large sample, the absolute contribution of CPLRs was much higher for poor households (landless and marginal farmers) as compared to others in the sample villages. On the other hand, Nadkarni *et al.* (1989) found that the non-poor derived more absolute benefits from the forests of the Western Ghats than the poor. Gowda and Savadatti (2004) made the same claim in the context of villages in Dharwad district of Karnataka, and Pasha (1992) observed the same trend even in villages in the drier regions of that state. In a more recent cross-country study using the NSSO data set, Chopra and Dasgupta (2008) found that NTFP incomes benefit the poor much less than the non-poor, raising questions about CPR-based livelihoods for the poor. Nonetheless, the NSSO data also suggest that a higher percentage of agricultural labourers depend on CPLRs than other categories of farmers in almost all regions of the country. The greater dependence on CPLRs holds true for fuelwood, fodder, and NTFPs. In the case of NTFPs, the greater dependence on NTFPs by agricultural labourers is most stark in less forested areas such as the Upper Gangetic belt, suggesting that they are derived more from private land and non-forest common land (Menon and Vadivelu 2006).

We believe that the intra-village variation in relative and absolute dependence is influenced by which kind of CPLRs one is talking about, their relationship with agriculture, and the nature of rights granted. In forested regions, if forests provide inputs to agriculture (such as manure and mulch for betelnut orchards), then the stratification in landholding is often reflected in the contribution of CPLRs to income, especially when (as in the Western Ghats case studied by Nadkarni *et al.* 1989) forest rights are granted in proportion to landholding. On the other hand, in semi-arid regions, grazing of livestock is one key form of dependence on CPLRs, and this is practised more by the poor than the richer households (who focus more on irrigated agriculture). Since the definition of CPLR studied and that of poor versus rich households varies significantly from study to study, our hypothesis remains to be properly tested. Within particular regions also, the relationship is probably more complex, as indicated by more recent detailed studies (Narain *et al.* 2008a, 2008b).

What does intra-village variation imply for forest and common land policy in general and the demand for decentralized governance of forests in particular? It appears to cut both ways. On the one hand, the greater dependence of the poorer groups on CPLRs (to the extent that it exists) implies that regeneration of CPLRs through participatory management would have major poverty-alleviation outcomes. This has been the argument made by most analysts (Jodha 1986; Reddy and Chakravarty 1999). For instance, Jodha concludes that 'the situation calls for greater attention to CPRs as a part of the anti-poverty strategy'.

On the other hand, uneven dependence can weaken the argument for local governance, as it reduces the chances of successful collective action (Arnold and Stewart 1991). This is particularly true if the elite are the ones disinterested in the resource, as their non-participation can quickly undermine the chances of successful collective action. But even worse, decentralized governance (in the absence of countervailing measures) tends to put power in the hands of the village elite, who may then use the CPLRs in ways that benefit themselves by way of cash income (say, from eucalyptus plantations) rather than benefiting the poor and the women (say, by growing firewood species or grasses). This phenomenon has been well documented in so-called 'successful' JFM cases (Lele *et al.* 2005). Chapter 10 in this volume describes in more depth the challenges posed by intra-village variation in dependence as well as other inequities in social and economic power for CPLR management. But, as the authors of that chapter also argue, this does not constitute an argument for neglecting the commons and depending only upon development strategies that target the development of private assets alone, or the privatization of the commons (see the next section). What it might necessitate is further decentralization below gram panchayat and even village levels to hamlets where dependence is perhaps more homogeneous. It may also require countervailing rights to landless households alone (such as rights to timber) beyond commonly shared rights (such as firewood collection).

Declining Dependence?

A bigger challenge to the demand for decentralized governance of CPLRs comes from arguments rooted in the mainstream development literature. The core argument is that dependence upon the commons for firewood, grazing, or NTFP collection is always a second-best option, as the products collected from CPLRs are generally what economists call 'inferior goods'. As rural communities modernize, they will start using kerosene or LPG as fuel, start practising stall-feeding instead of open grazing for their livestock, start growing their own high-yielding varieties of fodder rather than collecting it from the commons, and start cultivating second agricultural crops (through irrigation) thereby obviating the need for a different source of livelihood in the summer. Similarly, even if the collection of certain special products such as medicinal plants or fruit from the wild is lucrative today, the future of all NTFPs is domestication, and cultivation on private lands (Homma 1996), because that increases returns to labour and hence profitability.

And eventually, development means industrialization, which is bound to reduce the importance of livelihoods directly based on biotic resources, such as grazing of livestock or collection of NTFPs. In other words, the importance of forests in particular and CPLRs in general in both production (income-generation) and reproduction (for example, firewood collection) would be expected to decline (Byron and Arnold 1999).

The explanation of this phenomenon is relatively straightforward within the conventional development economics paradigm. Sunderlin *et al.* (2003) argue that while NTFPs may provide a safety net and also be a complement to single-crop agriculture, they 'can also be a poverty trap. Rural people rely on NTFPs because they are poor, but it is also possible they are poor because they rely on NTFPs and activities for which remuneration is low'. The remuneration from forest products is often low because easily available products such as firewood have low value, and high-value products often have poorly developed markets and patchy distributions in the forest (Belcher and Schreckenberg 2007). This appears to fit with studies showing that the increase in incomes resulting from even 'successful' JFM activities is often fairly marginal (Ravindranath *et al.* 2000).

Apart from these inherent limitations of CPLR products, there is some evidence to suggest a declining level of material dependence on CPLRs amongst local communities. Kiran Kumar *et al.* (2007) compared dependence in a village with canal irrigation with a village without, and found that CPLR dependence was much higher in the latter. Lele (2001a) shows that in the Western Ghats of Karnataka, dependence on forests is much lower where large tracts of forests have been converted into coffee and other plantations. The NSSO 54th round data of 1998 show shifts away from public and common land to dependence on resources from privately owned but seasonally open-access land when one goes from less agriculturally developed regions to regions like Punjab and Haryana. Sarkar (2008) has pointed to the shift to LPG for cooking even in the heavily forest-dependent villages of the middle Himalaya.

Economists explain this phenomenon by characterizing firewood and grazed fodder as 'inferior goods', that is, those goods consumed because there is no alternative. As Byron and Arnold (1999) argue, such low-value products are likely to be abandoned if other income-generating activities are made available. Such alternatives become available when, for instance, agricultural intensification enables double cropping and, therefore, higher availability of crop residue, green fodder, and less labour for grazing outside land. In addition, the public distribution system has increased access to LPG and kerosene, which in most cases are superior to firewood from the user's point of view.

Perhaps the most important challenge is not a decline in material dependence but a shift in cultural perceptions. Baviskar (2012) argues that agrarian India has had 'a "loss of nerve" [that is] the moral core of peasant economy—its normative sense of itself as the right or the good life—is now dead. Many small peasants observe that there is no future in farming any more.' Given that forest-based livelihoods or pastoral livelihoods were already considered inferior to sedentary agriculture, there is a possibility that even when such livelihoods become economically viable, rural households will not be interested in continuing or expanding them.

Explaining Declining Dependence and the Role of 'Push' Factors

In light of the above, understanding whether, and to what extent and under what circumstances—this 'decline in dependence' or 'loss of interest/nerve' is occurring becomes an important matter. Investigating this empirically is difficult, because what can be measured is only level of use, not the reasons for lower or higher use, and translating the level of use into 'imputed income' and 'level of dependence' has always been fraught with uncertainties.

We do not present new data on this matter. Rather, we wish to examine the possible causes for what might indeed be a decline in the percent contribution of CPLRs to rural livelihoods, even those of the poor.[8] The question we ask is whether the perceived decline can be explained only in terms of the 'pull' of modernization and urbanization (both material and cultural) or is it better explained as the result of a 'push', where communities, and especially the rural poor, have been *forced* to move away from the use of CPLRs. While the possibility of a pull cannot be denied, nor can its magnitude be easily estimated, there is ample evidence of a push that goes back in history but has taken a variety of forms. The two predominant forms that we discuss below are state policies that have physically pushed off or excluded local communities from the use of CPLRs, and other policies that have kept the level of benefits from CPLRs lower than it could be, thereby providing an economic push towards other forms of livelihood.

Taking Away the Commons: Colonial and Post-colonial Developments

The colonial period was undoubtedly a watershed of sorts for CPLRs in India. Although in the pre-British period, rulers often plundered forests to colonize areas or reserve specific species for royal use, the British in the second half of the nineteenth century went much further and took over outright large tracts of forests. While the process of such a takeover was uneven and took different forms in different places, large-scale abrogation of rights was common and remaining rights were converted into privileges. The Indian Forest Act, 1878, was a watershed as it resulted in the reservation of most forests and state-ownership over them. Local use of forests for fuelwood, fodder, timber, non-timber forest produce, and even agriculture were significantly curtailed. Forest-dependent communities, who rarely distinguished between forests and non-forest use, often practising shifting cultivation in forests, were branded as encroachers by the colonial state (see Chapter 8 in this volume).

State takeover of forests was part of a wider process of converting 'waste' into 'productive' uses and took two predominant forms: scientific (read monocultural) cultivation of industry-relevant forest products on the one hand and conversion of forests and other waste lands to non-forest uses on the other. While the reservation of forests was primarily aimed at securing and augmenting timber supply, the conversion of forests and non-forest land into plantations of coffee, tea, and cardamom in the Western Ghats and in Assam and Darjeeling was aimed more at revenue-generation. There were also cases of non-forest CPLRs being put under the plough for foodgrain production (Satya 2004). As Chakravarty-Kaul (1996) has detailed, the act of 'claiming waste' did not imply that lands were not used locally, but that the state wanted these lands for other uses. Productive use had a social dimension to it too. Philip (2004) argues that, in the context of the Western Ghats of the Nilgiris, takeover was a way not only to civilize nature, which was unruly and in need of taming, but also the people who inhabited this nature.

The immediate post-colonial period was much of the same. The 1952 Forest Policy was a continuation of colonial forest policy. State-ownership of forests remained as part of a bigger vision that saw the state at the helm of development and forests as a critical raw material necessary for industrial development. A detailed analysis of the forests lost to developmental projects is given in Chapter 7 in this volume. Suffice to say that approximately 30 million people had been displaced till the early 1990s because of development projects, mainly dams but also mines and industries. Many of these projects, especially dams and mines, were in forested areas and in areas with large adivasi populations.

At the same time, releasing of forest land for cultivation continued not so much for revenue purposes (as in the British period) as for populist reasons, as the easy way out in the absence of meaningful land reforms in most states, and in the name of meeting food production goals, such as under the Grow More Food campaign. Similarly, large areas of grazing land were also released for cultivation, supposedly (but not always) to landless households. That this was happening while the rights of pre-existing forest-dwellers, particularly the tribal communities in central India, were not recognized is just one of the many bitter ironies of the post-colonial period.

While the passing of the Forest Conservation Act (FCA) in 1980 brought a brief hiatus and certainly slowed down the release of land for agriculture, the release of forest land for dams, mining, roads, transmission lines, and other developmental projects, now often in the private sector, has gathered momentum since the 1990s. While the FCA provides a means for centralized tracking (not so much actually regulating) the conversion of forest land (see Chapter 7 in this volume for more details), in parallel large areas of non-forest CPLRs have been transferred by the states for developmental projects, including industrial development, quarrying, defence projects, and now Special Economic Zones (SEZs).[9] While comprehensive documentation and estimates are missing, Jodha (1987; 1989) estimated the loss of non-forest CPLRs in the 1950s–80s period. What Jodha also documents and subsequent studies have also highlighted (Damodaran 1987; Iyengar 1989) is that often this loss (especially post-1980) is not overt but covert, through the encroachment of CPLRs, typically by the village elite.[10] Kashwan and Lobo have described this process in more detail in Chapter 10 in this volume. The net result of state giveaways and encroachments has been a drastic curtailment in de facto accessible CPLRs in virtually all peninsular states.

Denying access, even while enhancing forest production, has been another, perhaps more ironic, dimension of this process of CPLR appropriation. The trend of replacing natural forests with timber plantations to which local communities were denied access, began in the British period and continued apace afterwards. In fact, in the 1980s, the scope was expanded to non-forest CPLRs—euphemistically named Social Forestry projects—and were taken up in non-forest land, typically grazing land, thereby denying villagers and pastoral nomads their grazing rights while producing softwood for industries. Farm forestry components of these projects further expanded this practice to private farmlands—lands that were otherwise seasonal CPLRs, as mentioned earlier. The leasing of thousands of acres of so called 'wasteland' to private companies to grow teak, eucalyptus, and acacia in Tamil Nadu (the Sterling Tree Magnum story: Pandian 1996) and similarly, policies elsewhere using the term 'wasteland' to justify the transfer of CPLRs to corporate houses for industrial plantations and agriculture (Bharwada and Mahajan 2006) is part of the same process.

A third form of alienation of the local community from the commons has emerged in the post-1970s, namely, the takeover of the local commons citing environmental reasons (in effect, global commons arguments). Even if conservation as such is seen as a necessary public goal, the ecological justification for complete stoppage of human use of and presence in forests is weak and untested (see Chapter 4 in this volume) and in practice, the process has been much more socially disruptive than just that. Given that even the deepest and remotest parts of this subcontinent have historical human presence and use, and given that the wildlife loss is driven by habitat loss, not habitat use, and much of this habitat loss has been for developmental purposes, the onus should have been on the state to devise a much more nuanced strategy for integrating local use with conservation. Instead, the state has declared a large number of protected areas (PAs), where the Wildlife Protection Act (according to the Supreme Court interpretations) requires almost complete elimination of human presence, but the process of settling the rights of various communities living in these PAs is in most cases still incomplete.[11]

The most recent example of the use of extreme conservationist approaches to usurp CPLRs is the declaration of 39 critical tiger habitats (CTHs)—constituted without following due process as laid down in the amended Wildlife Protection Act, 2006 itself and against the spirit of co-existence suggested by the Tiger Task Force, whose report led to the creation of CTHs (Bijoy 2011; Taghioff and Menon 2010). Again, the bitter irony is that while local communities, who are both long-term residents and largely socially and economically marginalized, are being relocated in the name of conservation, the spaces are not really maintained inviolate as tourism is permitted to a significant extent, and tourism revenues are either pocketed by the state or by tourism operators, being individuals and groups with substantial capital and political reach. It is, therefore, not surprising that tourism may further alienate local communities from their erstwhile commons (Baviskar 2012), as they resent the loss of livelihoods and the denial of new opportunities.

Finally, the past decade or so has seen a new global environmental claim on the local commons, namely, for mitigating climate change either through plantations of biofuels or for carbon sequestration. While the latter is still largely in its planning stage, the former is well under way. The Indian government set up a National Mission on Biofuels with biofuel authorities at both central and state levels. *Jatropha curcas* and *Pongamia pinnata*, both trees with oil-bearing seeds, were chosen with a target area of over 30 m ha of wasteland. Wasteland was chosen because the opportunity costs of using wasteland were considered lower than for more 'productive' lands. Hence, several states have begun to replicate the Social Forestry story by taking over or leasing out so-called wasteland for biofuel plantations. In Chhattisgarh, the FD and the Forest Development Corporation have since 2007 aggressively promoted jatropha on all types of lands, especially grazing and forest land (Lahiri 2008). The same has been the case in Rajasthan and Tamil Nadu, in the latter case as part of its watershed development programme (Ariza-Montobbio *et al.* 2010).

These 'low opportunity cost' wastelands are important commons for village communities. Many of these lands in Rajasthan, for example, are available for open grazing at least for two to three months. People also collect fuel wood from the commons and in some regions they are part of the traditional route taken by pastoralists.

Devaluing the Commons

Not only did the colonial state take over large tracts of forests for its exclusive use, but several policies initiated then and subsequently, have led to a serious decline in the economic returns per unit area of CPLRs that the local community can derive. To begin with, the colonial state significantly limited customary rights over produce from even those forests that were left open to villager access through a series of measures. First, all major timber species were declared as state property. Second, even bamboo was designated as a timber species, and thereby free access to it was denied. Third, most of the valuable non-timber products were also brought under state control, so as to maximize revenue to the colonial state. And the post-colonial state did nothing to redress this attenuation of local rights. In fact, the 1960s saw a spate of state takeovers of valuable NTFPs such as tendu patta (*Diospyros melanoxylon*), sal (*Shorea robusta*) seeds and leaves, and mahua (*Madhuca indica*)—species that were often critical to tribal livelihoods in central India (see Lele *et al.* 2010 for details). Even the recent (2004) decision to 'de-nationalize' NTFPs in Odisha kept the most valuable NTFP, namely, tendu patta, out of its purview.

Simultaneously, the colonial state systematically denied local communities the right to *manage* their CPLRs (except in a few cases like the Van Panchayats of Kumaon, where these rights were re-created after much protest). The bureaucratic mechanisms put in place for regulating the use of these lands were limited and impossible to sustain for long, especially post-independence. Thus, the land set aside for use by villagers became de facto open access.

When combined with the overall reduction in accessible areas, this was bound to lead to degradation of these commons, and it did. The productivity of these lands, both for grass and trees, declined dramatically, and when combined with the fact of reduced areas (due to state takeover, state hand-out, and encroachment) on the one hand and increased human populations on the other, the return per household from these CPLRs declined sharply.

Finally, the state has not only controlled the resource, but also access to product markets. Thus, even when full rights to harvest were conceded, such as when NTFP collection rights have been given to tribal cooperatives, the functioning of these cooperative societies has been controlled and manipulated by government officials in such as way as to unfairly limit economic returns to the collectors, with official and unofficial rent-seeking taking away almost half of the profits (Lele and Rao 1996). The movement towards reducing the margins retained by the state or its various agencies and intermediary corporations has been slow and haphazard (Lele *et al.* 2010). Even in a situation where the state has officially given up its claims to any royalty from any NTFP, such as in Madhya Pradesh, the actual price realized by the collectors leaves a lot to be desired because of the manner in which the state controls the entire process (Lele *et al.* 2012).

JFM was supposed to mark a major shift in state policy on this issue through the (re-) creation of material incentives for local communities to participate in forest protection and management. On paper, this was supposed to happen because JFM would give communities a share in hitherto state-controlled timber, and full rights to all commercial NTFPs. In practice, the JFM programme has failed to achieve a sub-stantial increase in the returns to local communities almost by design. Firstly, in most cases, JFM was only taken up on degraded land, so the returns have been slow in coming, and the opportunity costs, in terms of denial of grazing and firewood collection opportunities, have been severe (Lele *et al.* 2005; Sundar *et al.* 2001). Furthermore, rights to the most valuable commercial NTFPs, including bamboo, have stayed out-side the purview of JFM. Even in the best case of West Bengal, the share in the final harvest of sal produce has not been transferred to the forest protection committees in many cases (Banerjee 2007). Rights transferred on paper are thus not translated into real incomes. Many ex-ante studies had estimated that the returns from JFM would be sub-stantial (for example, Hill and Shields 1998), but their calculations have gone awry mainly because JFM never got implemented in the way they visualized.

The Potential of CPLRs

In short, if CPLRs appear to be making only limited contributions to rural livelihoods and if then the idea of rural development based on the simultaneous regeneration of both CPLRs and agriculture seems improbable, 'push' factors play a major role in bringing this about. Loss of rights, attenuation of rights, open-access situations, and rent-seeking state policies have all contributed to this situation. But will the handing over of substantial rights of collection, management, and marketing of forest and non-forest produce to local communities actually generate incomes at such a level as to substantially enhance rural livelihoods and help households and communities meet their developmental aspirations, at levels similar to those achieved by say, Green Revolution agriculture?

In some ways, this question remains hypothetical. The absence of any example where such rights have been unambiguously and securely granted (unlike the case of forest-user groups in Nepal after 1992) makes it impossible to answer this question properly. Nevertheless, there is substantial anecdotal evidence to suggest that this might be possible. The high potential for increased income from forest products can be seen from the rare cases where rights are unambiguously transferred and associated institutions have functioned reasonably well. For instance, the Soligas of the Biligirirangan Hills of Karnataka had (till recently) exclusive NTFP harvesting rights to large patches of forest, and these NTFPs had to be sold through their own co-operative societies called LAMPS. In one LAMPS, when the community could be mobilized and pressure brought upon its office-bearers to conduct auctions transparently and manage accounts honestly, the returns to Soliga NTFP collectors were typically 50 per cent higher than when the society is malfunctioning—with no change in access conditions, technologies of processing, or market conditions (Lele *et al*. 2004). Value-added processing added another 20 per cent to the returns. Another example is the traditional individually controlled woodlots in coastal Karnataka, which are playing a 'banking' function while also providing inputs to agriculture (Srinidhi and Lele 1999). A third, more substantial but more controversial example is, of course, that of the timber rights that various villages, clans, and individuals in the Northeastern states enjoyed till recently. While there was mixed evidence as to the long-term sustainability of the logging regimes followed, it is clear that the returns were very substantial for quite some time (Nathan 2000). Instead of improving the regulation for sustainability and offsite impacts, the Supreme Court's 1996 decision to clamp down completely on logging has had serious livelihood impacts (Nongbri 2001) and subsequent relaxation through bureaucratically approved working plans has not really addressed the core issues.

Will this increased economic return be enough to ensure local-level collective action? Can the cultural hegemony of urban lifestyles that Baviskar talks about be offset by improved material benefits from traditional occupations such as NTFP collection or forest-livestock-agriculture-based livelihoods? There is no easy answer to these questions in empirical terms. But there is a normative aspect to this discussion to which we now turn.

Beyond Economic Dependence: The Normative Case for Local Control

We began this chapter from where many of the proponents of decentralized governance have begun, namely, that villagers are critically dependent upon forests, and therefore, they must be provided adequate access and management rights over them. The assumption of dependence then makes it plausible that village communities will manage the resource sustainably. In this discourse, as Menon (1999) had pointed out, the ultimate normative concern is an environmentalist one, namely, sustainable resource-use. If poorer communities turn out to be more dependent on CPLRs, then one can in addition invoke the discourse of poverty alleviation to justify decentralized governance, although one has to then grapple with the danger of elite capture. In both cases, decentralized governance remains a means to an end or ends. If then it turns out that material dependence is actually waning rapidly due to economic development, or that the CPLRs can only provide some insurance but cannot be the engine of rural development, then the chances of successful collective action seem to drop and the case for decentralized governance is correspondingly weakened.

Our examination of the evidence regarding dependence and the factors influencing it has highlighted several points. First, CPLRs are not just forests, but a range of resources used in multiple ways by rural communities as an integral part of their livelihoods. The discussion, therefore, needs to move beyond forest dependence and governance to CPLR dependence and governance.[12] Second, there is ample evidence that CPLRs continue to contribute significantly to the income of the rural poor (even if not the major portion). While it is possible that this contribution has been over-estimated in some studies, there is also substantial evidence that the importance of CPLRs goes beyond their percentage or absolute contribution to incomes: they provide livelihoods during critical lean seasons, help build agricultural productivity, and have cultural significance. Third, there is some (largely anecdotal) evidence that the dependence on CPLRs may be declining and that some of this is because households actually choose (or will soon choose) to move to other occupations. However, there is also substantial evidence that much of the relative 'unimportance' of CPLRs or their perceived inability to contribute substantially to rural development is probably the result of a systematic reduction in accessible areas, attenuation of harvesting rights, degradation due to open-access situations, and other policies that constrain the economic potential of these resources. CPLRs could, under the right conditions, make a major contribution to rural development and poverty alleviation, while being sustainably managed. It is important to also remember that rights are not only a means to improved livelihoods, but a recognition of the cultural value associated with certain environment–community relationships.

But at another level, one may question the starting-point itself. One may ask whether the likelihood of 'good' outcomes, whether in sustainability or in poverty-alleviation terms, should be a necessary precondition for decentralizing governance of CPLRs or of the immediate environment of rural communities. One may argue that the ability to have a major say in how one's immediate environment is managed is the democratic right of each group of citizens, whether rural or urban, whether they depend upon that environment in material terms or not, whether they have cultural ties to it or not, whether they have traditional knowledge and a traditionally conservationist culture or not. After all, the inability of 'natives' to manage their own affairs peacefully and successfully was for long the justification given by colonial powers for their continued presence in and control over the colonized nations. The clinching argument against colonialism was not that colonized peoples knew to manage their affairs better, but that colonialism was morally repugnant. By the same token, a certain level of democratic control over one's environment can be seen as morally necessary, even if not guaranteeing any better outcomes.

To summarize, the relationship between local commons and communities is an important one, but also complex, variable, and shifting over time, not cast in stone. And the history of local commons in India is the history of state takeover, neglect, and antipathy to the point where the inability of local communities to derive high levels of benefits in a sustained manner from these resources has become a self-fulfilling prophecy. The current development paradigm may indeed provide an irresistible, if illusory, attraction. But communities will continue to reconstruct and relate to the local commons in different ways, and the commons in turn, if conceived of broadly, addressed in tandem with private resources, and regenerated imaginatively, may play an important role in sustainable rural development.

Notes

1. The other pillars are that local communities have a common interest in the forest, that they are 'natural' conservationists, and that they have superior ecological knowledge compared to the forest bureaucracy.

2. Indeed, this separation of the right to use or access from the right to exclude others or manage the resource has repercussions for the magnitude and sustainability of the benefits derived from these lands, as we shall see later.

3. Heavily forested areas include the Himalayas, Western Ghats, the adivasi belt of central India, and the Northeast predominantly.

4. Singh *et al.* (1996) studied eight semi-arid villages in Punjab and concluded that the landless derived 27 per cent of their income from CPRs and cultivating households 22 per cent. In the four semi-arid villages of Dharwad district studied by Gowda and Savadatti (2004), between 22 and 25 per cent of the total income was derived from CPRs. See also Jodha (1986, 1990).

5. NSS 54th round estimates with regard to a disaggregated estimate of different CPR usages pertained to the 'collection' of CPR products. It is important to note, as Menon and Vadivelu (2006) have, that the use of CPRs also include activities such as grazing that does not involve collection. Hence, total estimates of use are likely to be underestimated and so, too, the contribution of fodder to total CPR use.

6. These studies also illustrate the significant contribution of livestock to agricultural GDP and the role of goats and sheep in meat production.

7. Based on fieldwork conducted by Ajit Menon in Gudalur during 2008–10.

8. Note, however, that large-sample studies such as FES (2012) continue to show a high level of dependence (20–40 per cent) across a number of states.

9. For instance, in Raigarh district, Maharashtra, where Reliance is building an SEZ, more than 12,000 ha of dali lands are under threat. Some of these

areas are forest land, at times used as commons but in other cases cultivated by adivasis (Asher 2006).

10. This is not to be confused with the so-called 'forest encroachment' that is, the outcome of redesignation of historically cultivated lands as forest land under a flawed forest settlement process, which was in part the genesis of the Forest Rights Act, 2006. Chapter 3 in this volume discusses this in more detail.

11. For instance, the Tiger Task Force estimated that a total of 1,01,077 people lived in the core area of so-called critical tiger habitats and 3,80,535 in the core and buffer areas together (Tiger Task Force 2005). This does not include the number of people who use the tiger reserves as transient pastoralists.

12. This point is reiterated in Chapter 2 in this volume.

References

Agarwal, A. and S. Narain. 1989. 'Towards Green Villages: A Strategy for Environmentally-sound and Participatory Rural Development'. New Delhi: Centre for Science and Environment.

Agrawal, A. 1998. *Greener Pastures: Politics, Markets and Community Among a Migrant Pastoral People*. Durham: Duke University Press and New Delhi: Oxford University Press.

Ariza-Montobbio, P., S. Lele, G. Kallis, and J. Martinez-Alier. 2010. 'The Political Ecology of *Jatropha* Plantations for Biodiesel in Tamil Nadu, India', *Journal of Peasant Studies*, 37 (4): 875–97.

Arnold, J.E.M. and W.C. Stewart. 1991. 'Common Property Resource Management in India'. Tropical Forestry Paper no. 24, Oxford: Oxford Forestry Institute.

Asher, M. 2006. 'Don't Call it Wasteland', *The Times of India*, 12 October; http://timesofindia.indiatimes.com/home/opinion/edit-page/Dont-call-it-wasteland/articleshow/2151679.cms?referral=PM.

Banerjee, A. 2007. 'Joint Forest Management in West Bengal', in O. Springate-Baginski and P. Blaikie (eds), *Forests, People & Power: The Political Ecology of Reform in South Asia*, pp. 221–60. London: Earthscan.

Baviskar, A. 1999. *In the Belly of the River: Tribal Conflicts over Development in the Narmada Valley*. New Delhi: Oxford University Press.

———. 2012. 'India's Changing Political Economy and its Implications', in Anonymous (ed.), *Deeper Roots of Historical Injustice*, pp. 33–45. Washington, DC: Rights and Resources Initiative.

Beck, T. 1994. 'Common Property Resource Access by Poor and Class Conflict in West Bengal', *Economic and Political Weekly*, 29 (4): 187–97.

Beck, T. and M.G. Ghosh. 2000. 'Common Property Resources and the Poor: Findings from West Bengal', *Economic and Political Weekly*, 35 (3): 147–53.

Belcher, B. and K. Schreckenberg. 2007. 'Commercialisation of Non-timber Forest Products: A Reality Check', *Development Policy Review*, 25 (3): 355–77.

Bharwada, C. and V. Mahajan. 2006. 'Quiet Transfer of Commons', *Economic and Political Weekly*, 41 (4): 313–15.

Bijoy, C.R. 2011. 'The Great Indian Tiger Show', *Economic and Political Weekly*, 46 (4): 37.

Byron, N. and M. Arnold. 1999. 'What Futures for the People of the Tropical Forests?' *World Development*, 27 (5): 789–805.

Chakravarty-Kaul, M. 1996. *Common Lands and Customary Law: Institutional Change in North India over the Past Two Centuries*. New Delhi: Oxford University Press.

Chopra, K. and P. Dasgupta. 2008. 'Nature of Household Dependence on Common Pool Resources: An Empirical Study', *Economic and Political Weekly*, 43 (8): 58–66.

Damodaran, A. 1987. 'Structural Dimensions of Fodder Crisis: A Village Study in Karnataka', *Economic and Political Weekly*, 22 (13): A-16 to A-22.

DeFries, R. and D. Pandey. 2010. 'Urbanization, the Energy Ladder and Forest Transitions in India's Emerging Economy', *Land Use Policy*, 27(2): 130–8.

Fernandes, W., G. Menon, and P. Viegas. 1988. *Forests, Environment and Tribal Economy: Deforestation, Impoverishment and Marginalization in Orissa*. New Delhi: Indian Social Institute.

Foundation for Ecological Security (FES). 2012. 'A Commons Story: In the Rain Shadow of Green Revolution'. Anand: Foundation for Ecological Security.

GoI. 1988. 'National Forest Policy'. New Delhi: Ministry of Environment and Forests.

Gowda, M.N. and P.M. Savadatti. 2004. 'CPRs and Rural Poor: Study in North Karnataka', *Economic and Political Weekly*, 39 (33): 3752–7.

Guha, R. 1983. 'Forestry in British and Post-British India: A Historical Analysis (Parts III and IV)', *Economic and Political Weekly* 18: 1940–7.

———. 1996. 'Savaging the Civilised: Verrier Elwin and the Tribal Question in Late Colonial India', *Economic and Political Weekly*, 31 (35/37): 2375–89.

Hill, I. and D. Shields. 1998. 'Incentives for Joint Forest Management in India', World Bank Technical Paper no. 394. Washington, DC: The World Bank.

Homma, A.K. 1996. 'Modernisation and Technological Dualism in the Extractive Economy in Amazonia', in M. Ruiz Pérez and J.E.M. Arnold (eds), *Current Issues in Non-Timber Forest Products Research*, pp. 59–82. Bogor: Center for International Forestry Research.

Iyengar, S. 1989. 'Common Property Land Resources in Gujarat: Some Findings about their Size, Status and Use', *Economic and Political Weekly*, 24 (25): A67–A77.

Jodha, N.S. 1985. 'Population Growth and the Decline of Common Property Resources in Rajasthan, India', *Population and Development Review*, 11 (2): 247–64.

———. 1986. 'Common Property Resources and Rural Poor in Dry Regions of India', *Economic and Political Weekly*, 21 (27): 1169–81.

———. 1987. 'A Case Study of the Decline of Common Property Resources in India', in P. Blaikie and H. Brookfield (eds), *Land Degradation and Society*, pp.196–207. New York: Methuen.

———. 1989. 'Depletion of Common Property Resources in India: Micro-level Evidence', *Population and Development Review*, 15: 261–83.

———. 1990. 'Rural Common Property Resources: Contributions and Crisis', *Economic and Political Weekly*, 25 (26): A65–A78.

Kapur, D. 2010. 'Dependence of Livestock Rearers on Common Lands: A Scoping Study'. Report for the South Asia Pro-Poor Livestock Policy Programme (SA PPLPP). Anand: National Dairy Development Board and Food and Agriculture Organization (FAO).

Kumar K., A.K., S. Lele, and P. Shivashankar. 2008. 'Impact of Irrigation-led Agricultural Development on Use of Common Lands in Dry Regions of Karnataka, India', in Anonymous (ed.), *Insights from the Field: Studies in Participatory Forest Management in India*, pp. 110–34. New Delhi: Resource Unit for Participatory Forestry (RUPFOR), Winrock International India.

Lahiri, S. 2008. 'Colonizing the Commons: It is Jatropha Now!' *Mausam*, 1 (1): 14–19.

Lele, S. 2001. 'Linking Ecology, Economics, and Institutions of Village-level Forest Use in the Karnataka Western Ghats', Project Report. Bangalore: Institute for Social and Economic Change and Pacific Institute for Studies in Development, Environment and Security.

Lele, S. and R.J. Rao. 1996. 'Whose Cooperatives and Whose Produce? The Case of LAMPS in Karnataka', in R. Rajagopalan (ed.), *Rediscovering Cooperation*, vol. II, pp. 53–91. Anand: Institute of Rural Management Anand.

Lele, S., A.K. Kiran Kumar, and P. Shivashankar. 2005. 'Joint forest planning and management in the Eastern plains region of Karnataka: A rapid assessment. CISED Technical Report'. Bangalore: Centre for Interdisciplinary Studies in Environment and Development.

Lele, S., K.S. Bawa, and C.M. Gowda. 2004. 'Ex-Post Evaluation of the Impact of an Enterprise-based Conservation Project in BRT Wildlife Sanctuary, India'. Paper presented at 'The Commons in an Age of Global Transition: Challenges, Risks and Opportunities', Eighth Biennial Conference of the International Association for the Study of the Commons, organized by the Institute of Social Research, National Autonomous University of Mexico, at Oaxaca, Mexico, from 9–13 August.

Lele, S., M. Pattanaik, and N.D. Rai. 2010. 'NTFPs in India: Rhetoric and Reality', in S.A. Laird, R. McLain, and R.P. Wynberg (eds), *Wild Product Governance: Finding Policies that Work for Non-Timber Forest Products*, pp. 85–112. London: Earthscan.

Lele, S., V. Ramanujam and J. Rai. 2012. 'Tendu Leaf Procurement, Rural Livelihoods and Forest Governance in Madhya Pradesh'. (unpublished manuscript). Bangalore: Ashoka Trust for Research in Ecology and the Environment.

Menon, A. 1999. 'Common Property Studies and the Limits to Equity: Some Conceptual Concerns and Possibilities', *Review of Development and Change*, 4 (1): 51–70.

Menon, A. and A. Vadivelu. 2006. 'Common Property Resources in Different Agro-climatic Landscapes in India', *Conservation and Society*, 4 (1): 132–54.

Nadkarni, M.V. 1990. 'Use and Management of Common Lands: Towards an Environmentally Sound Strategy', in C.J. Saldanha (ed.), *Karnataka State of Environment Report*, pp. 31–53. Bangalore: Centre for Taxonomic Studies, St. Joseph's College.

Nadkarni, M.V., S.A. Pasha, and L.S. Prabhakar. 1989. *The Political Economy of Forest Use and Management*. New Delhi: Sage Publications.

Narain, U., S. Gupta, and K. van't Veld. 2008a. 'Poverty and the Environment: Exploring the Relationship between Household Incomes, Private Assets, and Natural Assets', *Land Economics*, 84 (1): 148–67.

———. 2008b. 'Poverty and Resource Dependence in Rural India', *Ecological Economics*, 66 (1): 161–76.

Nathan, D. 2000. 'Timber in Meghalaya', *Economic and Political Weekly*, 25 (4): 182–6.

Nongbri, T. 2001. 'Timber Ban in North-East India: Effects on Livelihood and Gender', *Economic and Political Weekly*, 36 (21): 1893–1900.

Pandian, A. 1996. 'Land Alienation in Tirunelveli District', *Economic and Political Weekly*, 31 (51): 3291–4.

Pasha, S. 1992. 'CPRs and Rural Poor: A Micro Level Analysis', *Economic and Political Weekly*, 27 (46): 2499–503.

Philip, K. 2004. *Civilising Natures: Race, Resources and Modernity in Colonial South India*. New Delhi: Orient Longman.

Qureshi, M.H. and S. Kumar. 1998. 'Contributions of Common Lands to Household Economies in Haryana, India', *Environmental Conservation*, 25 (4): 342–53.

Ravindranath, N.H., K.S. Murali, and K.C. Malhotra (eds). 2000. *Joint Forest Management and Community Forestry in India: An Ecological and Institutional Assessment*. New Delhi: Oxford and IBH.

Reddy, S.R.C. and S.P. Chakravarty. 1999. 'Forest Dependence and Income Distribution in a Subsistence Economy: Evidence from India', *World Development*, 27 (7): 1141–9.

Saberwal, V. 1999. *Pastoral Politics: Shepherds, Bureaucrats and Conservation in the Western Himalaya*. New Delhi: Oxford University Press.

Sarkar, R. 2008. 'Decentralised Forest Governance in Central Himalayas: A Re-evaluation of Outcomes', *Economic and Political Weekly*, 43 (18): 54–63.

Satya, L.D. 2004. *Ecology, Colonialism and Cattle: Central India in the Nineteenth Century*. New Delhi: Oxford University Press.

Shanmugaratnam, N. 1996. 'Nationalisation, Privatisation and the Dilemmas of Common Property Management in Western Rajasthan', *Journal of Development Studies*, 33 (2): 163.

Singh, C. 1999. *Natural Premises: Ecology and Peasant Life in the Western Himalaya 1800–1950*. New Delhi: Oxford University Press.

Singh, K., N. Singh, and R. Singh. 1996. 'Utilisation and Development of Common Property Resources—A Field Study in Punjab', *Indian Journal of Agricultural Economics*, 51 (1/2): 249–59.

Skaria, A. 1999. *Hybrid Histories: Forests, Frontiers and Wildness in Western India*. New Delhi: Oxford University Press.

Srinidhi, A.S. and S. Lele. 1999. 'Factors Influencing Sustainable Forest Management under Regulated Private Property: The Case of the *Haadis* of Dakshina Kannada'. Paper presented at the First INSEE Biennial Conference, organized by the Indian Society for Ecological Economics at Bangalore on 20–22 December.

Sundar, N., R. Jeffery, and N. Thin. 2001. *Branching Out: Joint Forest Management in India*. New Delhi: Oxford University Press.

Sunderlin, W., A. Angelsen, and S. Wunder. 2003. 'Forests and Poverty Alleviation.' (unpublished manuscript). Bogor: Centre for International Forestry Research (CIFOR).

Taghioff, D. and A. Menon. 2010. 'Can a Tiger Change its Stripes? The Politics of Conservation as Translated in Mudumalai', *Economic and Political Weekly*, 45 (28): 69.

Tiger Task Force. 2005. 'Joining the Dots: Report of the Tiger Task Force'. New Delhi: Ministry of Environment and Forests, Government of India.

SHARACHCHANDRA LELE
AJIT MENON

Epilogue

We began this volume by describing the forest sector in India as highly complex and currently undergoing enormous churning. Four years and eleven chapters after we began the project, the complexity and dynamism in the sector have only heightened. As we write this:

1. The Supreme Court has accepted the relevance of the Forest Rights Act (along with the Panchayats (Extension to the Scheduled Areas) Act, 1996 (PESA) and constitutional guarantees of tribal rights) in decisions regarding diversions of land held sacred by tribal communities, and asked the Odisha government to consult the gram sabhas of communities affected by the decision to mine the Niyamgiri Hills.

2. A report commissioned by the Ministry of Environment and Forests (MoEF) on the question of Net Present Value (NPV) to be charged when forests are converted to non-forest uses recommends that 50 per cent of this value be transferred to local communities whose rights are abridged due to such conversion.

3. Several hundred villages in eastern Maharashtra and more than 2000 villages in Odisha have received Community Forest Rights. Some villages in Gadchiroli and Gondia districts of Maharashtra for the first time harvested and marketed *tendu patta* (*beedi* leaves) on their own, exercising their rights under the Forest Rights Act (FRA) even though

the law establishing state control over tendu patta remains on the books and is implemented in the rest of the state.

4. Some conservationists have begun to come out in favour of the implementation of the FRA (Raman and Madhusudan 2013).

And yet, it is not clear whether these changes represent a real trend towards greater democratization. The bigger picture contains many trends that undermine the idea of democratizing forest governance. These include:

1. The Forest Service has closed ranks against the push for democratization and the FRA in particular, and refused to cooperate even with its own minister. Whereas, in the pre-FRA era, a few foresters might have gone out on a limb and made statements such as 'there is little doubt that the demarcation of forest lines was done in a very haphazard way' (Rangachari and Mukherji 2000: 106), the forester community now seems to have abandoned any constructive engagement with these questions.

2. Aid agencies such as the Japanese bank continue to support state forest departments (FDs) for conventional afforestation projects as if community forestry or the FRA did not exist. Many non-governmental organizations (NGOs) continue to work within the framework of JFM, blind to all its flaws.

3. Many international donors and actors have now lost interest in local forestry issues per se, and are narrowly focused on setting up payment schemes wherein Western countries buy carbon credits from developing countries. Centralized control of forests is, therefore, seen to be more convenient than a democratically decentralized one (Phelps *et al.* 2010).

4. National Park and tiger reserve authorities by and large continue to operate as if the FRA does not exist, and as if they had never heard of the term co-management.

5. The Supreme Court continues to believe that forests are national property, and continues to micro-manage decisions in the forest sector through the Godavarman case.

6. In many pockets in the central Indian tribal and forested belt, the conflict between the state, using military, paramilitary, police, and civilian fronts like *salwa judum*, and the Maoists continues unabated and has prevented any meaningful discourse on forest rights and

governance. Indeed, there are reports of Maoists cooperating with industries to prevent tribal communities from getting community forest rights (Pallavi 2012).

7. The larger discourse continues to focus on 'development as modernization/industrialization' and completely neglects (if not derides) ideas of forest-based livelihoods (Baviskar 2012). Even pro-poor programmes such as the National Rural Employment Guarantee Act (NREGA) are unable to integrate systematically the building of natural capital into their portfolio.

In this context, it is not at all clear what the next decade holds for the forest sector in India, and specifically whether the process of democratization will go further and what shape it will take. Nevertheless, we believe the fractiousness of the forest discourse and internal contradictions within the state over the forest question are signs of a continued ferment that will open up opportunities, albeit in unexpected ways. Some have argued that 'participation in forestry alone will not work until the larger state–society relationship is changed' (Vira 2005). We are hopeful that the forest sector could be at the vanguard of changing this state–society relationship. But for this to happen, a more nuanced approach to thinking about the forest question has to emerge in theory and practice. More attention is required to the meaning, structure, and context of democratization in the forest sector. Drawing upon the chapters in this volume and going beyond them, we highlight two crucial areas that need greater attention.

Democratization is More than Devolution[1]

In a recent overview piece, Ramachandra Guha says that 'in one essential respect the forestry debate hasn't moved on beyond the 1980s ... the overall framework of the debate is still determined—or confined—by the opposition of two narratives, one which privileges state control and management of forests, the other which offers the alternative of community control and management' (Guha 2012: 10). In this volume too, the tone of many of the chapters has been, either implicitly or explicitly, one of casting the community in opposition to the state. To an extent, this is understandable: although 40 years have passed since the villagers in Garhwal protested against the lack of harvesting rights, and 25 years since the first full-fledged forest policy statement of independent India made noises about participatory forestry, the FD and the Indian Forestry Service (and by extension, the IFS-dominated MoEF at the centre) continues to loom large on the forest landscape.[2] There has been hardly any change in the legal or actual power wielded by foresters, nor any significant shift in their mindsets.[3] Breaking the foresters' stranglehold on forest policy, forest management, forest funding and forest research, and transferring substantial rights and power to local communities, therefore, becomes the primary practical goal of most social/tribal activists.

Without perhaps always meaning to, this ends up setting up a state–community dichotomy and opposition, with the state being the 'bad guy' and the community being the 'good guy'. When one further attaches the labels 'tribal', 'poor', and 'forest-dweller' to the community, the polariszation becomes even sharper. Sections of academia have contributed to this simplification by focusing on the concept of collective action: the idea that, if left to themselves, small local communities with a shared interest in the forest can and will come together to protect and manage it. But neither the pragmatic nor the academic simplifications is logically tenable in the context of forest governance.

Forests are by their very nature a complex socio-ecological entity, generating tangible products and intangible services to multiple stakeholders at different scales through a combination of ecological and social processes. They not only produce tangible products such as timber, bamboo, fodder and wild honey, but they also moderate the hydrological behaviour of watersheds, provide habitat for wildlife, and sequester CO_2.[4] And these benefits accrue to different groups in society. While the tangible products may benefit villagers or logging contractors, hydrological regulation benefits downstream water-users, climate change mitigation due to CO_2 sequestration benefits the entire world, and wildlife is also seen today as a global heritage of humankind. Even tangible products may, as a result of social arrangements, 'flow' to communities that are not local but come from afar, such as pastoral nomads. Decisions about one aspect, such as logging for timber, have repercussions for all other goods and services. Moreover, forests are slow-growing entities, so decisions taken today will have repercussions over a long period of time. Thus, leaving decision-making about forests to any one beneficiary group *alone*, whether the global wildlife conservation community or the local firewood user, would be likely to jeopardize the legitimate stakes of others.

Admittedly, the notion of 'regional', 'national', or 'global' (service) benefits of forests has in the past been used simply as an excuse (by the British and by the post-independence Indian state) to grab control of the tangible benefits (such as timber or the land itself). The bogey of forests being 'degraded' by local use has been raised at several points to justify high-handed takeovers and exclusion of legitimate users, as Nitin D. Rai, Kundan Kumar, and others have argued in this volume (see also Lele 2000). And, as Shomona Khanna has pointed out in chapter 6, the Supreme Court itself has used the 'national interest' argument to justify the centralization of power in its own hands. But to reject any legitimate stake for non-local communities would be as problematic as the idea that forests are simply a national-level public good in which local uses and users are an illegitimate presence. It would be akin to saying that, in a river basin, only users living on river banks have rights to water, and they have neither to consider the needs of those away from the banks or those downstream of themselves. The question, therefore, is not whether non-local stakes are legitimate, but rather how they can be given a voice in decision-making about forests that is not disproportionate, not driven by the greater power they wield by virtue of being economically and politically powerful, as the conservation groups and carbon-sequestration advocates generally are.

Furthermore, the tension is not just between local, national, and global stakeholders. As Kashwan and Lobo point out, there is ample evidence of competing stakes *within* the local, whether between graziers and firewood collectors, collectors of non-timber forest products (NTFPs) versus those interested in the timber, and so on: differences that get exaggerated due to their correlation with caste, class, or gender divisions. Study after study of bottom-up community forestry has shown how easily women, pastoralists, or NTFP collectors can get sidelined (Bhatta 2002; Sarin 1995; Sarin *et al.* 2003). Democratic governance requires that all these stakeholders get a fair voice in decision-making irrespective of their social position and location, while allowing for a hierarchy of stakes that varies with the context. This cannot be easily achieved by collective action institutions, because these institutions tend to emerge in the context of commonly shared interests and focus on overcoming problems of free-riding and achieving economies of scale. What are required here are regulatory institutions that focus on equity and fairness (Lele 1999). In this context, casting the state in opposition to the community is of little analytical value. Institutions that have a regulatory role and the membership of which is citizenship based, whether gram sabhas or provincial governments, are all forms of a state. Governance requires the allocation of power to take decisions, while simultaneously trying to guard against its misuse. In any allocation of power, questions of voice for marginalized sections, and of transparency and accountability will arise as much for village-level bodies as they arise for higher level ones.

Deepening and broadening the idea of democratic forest governance will, therefore, require thinking in terms of a multi-layered governance system, albeit one in which the allocation of power and responsibility, the scale and the structures of transparency and accountability would have to be very different from what they are right now (see Lele 2004; Lele 2011, for a detailed discussion). The contents of such restructuring are only recently beginning to be discussed, as in the report of the Joint Committee on the FRA (Joint Committee 2010, Chapter 8), but much of the debate continues to be polarized between foresters wanting to hang on to their traditional role (of forest planter, police, regulator, planner, policy-maker, scientist all combined) and activists wanting to simply do away with the FD.

In the case of identification of conservation-priority areas, as Pathak *et al.* have discussed in their chapter, the CWH/CTH provisions have the potential to democratize the procedure for identification and ensure some due process before deciding upon relocation of local communities. More exploration is, however, required on what democratizing the actual governance of conservation-priority areas means, on how local and off-site voices can be represented in this governance, on how indigenous ecological knowledge can be integrated meaningfully with modern ecological knowledge, and what claims local communities should have on the stream of economic benefits that even conservation areas generate (whether through tourism or through external support for conservation work).

Similarly, giving local communities a veto in the conversion of forests to (say) mining is only one step in the democratization of that process. Questions such as who constitutes the local community and therefore holds veto power,[5] who represents the community when the community is too large, what if some want a development project and others do not (or in other words, is their power limited to saying 'no' but not 'yes'), and can communities enter into profit-sharing with mining companies—all need to be studied and debated.

Furthermore, the changes required are not just structural but also behavioural.[6] On the one hand, the most urgent behavioural change required is in the mindset. The challenge of forest governance is not just one of overcoming the resistance of a deeply and long-entrenched bureaucracy to redress the historic injustice of non-recognition of local stakes. It also involves accepting that undemocratic behaviour is not the exclusive reserve of the bureaucracy but is deeply entrenched in a society where feudalism, colonialism, and post-colonial centralist tendencies (now heavily influenced by corporate interests) have largely prevented the fostering of a truly democratic spirit at all levels in our society, from the household to the nation, from rural to urban, from marginalized to mainstream sections. Again, this cannot be an excuse for withholding progressive structural changes or for inserting arbitrary bureaucratic safeguards (such as when divisional forest officers are given unilateral power to dissolve village forest committee, or registrars of cooperative societies are empowered to take over cooperative societies) as has been the case in the past. But it challenges us to look at questions of behaviour, leadership, and ethos that are often considered too 'managerial' or in the realm of 'social engineering' by mainstream social scientists.

Finally, the importance of regional diversity cannot be lost sight of. Given that the colonial forest policy operated at the national level (even as it allowed for provincial variations) and that post-independence policies further homogenized the process, obliterating nuances that even the British had allowed (Lele and Srinidhi 1998), the entire debate on forest governance, from Chipko and the FCA to the draft forest bills, national forest policy, JFM, and Godavarman has been at the national level. Such a centralized system required a centralized response, which the FRA has provided to an extent. But it will be impossible to interpret a broad national-level legislation like the FRA, or amendments to the Wildlife Protection Act (WLPA), or the FCA in ways that are sensitive to local context and history. To give only one example: the Western Ghats present an extreme case where communities are largely non-tribal and the economy is relatively prosperous, and where substantial forest rights were granted to forest-dependent households, at the individual rather than community level (Srinidhi and Lele 2001), but in a discriminatory manner (favouring landed versus landless, and so on). Defining the contours of democratic forest governance is much more challenging in this context than in a central Indian tribal forest-dweller context. Equally challenging is the task of visualizing what democratization means in the complex situations that prevail in Northeastern India, where forests are much more within local control, but local does not necessarily mean democratic nor does it always translate into sustainable forms of forestry. Taking the democratization discourse to the state level rather than hope for one-size-fits-all solutions is, therefore, going to be essential.

Democratization is Not Enough: What are Forests For?

It is a fact of life that few people are interested in the quality of governance unless it affects them in some material way. It then follows that we cannot discuss the process of forest governance without asking how it relates to the material and cultural conditions of the stakeholders. The assumption in much of the literature, including several of the chapters in this volume, has been that forest-fringe communities are heavily forest-dependent and hence are automatically interested in questions of forest governance. But, as Menon *et al.* have warned in their chapter, one may not want to take this dependence (and hence this interest) for granted. The pace of industrialization and urbanization is such that both the material dependence upon and cultural links to forests might be diminishing rapidly in younger generations in many parts of the country. Without concluding that 'the moral core of the peasant economy—its normative sense of itself as the right or good life—is dead' (Baviskar 2012),[7] we would argue that the discourse on forest governance must also engage with the question of substantial improvements in forest-based livelihood opportunities, while ensuring that the goal of sustainability is not ignored.

Livelihood enhancement in today's increasingly neoliberal and technologically specialized world will require substantial engagement with the market so as to generate cash income. This is something that many activists and social movements have stayed aloof from, preferring to believe in the myth of the tribal economy as a subsistence economy. What has actually existed is a system of forest product extraction and sale wherein the state, through the FD, or its favoured interest groups, such as timber contractors, engage the forest-dwellers' labour for timber logging and NTFP harvest at lower than minimum wages while extracting all the surplus value. What forest-dwellers would probably like to see is that this system is replaced by a system wherein they continue to harvest forest produce but capture most of its value. The governance question here is how to ensure that gram sabhas, which are citizenship-based institutions that have to include all villagers, do not capture the economic gains that should accrue to the actual NTFP collectors, who may often be in a minority in the village; in other words, that gram sabhas do not replace the state in surplus extraction. The developmental question here is how to enable NTFP collectors, who often belong to the weakest sections of society, to engage with the market without draining the state treasury in the long run[8] or repeating the mistakes of the multitude of highly paternalistic and bureaucratic schemes such as LAMPS and fully state-controlled cooperatives that various states have launched in the past (Lele and Rao 1996; Lele *et al.* 2012).

Linked indirectly to the question of livelihood enhancement is the question of the fiscal policy on forests. Modern states function primarily through budgetary allocations. The contradictions between more decentralized forest governance and highly centralized fund generation and allocation are an area of major concern. The CAMPA fund mess, as described by Kohli and Menon in Chapter 7 in this volume, is just one such issue related to the fiscal complexity surrounding forest and environmental governance in the country. Can 'payment' schemes, where local forest-dependent communities are paid subsidies in proportion to their 'delivery of forest ecosystem services' that were advocated in India as far back as 1994 (Gadgil and Rao 1994) actually work? Under what conditions? Can the new carbon-based payment schemes do the trick? Political objections to REDD+ apart (Lele 2013), economic analyses show that the returns per household from payments for carbon sequestration are a pittance (Chetan Agarwal, personal communication). Payments for hydrological services are even more complicated, as the relationship between forest cover and hydrology is much more complicated (Lele 2009). Eco-tourism is often seen as the more tangible alternative to these ephemeral payment schemes, although the challenges here are similar to those in the NTFP sector—entrenched external interests and limited skills and capital in the hands of the forest-dwellers.

Ensuring that livelihood enhancement is ecologically sustainable will be even more challenging. Traditional knowledge may not have all the answers (Corbridge and Jewitt 1997). Knowledge that may have been adequate in the context of meeting subsistence needs of much smaller populations may be inadequate when it comes to anticipating the impacts of simultaneous commercial-scale harvest of bamboo, cane, tendu leaves, and a host of other products, not to mention timber, that have market value. The forestry discourse has to engage with questions of knowledge production: how can publicly funded scientific institutions generate socio-ecological knowledge that is truly useful and relevant to forest-dependent communities? The capture and control of forestry research in the country by the forest service is an area that is almost as much in need of reform as the control of the forest landscape itself.

***

This volume attempts to provide a bird's-eye view of the forest question in India today. Ranging from abstract questions of what are forests and what is participation to the nitty-gritty of the FRA and the socio-legal complexities of the Godavarman case, the contributors to this volume have sought to capture and elucidate some of the key dimensions of the churning that is taking place in the country today around how we can democratically govern our forests. In this brief epilogue, we have sought to point to other crucial debates that are emerging in this area. We are hopeful that these debates will contribute in some ways to progressive changes in the forest sector, while also providing a rich source of insights into the wider society–nature relationship as it is evolving today.

Notes

1. The terms 'devolution' and 'decentralization' have been used in contradictory and confusing ways. Here, we use the term 'devolution' to mean transfer of power and responsibility to lower levels by legislation (Pomeroy and Berkes 1997), something that the FRA of 2006 attempts.
2. Guha's mentioned piece itself talks mostly about the IFS!
3. Jayanthi Natarajan, former minister of MoEF, indicated in a conversation that the top forest official in the country refused to abide by her instructions on the implementation of the Forest Rights Act!

4. They also produce some 'dis-services' such as providing habitat for malaria-bearing mosquitoes or for wildlife that preys on nearby agricultural crops.

5. A debate that emerged recently when the Odisha government limited the Supreme Court-mandated referendum on the Niyamgiri mining question to only 12 palli sabhas.

6. And, of course, the dialectic between the two. Much of the environmental behaviour we see today, whether in the form of rampant consumerism by individuals or the single-minded pursuit of economic growth by producers, is the product of a neoliberal ideology that has established itself in society.

7. We would, however, hazard that unless sustainability constraints are first imposed on urban India, the demand of ecological sustainability from rural households, farmers or forest-users will become increasingly untenable.

8. Which schemes such as the recently announced 'Mechanism for marketing of Minor Forest Produce (MFP) through Minimum Support Price (MSP) and development of value chain as a measure of social safety for MFP gatherers' (http://pib.nic.in/newsite/erelease.aspx?relid=97552) are bound to do.

References

Baviskar, A. 2012. 'India's Changing Political Economy and its Implications', in Anonymous (ed.), *Deeper Roots of Historical Injustice*, pp. 33–45. Washington, DC: Rights and Resources Initiative.

Bhatta, B. 2002. 'Access and Equity Issues in High Mountain Regions: Implications of a Community Forestry Programme'. In *Policy Analysis of Nepal's Community Forestry Programme: A Compendium of Research Papers*, pp. 1–32. Kathmandu: Winrock International-Nepal.

Corbridge, S., and S. Jewitt. 1997. 'From Forest Struggles to Forest Citizens? Joint Forest Management in the Unquiet Woods of India's Jharkhand', *Environment and Planning, A* 29 (12): 2145–64.

Gadgil, M., and P.R. Sheshagiri Rao. 1994. 'A System of Positive Incentives to Conserve Biodiversity', *Economic and Political Weekly*, 29 (32): 2103–7.

Guha, R. 2012. 'The Past and Future of Indian Forestry', in Anonymous (ed.), *Deeper Roots of Historical Injustice*, pp. 1–12. Washington, DC: Rights and Resources Initiative.

Joint Committee. 2010. 'Manthan: Report of the National Committee on Forest Rights Act', New Delhi: Ministry of Environment and Forests and Ministry of Tribal Affairs, Government of India.

Lele, S. 1999. 'Institutional Issues in (J)FM(& R)', in Anonymous (ed.), *National Workshop on Joint Forest Management*, pp. 19–29. Ahmedabad: VIKSAT, Gujarat Forest Department and Aga Khan Foundation.

————. 2004. 'Beyond State–Community Polarisations and Bogus "Joint"ness: Crafting Institutional Solutions for Resource Management', in M. Spoor (ed.), *Globalisation, Poverty and Conflict: A Critical 'Development' Reader*, pp. 283–303. Dordrecht, Boston and London: Kluwer Academic Publishers.

————. 2009. 'Watershed Services of Tropical Forests: From Hydrology to Economic Valuation to Integrated Analysis', *Current Opinions in Environmental Sustainability*, 1 (2): 148–55. doi: 10.1016/j.cosust.2009.10.007.

————. 2011. 'Rethinking Forest Governance: Towards a Perspective Beyond JFM, the Godavarman case and FRA', in Anonymous (ed.), *The Hindu Survey of the Environment 2011*, pp. 95–103. Chennai: N. Ram.

————. 2013. 'Buying Our Way Out of Environmental Problems?' *Current Conservation* 6 (1): 9–13.

————. 2000. 'Degradation, Sustainability or Transformation?' *Seminar*, 486: 31–7.

Lele, S. and A.S. Srinidhi. 1998. 'Indian Forest Policy, Forest Law, and Forest Rights Settlement: A Serious Mismatch', in *International Workshop on Capacity Building in Environmental Governance for Sustainable Development*. Mumbai: Indira Gandhi Institute for Development Research.

Lele, S. and R. Jagannath Rao. 1996. 'Whose Cooperatives and Whose Produce? The Case of LAMPS in Karnataka', in R. Rajagopalan (ed.), *Rediscovering Cooperation*, pp. 53–91. Anand, Gujarat: Institute of Rural Management.

Lele, S., V. Ramanujam, and J. Rai. 2012. *Tendu Leaf Procurement, Rural Livelihoods and Forest Governance in Madhya Pradesh*. Bangalore: Ashoka Trust for Research in Ecology and the Environment.

Pallavi, A. 2012. 'Don't Say Bamboo', *Down To Earth*, 21 March.

Phelps, J., E.L. Webb, and A. Agrawal. 2010. 'Does REDD+ Threaten to Recentralize Forest Governance?', *Science*, 328 (5976): 312.

Pomeroy, R.S., and Fikret Berkes. 1997. 'Two to Tango: The Role of Government in Fisheries Co-management', *Marine Policy*, 21 (5): 465–80.

Raman, S. and M.D. Madhusudan. 2013. 'Development Minus Green Shoots', *The Hindu*, 12 February.

Rangachari, C.S. and S.D. Mukherji. 2000. 'Old Roots New Shoots: A Study of Joint Forest Management in Andhra Pradesh, India', Kinsuk Mitra and Doris Capistrano, (eds), *Contemporary Issues in Indian Forestry*. New Delhi: Winrock International and Ford Foundation.

Sarin, M. 1995. 'Regenerating India's Forests—Reconciling Gender Equity with Joint Forest Management', *IDS Bulletin*, 26 (1): 83–91.

Sarin, M., N.M. Singh, N. Sundar, and R.K Bhogal. 2003. *Devolution as a Threat to Democratic Decision-making in Forestry? Findings from Three States in India*. London: Overseas Development Institute.

Srinidhi, A.S. and S. Lele. 2001. *Forest Tenure Regimes in the Karnataka Western Ghats: A Compendium*. Bangalore: Institute for Social and Economic Change.

Vira, B. 2005. 'Deconstructing the Harda Experience: Limits of Bureaucratic Participation', *Economic and Political Weekly*, 40 (48): 5068–75.

Index

Editors and Contributors

Arshiya Bose is a member of Kalpavriksh and a doctoral student at the Department of Geography, University of Cambridge. Her research interests include the political ecology of conservation of biodiversity on coffee plantations in the Western Ghats, India, and in particular, the role of market-based incentives as the primary mover of behaviour and land-use practices. She can be reached at arshiyabose@gmail.com.

Neema Pathak Broome is a member of Kalpavriksh and coordinates the Conservation and Livelihoods Programme in the organization. She also coordinates activities of the Consortium supporting Indigenous Peoples, and Local Community Conserved Areas and Territories (ICCA) in South Asia. Her interests lie in researching, documenting, understanding, facilitating, and advocating for processes towards decentralized, equitable, diverse, and context-specific forms of biodiversity governance. She can be reached at neema.pb@gmail.com.

Dhrupad Choudhury is the Regional Programme Manager of International Centre for Integrated Mountain Development's (ICIMOD) Regional Programme on Adaptation to Change. Dhrupad has a special interest in traditional practices in natural resource management, access and control regimes, and livelihoods of ethnic communities in Northeast India. His present work includes community-based adaptations to change and managing transformation in shifting cultivation areas. He can be reached at dchoudhury@icimod.org.

Shiba Desor is a member of Kalpavriksh, based in Pune. She works on issues of conservation and livelihoods with a focus on policy and practice related to community-based governance of forests. She is interested in gaining a deeper understanding of human perceptions of, and relationships with, their surrounding ecosystems. She can be reached at desor.shiba@gmail.com.

Prakash Kashwan is Assistant Professor in the Department of Political Science at University of Connecticut, Storrs. His research and teaching focuses on international environmental policy and politics, institutional analysis, property rights, and research methods. At the time of the writing of his contribution to this book, he was a doctoral student at Indiana University, Bloomington, USA. He can be reached at prakash. kashwan@uconn.edu.

Shomona Khanna is an advocate practising in the Supreme Court of India and the Delhi High Court. She has worked on numerous civil and democratic rights issues both as a lawyer and a writer. Her main areas of interest are rights of indigenous peoples under Indian laws, and also rights of other marginalized communities such as dalits, women, and victims of state violence. She can be reached at shomona@gmail.com.

Kanchi Kohli is a researcher, writer, and campaigner working on environmental, forest, and biodiversity governance in India. Her work seeks to draw empirical evidence from sites of conflict and relates to the power politics emanating from the legal and policy realm. In her current profile, she seeks to work with the strength of various networks and organizations, in an independent capacity. She can be reached at kanchikohli@gmail.com.

Ashish Kothari is Founder-member of the Indian environmental group Kalpavriksh, and has taught at the Indian Institute of Public Administration, coordinated India's National Biodiversity Strategy and Action Plan process, and chaired an IUCN network dealing with protected areas and communities. Active in several civil society movements, Ashish has authored or edited over 15 books, and is currently working on globalization and its alternatives. He can be reached at chikikothari@gmail.com.

Kundan Kumar completed his PhD in Resource Development from Michigan State University and is Assistant Professor at Faculty of Forestry, University of Toronto. His research interests are in forest and land tenure, social and environmental movements, environmental justice, and political ecologies of forested landscapes. He is currently doing research on India's Forests Rights Act. He can be reached at kumarkun@gmail.com.

Sharachchandra (Sharad) Lele is a Senior Fellow at the Centre for Environment and Development at the Ashoka Trust for Research in Ecology and the Environment (ATREE), Bangalore. His research interests include conceptual issues in sustainable development and sustainability, and interdisciplinary analyses of forest and water resource management across South Asia. He was a member of the MoEF–MoTA Joint Committee on the Forest Rights Act. He can be reached at slele@atree.org.

Viren Lobo is Director of Society for Promotion of Wastelands Development (SPWD) and has been engaged in action research on various issues related to livelihoods, governance, and ecology of natural resources. He has worked through and facilitated the development of various networks dealing with aspects of natural resources, ecology, and livelihoods. He has written extensively on food security, common land management, climate change, forest rights, and the relevance of biodiversity in the life support systems of rural communities. He can be reached at vlobo62@gmail.com.

Ajit Menon is Associate Professor at the Madras Institute of Development Studies (MIDS), Chennai. His research includes both conceptual and empirical studies of the political economy of the environment with special reference to forestry and fisheries in Tamil Nadu. He is also interested in the wider debates around capitalism and neoliberalism and their manifestations within development and environment policy in India. He can be reached at ajit1112@gmail.com.

Manju Menon is the Programme Director of the Namati-CPR Environmental Justice Program, and a member of Kalpavriksh. Her research interests include environmental law and implementation analyses, policy and institutional arrangements for environmental governance, monitoring and compliance. She is currently a PhD candidate at the Centre for Studies in Science Policy (CSSP), Jawaharlal Nehru University, New Delhi. She can be reached at manjumenon1975@gmail.com.

Nitin D. Rai is a Fellow at the Ashoka Trust for Research in Ecology and the Environment (ATREE), Bangalore. His research in political ecology incorporates situated knowledge and ecological science to understand the implications of conservation and development for people and landscapes. He can be reached at nitinrai@atree.org.

Sagari R. Ramdas is Director of ANTHRA, Hyderabad, a resource group that works with peasant, pastoralist, and adivasi communities, particularly women, on environment and social justice concerns in the context of food sovereignty. Her research, carried out with people's alliances, looks at ways in which policies and programmes that span multiple sectors impact food-farming systems, and how communities are organizing themselves to resist and assert their rights over resources. She can be reached at sagari.ramdas@gmail.com.

Madhu Sarin has worked for over 30 years on community-based natural resource governance by gender-equal and democratic community institutions. Over the last few years, she has advocated legal recognition of community forest rights and authority for genuine democratization of forest governance in India. She has played a significant role in drafting the recently enacted Forest Rights Act, 2006. She can be reached at madhu.sarin1@gmail.com.